SILICICLASTIC SHELF SEDIMENTS

Based on a Symposium

Sponsored by the Society of

Economic Paleontologists and Mineralogists

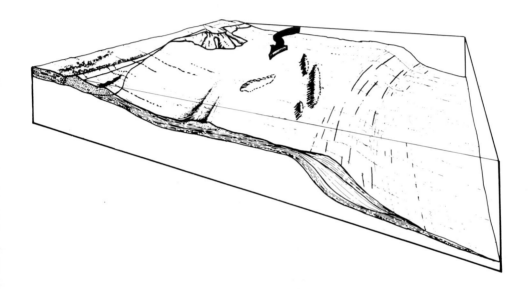

Edited by

Roderick W. Tillman, Cities Service Research, Tulsa, Oklahoma,

and

Charles T. Siemers, Sedimentology, Inc., Boulder, Colorado

Tulsa Oklahoma, U.S.A.

May, 1984

A Publication of

The Society of Economic Paleontologists and Mineralogists

a division of

The American Association of Petroleum Geologists

PREFACE

This volume is a collection of papers which, for the most part, were included in a symposium on Shelf Sandstone deposits sponsored by the Society of Economic Paleontologists at the Annual Meeting in Denver in 1980. Six of the papers presented at the symposium are included in this volume; additional papers were added to give coverage to Recent shelf deposits and to increase the number of ancient examples.

This volume is divided into three sections. The sections are arranged in a sequence proceeding from youngest to oldest. To date, the maximum amount of data available for making interpretations on shelf-sand depositional systems is from Mesozoic rocks, and this bias is reflected in this volume.

A variety of techniques are useful in documenting shelf depositional process and sand body geometries. Among these are sedimentary structures observed in outcrops and cores, biogenic data including trace fossils, micro- and macro-fauna, detailed seismic sections, and detailed subsurface correlations. These types of data are all readily available, at least locally, for Mesozoic rocks. In studying older shelf sandstones such as those from the early Paleozoic and Precambrian, some techniques cannot be used; trace fossils are rare to absent, and it is difficult to establish the relationships of the shelf sand-bodies to the shoreline of the broad shallow seas that were common at that time. Most papers that have been published on modern shelf sandstones have provided little or no information on the internal features of the sand bodies. The inability or lack of interest in obtaining long undisturbed cores (10 meters or more in length) from modern offshore sedimentary environments has contributed to this problem.

MODERN SHELF SAND-RIDGES

Considerable oceanographic work has been done on the New Jersey Shelf during the last quarter century, however, it has only been in recent years that detailed seismic and vibracore data have been coordinated to yield detailed process-oriented depositional models.

Information from studies such as these by Stubblefield et al. in this volume can be used to better understand the process of formation of ancient shelf sandstones. Since the Holocene is in general a period of rising sea level, much controversey exists in areas such as the New Jersey Shelf as to whether sediments in the are are "reworked", "palimpsest" or derived from external sources. One of the problems addressed in the first paper of this volume by Stubblefield et al. is, "What would occur during short periods of stillstand during an overall transgression"? Could Holocene shoreline sands be deposited during a stillstand and be preserved wholly, or in part, even after being transgressed? Swift et al. in the discussion of Stub-

blefield et al. argue that the features observed on the New Jersey mid-shelf are not preserved shoreline feature but instead are inner shelf-ridge fields. According to Stubblefield in his Reply, if some of the present sand ridges such as those on the New Jersey mid-shelf are the remnants of barrier islands formed during transgression as he proposes, they might be expected to have lagoonal muds and sands stratigraphically below them. If the sand ridges were formed in an offshore area independent of barrier island-forming processes as suggested by Swift, the muds below or interbedded with the sand-ridges should have been formed on the shoreface or the shelf. Arguments for and against these and other scenarios are presented by these authors.

CRETACEOUS SHELF SEDIMENTS OF THE ROCKY MOUNTAINS

Swift and Rice in their paper dealing with Sand Bodies on Muddy Shelves compare the evidence from modern shelf sediments with those of the Cretaceous of the Rocky Mountains. This paper bridges the gap between studies of Recent and ancient shelf deposits. They suggest that in the Western Interior seaway sandy-mud bottom-sediment-loads may have been transported great distances along the shelf, especially during major storm events. They go on to discuss how, through a series of events, sand may be winnowed and deposited on slight topographic highs. Once deposition begins they suggest how a coarsening-upward sandstone sequence might be deposited.

Two papers on the Shannon Shelf sandstones discuss different aspects of this unit. Both papers stress the importance of storms in forming shelf sandstones. Shurr emphasizes the regional to sub-regional aspects of what he terms the "hierarchy" of sandstone bodies, while Tillman and Martinsen emphasize processes of deposition, facies recognition, and the effects of local topography on shelf sandstone distribution. Shurr discusses five hierarchical elements diminishing in size from the largest scale element, "lithosome" through a series of units termed: "sheet", "lentil", "elongate lens" and "facies package". He emphasizes the differences in the type and duration of the processes that are responsible for formation of each of these hierarchical elements. The scale and characteristics of each of these "elements" are discussed in terms of Shannon outcrops and subsurface data in Southeast Montana.

Tillmann and Martinsen discuss the Shannon in The Powder River Basin of Wyoming. They put into context the oil prone facies of the shelf-ridge sandstones in terms of their distribution, geometry and processes of deposition. Eleven facies, associated with the up to 80-foot thick "shelf-ridge sandstone complexes", are discussed in terms of sedimentary features, processes of deposition (including nearly unidirectional transporting currents) and reservoir potential. The vertical sequence of facies they observe indicates a general upwards increase in energy within any bar complex. Surface and outcrop data are integrated in an effort to provide one or more models of shelf sand deposition on what they interpret as sandstones deposited 100 miles offshore on the Upper Cretaceous mid-shelf. Evidence concerning the effects of local tectonics in creating local topography, which in turn localizes sandstone deposition, is also discussed.

Rice, in his paper on the lowermost Upper Cretaceous Mosby Sandstone in central Montana, emphasizes the potential for long distance (1,000 miles) transport of sediments from deltaic source areas to their eventual resting place as shelf sandstones. He suggests that the Dunvagen Delta in west-central Alberta was the source for Cenomanian sediments that were deposited as far south as central and southern Montana. The thin Mosby Sandstone, which averages about 1.5 meters in thickness, contrasts sharply with the thick sandstones typical of the Shannon in Wyoming. In the Mosby, Rice observes coarsening-upward cycles containing individual sandstones which at the top are "gradational" with the overlaying shale, commonly contain hummocky bedding, and contain an overall sequence of structures which suggest upward decreasing energy levels.

Beaumont in his paper on the Lower Cretaceous Viking Formation in Alberta identifies "shoreward shingling" elongate sand bodies west of a series of "irregular-shaped bodies". The former, which includes Joffre and Joarcum Fields, he attributes to deposition on the shelf during an overall transgression, and the latter he interprets as drowned delta complexes. A common coarsening-upward sequence he observed in the shelf deposits is silty shale, rippled to bioturbated sandstone, glauconitic sandstone and polymictic pebble conglomerate. He also recognizes conglomerates within the sequences. Subsurface sections indicate that the Viking thickens westward, pinches out still further westward, and contains overlapping "sheets" of sandstone that contain elongate "cleaner" sandstone bodies. The presence of the elongate shingled shelf sandstone bodies shoreward (westward) of only slightly older delta sand bodies he terms "retrogradational" shelf sedimentation.

Kitely and Field discuss a series of formations in Colorado which grade from shoreline deposits in the west to interbedded shoreline and shelf deposits in the central part of Colorado to exclusively shelf deposits in eastern Colorado. Eustatic sea level changes, they believe, are the major cause for changes in environments of deposition. They attribute minor local transgressions and regressions to local tectonic uplifts and the consequent increase in sediment supply. Formations such as the Hygiene, Terry, and First Mancos sandstone are tied into Kauffman's (1977) sea level changes, T_7 and T_9. The R_9 regression, they believe, is a consequence of local uplift in northwest Colorado.

This and other similar uplifts are tied to deposition of delta front and shoreface sandstones. Using data from Recent sediments, they attempt to quantify flow rates for shelf units such as the Hygiene Sandstone. They estimate that the bottom current velocities which deposited the Hygiene sandstone ranged from 70 to 100 cm/sec.

CRETACEOUS SHELF SEDIMENTS OF THE SOUTHERN UNITED STATES

The seaward transport of sand from the shoreline on to the shelf is a topic discussed in Hobday and Morton's paper on the Lower Cretaceous Grayson Sandstone of northeast Texas. They take issue with one of the often quoted theories espoused by Hayes (1967) as a result of his Hurricane Carla studies. Hayes proposed that the hydrodynamic head created by storm washover into back-barrier bays would exit via washover channels and inlets, and within the flows transport coarse sediment across the shoreface and inner shelf as a density current. Hobday and Morton argue that the runoff from the lagoon studied by Hayes was negligible, that maximum bottom current velocities occurred significantly prior to the maximum surge currents through the inlets, and that the thickest and most extensive deposits of the graded sand units was off a portion of the coast where washover channels were entirely absent. They believe they can recognize deposits formed by strongly oblique flow from the shoreline onto the shelves in the Grayson Formation where it outcrops in north-central Texas. The vertical sequence of facies, sedimentary structures, and trace fossils in the Grayson are discussed and evidence is presented that the major process of sand deposition for the sandstone was infrequent storms.

Kurten Field, which includes more than 103 wells producing from the Woodbine-Eagle Ford Formations in south Texas, consists of five sandstone units and is interpreted by Turner and Conger to be a stratigraphic and diagenetic trap. The sandstones are interpreted, on the basis of sedimentary structures, trace fossils, and microfauna, to be offshore bars which were generated by river-mouth by-passing, storm-generated sheet flows and longshore currents. Their study of the field included, in addition to subsurface log studies, analysis of five slabbed cores and 59 thin sections. Most of the sandstones are interpreted to have been deposited by processes in which there was a decrease in depositional energy upwards. Sheet flows or shelf turbidites are suggested as possible depositional mechanisms for Kurten Field sandstones. Porosity in many of the Woodbine sandstones is secondary and is the result of diagenesis. They found that sandstones containing clay are more porous than beds that were initially clean sandstones.

PRECAMBRIAN SHELF SANDSTONES

High energy shelf deposits from the Precambrian Wishart Formation in the Labrador trough are described by Simonson. He suggests that upward-coarsening 12 m thick cycles resulted from both storms and tides that were important on the shelf during the Proterozoic. Five facies are defined and their areal distribution is discussed; three of the facies form 90 percent of the aggregate thickness of the only sparsely shaley Wishart Formation; the other two facies are not part of what he describes as upward-thickening cycles. In addition to presenting detailed descriptions and interpretations of the shelf sediments, the difficulties arising from not being able to recognize a shoreline are discussed. This problem is common in many studies of shelf sediments in the early Paleozoic and Precambrian, and the problem remains largely unsolved.

ACKNOWLEDGMENTS

This volume represents the efforts of a dedicated group of authors from relatively diverse fields of endeavor. Their cooperation in meeting deadlines and their willingness to make changes which were suggested by the editors to aid in the overall consistency of the volume are appreciated.

We acknowledge with thanks the technical reviews of papers in this volume by Roger Slatt, Jim Rine, Mary Walker, Karen Porter, Mike Boyles, Jim Ebanks, Charles Nittrouer, Don Swift, and Jack Etter.

RODERICK W. TILLMAN
CHARLES T. SIEMERS

CONTENTS

RECOGNITION OF TRANSGRESSIVE AND POST-TRANSGRESSIVE SAND RIDGES ON THE NEW JERSEY CONTINENTAL SHELF

W.L.STUBBLEFIELD[1], D.W. MCGRAIL[2], AND D.G. KERSEY[3],[4]

[1]National Oceanic and Atmospheric Administration, National Sea Grant College Program, 6010 Executive Boulevard, Rockville, Maryland 20852, [2]Department of Geological Oceanography, Texas A&M University, College Station, Texas 77843, and [3]Department of Geology, Texas A&M University, College Station, Texas 77843

ABSTRACT

A new perspective regarding the genesis of the sand ridges on the storm-dominated New Jersey continental shelf has been developed by synthesis of two previous models. One model suggests the ridges formed as barriers during the last marine transgression. The other model suggests the ridges are post-transgressive features forming at the base of the shoreface. The bathymetric and textural data on the New Jersey nearshore and mid-shelf suggest two different groups of ridges: one group formed as barriers whereas the other group represents a post-transgressive origin. These ridge groups have distinctive orientations, wavelengths, morphology, and sediment textures. Intersecting the coast at 15° to 30° are coast-oblique ridges. Seaward of these is a 30-km-wide complex of ridges that are approximately coast-parallel. Farther seaward is a group of ridges parallel to the nearshore ridges. Cluster and spectral analyses produce similar groupings.

Vibracores through the nearshore ridges reveal a medium to medium-coarse mean grain size with numerous occurrences of coarsening and fining (high-energy events). The vibracores through the mid-shelf coast-parallel ridges indicate two depositional zones. The upper approximately 1 m is similar to the nearshore ridges. Beneath this upper zone the grain size is fine to medium-fine, with fewer high-energy events. Also unique to the mid-shelf ridges are zones of interlayered mud and fine sand.

The New Jersey mid-shelf coast-parallel ridges are interpreted as degraded barriers that were submerged by rising sea level. The nearshore and outer shelf coast-oblique ridges formed as post-transgressive, shoreface-connected ridges. Thus, the degraded barriers and seaward intersecting ridges are an earlier Holocene analog to the present barrier and shoreface-connected ridge complex.

INTRODUCTION

Early theories of ridge origins on the modern shelves include Agassiz's (1894) study of the Bahama Bank, van Veen's (1935) work in the North Sea, and those of Off (1963) suggested that the ridges formed by alternating bands of slower-moving water interspersed with faster-moving waters. Van Veen (1935) suggested different paths of tidal ebb and flow currents. Many of the ridge features described by Off as having been formed by tidal currents have been studied by other workers who concurred with his hypothesis, e.g., Georges Bank (Jordan, 1962; Stewart and Jordan, 1964; Smith, 1969) and the North Sea (Belderson et al ., 1971; Houbolt, 1968; Caston, 1972; McCave, 1972). However, tidal currents on many shelves are substantially less than the 50 to 250 cm/sec suggested by Off as being necessary for ridge formation. For example, the tidal currents on the New Jersey shelf are generally less than 20 cm/sec (Redfield, 1956). Consequently, other models have been invoked to explain the origin of the ridges on continental shelves that have low tidal velocities.

On the Long Island shelf, McKinney and Friedman (1970) contended that the Holocene shelf topography reflects fluvial drainage topography of a Pleistocene coastal plain that has not been obscured by recent sediment flex. Similarly, Swift et al. (1972) interpret portions of the bathymetry on the northern New Jersey and inner Long Island shelf as bearing the imprint of subaerial tributaries.

Many workers, however, have interpreted the shelf topography of the U.S. Middle Atlantic Bight and adjacent coastal waters as being indicative of overstepped coastlines. In their interpretations, coastal features such as subaerial beach ridges or barrier islands, which date from either the late Pleistocene regression or the more recent Pleistocene-Holocene transgression, were partially preserved. Workers using bathymetry as their criteria for this interpretation include: Emery's (1967), Shepard's (1973), and Uchupi's (1970) investigations of the U.S. East Coast: McClennen and McMaster's (1971) and Kraft's (1971) study of the New Jersey and Delaware shelves; McMaster and Garrison's (1967) work on the Rhode Island shelf; and Riggs and Freas's (1968) investigation in the Southern Atlantic Bight.

McClennen (1973) used high-resolution and vibracore data to obtain a similar interpretation. His study indicated that the modern sand ridges on the New Jersey continental shelf were built on Late Wisconsinan or older sediments. The dipping seismic reflectors within these sand ridges and the inhomogeneity of the sediments prompted McClennen to suggest that the origins of the ridges were related to nearshore processes of barrier-beach and lagoon formation.

Sanders and Kumar (1975a, 1975b) have suggested that certain barriers off Fire Island, New York were preserved by in-place drowning. These authors sug-

[4]Present Address: Reservoirs Inc.. Houston, Texas 77043

gested that as sea level rose the barriers remained in place until the wave breaker zone reached the top of the barrier. At that point the breaker zone jumped landward to the inner margin of the lagoon, thus drowning the barrier. With this mechanism the breaker zone skips across the shelf rather than universally progressing over the entire area.

A more recent origin for the sand ridges has also been proposed. In a detailed investigation of the Atlantic shelf of the United States, extending from Long Island to Florida, Duane et al. (1972) and Swift et al. (1972) concluded that the shelf topography reflects post-transgressive processes. With the exception of an imprint of fluvial drainage in isolated areas, these authors suggested that most of the features formed at the foot of the retreating shoreface, either as shoreface-connected ridges or ridges associated with a retreating estuary or cape.

In their examination of shoreface-connected and similar nearshore ridges, Duane et al. (1972) observed a frequent 15°-30° intersection angle with the coastline. They used this intersection angle as an argument against the idea of the ridges being submerged coastlines. They suggested instead that the ridges formed in the nearshore by means of a storm-generated helical flow in the water column. Later, Swift (Swift and Ludwick, 1976; Swift et al., 1977; Swift and Field, 1981) questioned the applicability of helical flow in sand ridge development, choosing instead an asymmetrical shear stress distribution over a bottom irregularity. Swift and the others base their model on sand wave theory as applied to tidal or fluvial sand waves by Smith (1969, 1970).

Not all ridges, however, intersect the coastline at an oblique angle. One such example is on the New Jersey shelf, seaward of the nearshore ridges. These and similar ridges were interpreted by Swift et al. (1972) as having been initially formed by tidal currents at the mouth of an inlet. These tidal ridges were abandoned by the retreating shoreface and rotated to an approximate coast-parallel trend by storm-generated currents. According to Swift et al. (1972, 1978) these offshore ridges are superimposed on a broad, shelf-transverse bathymetric feature of subdued relief (shoal retreat massif). This broad feature marks the retreat path of a littoral drift depositional center on the sides of an estuary mouth. A similar offshore topographic high, also molded into a series of arcuate ridges, is thought to be associated with the retreat of a cape (Swift et al., 1972).

Thus, the sand ridges on certain continental shelves possessing weak tidal currents have been ascribed to differing formative processes. Suggested times of formation for these ridges vary from prior to transgression, to during transgression, to post-transgression. Unstated, but assumed for most of these models, is that all such ridges formed during one of these times, that is, prior to, during, or subsequent to the passage of the transgressive coastline. Only a few workers have alluded to the possibility that different ridges formed at different times. One such exception is Knebel and Spiker's (1977) use of seismic profiles to suggest that the ridges formed both during and after passage of the coastline. They based their interpretation for some of the ridges being formed by nearshore processes during times of lowered sea level on the truncation of internal seismic reflectors along the flanks of the ridges. For those ridges suggested as being formed by modern depositional processes, they based their interpretation on the presence of dipping internal reflectors, increased thickness of sand, and the close proximity of the sand ridges to large asymmetric sand waves.

Based on the author's interpretations of data collected on the New Jersey shelf, a case can be made to support Knebel and Spiker's (1977) contention that two types of ridges exist. We interpret the data as indicating that some of the ridges formed as barriers during the passage of the coastline, whereas other ridges formed as post-transgressive, shoreface-connected ridges.

METHODS

Field Methods

In order to address the question of ridge formation on the western Atlantic shelf, data were collected in a 37-km-wide corridor extending 65 km seaward from the New Jersey coast (Fig. 1). This area is bounded by latitudes 38° 45′ N and 39° 30′ N and longitudes 73° 35′ W and 74° 30′ W. Data utilized in this study consist of published bathymetric maps (Stearns, 1967), and 12 vibracores; four of these vibracores were previously discussed by Stubblefield et al. (1975). In this paper all data is expressed in metric units with the exception of the bathymetric maps in Figures 1 and 2. These figures are in fathoms (1 fathom = 6 feet = 1.83 meters).

The vibracores were collected from the crests, flanks and troughs of various ridge systems. The recovery length for individual cores ranged from 0.6 to 5.75 m. However, by varying the core sites on a ridge, it is expected that more than 5.75 meters of stratigraphic section were sampled. This is based on the assumption of lateral continuity of units within individual ridges. Pertinent vibracore data, including recognizable macrofauna, are in Table 1. Vibracore sites and the two grab sample transects used by Stubblefield and Swift (1981) are in Figure 1.

Laboratory Methods

In this study the inner and outer shelf boundary of 37 m - 40 m, as defined by Swift et al. (1972), will be followed. However, their inner shelf is subdivided into a nearshore and a mid-shelf province. Examination of Figure 1 indicates a ridge complex near the coastline. Seaward is an area of few ridges, and still farther seaward is another large complex of ridges. This intermediate zone, with few ridges, marks the seaward limit of the nearshore zone. The maximum water depth for

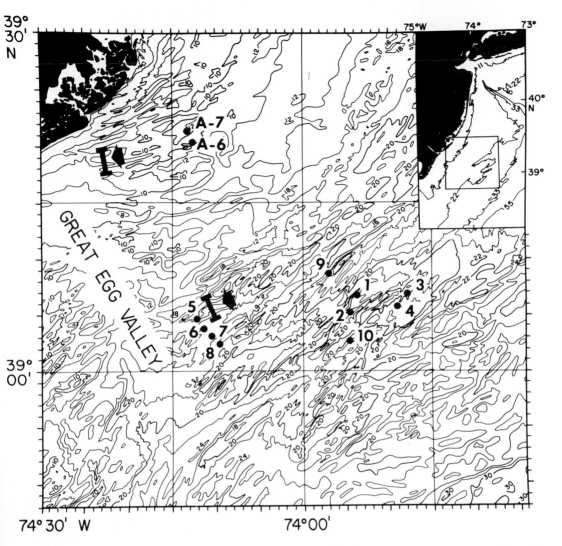

Fig. 1.—Index map of study area. Vibracore sites are marked with a solid circle; two grab sample transects are shown with solid line. The bathymetric map (Stearns, 1967) is contoured in two-fathom intervals (2 fathoms = 3.66 meters).

the nearshore is 20 m and includes the shoreface with a maximum depth of 15 m. The mid-shelf province in turn extends from the 20 m to the 40 m contour (11-22 fathom contours of Fig. 1).

Bathymetry.—The complex nature of the bathymetry makes interpretation difficult. The major trends are apparent, but the more subtle features are partially masked by discontinuities and bifurcation in the ridges (Fig. 1). Particularly difficult to quantify are the length and width of the mid-shelf ridges. This problem was resolved by preparing a map of deepest closed contours to define each topographic high. As a result, individual highs are isolated, permitting easy discernment of both ridge trends and interrelationships. However, an erroneous impression may be created from this method

in that individual ridges appear totally isolated when in actuality they may be a part of a larger trend. To avoid possible error, the closed contour map was constantly used in conjunction with the bathymetric map. The sideslope angles and distance between ridges (wavelengths) were determined by means of a series of cross-shelf bathymetric profiles drawn normal to the shoreline (see Fig. 2 for profile locations).

Vibracores processing.— After removal from the core barrels, the vibracores for this study were maintained in an upright position and kept at approximately 5°C. In the laboratory, the cores were transversely cut into 70 cm lengths and radiographed. Each section was then longitudinally split, immediately described for color, structure, lithology, macrofauna content, and fi-

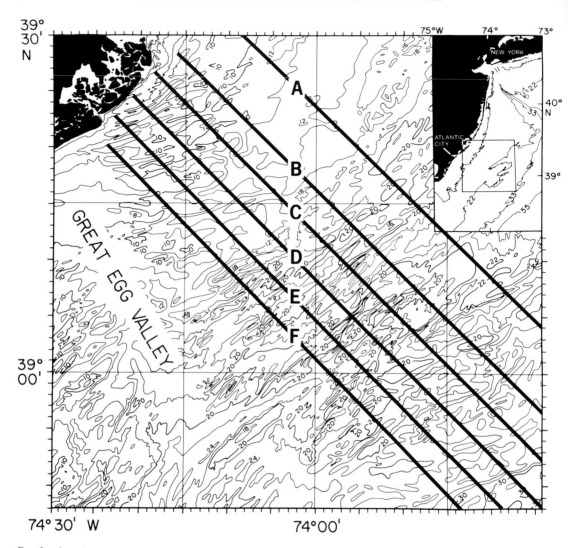

FIG. 2.—Location map of six bathymetric shore-perpendicular cross-shelf profiles. The contour interval is two fathoms.

nally photographed in color. The cores were sampled for grain size every 10 cm and at all discernible lithologic discontinuities.

Grain size analysis.—The sand-size fractions from the surficial and vibracore samples were analyzed with a Rapid Sediment Analyzer settling tube (Nelsen, 1976). The grain size distribution was determined at quarter-phi intervals.

Paleontology.—Both macrofauna and microfauna were collected from the grab samples and vibracores. Identifiable macrofauna include: the mollusks, *Mercenaria mercenaria, Crassotrea virginica, Ensis directus, Anomia simplex, Placopecten magellanicus,* and *Spisula solidissma;* the gastropod *Nassarius trivittatus;* and the sand dollar *Echinarachnius parma* (see Table 1 for partial fauna location). The microfauna consisted

principally of benthonic foraminifera with some planktonic foraminifera and nannofossils. Table 2 is a listing of all microfauna that were identified.

Spectral analysis.—Spectral analyses, in the form of Fourier transforms, were used to examine the spatial harmonics of the bottom topography in the cross-shelf profiles. Bathymetric input data consisted of a depth at 300 m intervals along each profile line. A least-squares method was used to remove the regional slope from the digitized data. This insured that the gradually changing depths associated with the shelf's slope did not introduce a bias to the data. The output data were: (1) amplitude in meters squared per frequency; and (2) frequency in cycles per 300 m. The frequency is a spatial term $(2\%\pi/L)$ and is a measure of the wavelength of the sand ridges. The frequency was later con-

TABLE 1.—VIBRACORE DATA

Core	Location	Water Depth (m)	Ridge Location On Shelf	Location on Ridge	Total Recovery (m)	Fauna Present
1	39°06.6'N 73°55.0'W	31.4	Middle Shelf	Crest	2.61	
2	39°05.1'N 73°55.8'W	44.8	Middle Shelf	Trough	2.4	*M. mercenaria* *C. virginica*
3	39°07.1'N 73°50.5'W	41.8	Middle Shelf	Lower Flank	4.23	*M. mercenaria* *Ensis directus* *M. mercenaria* *M. mercenaria*
4	39°20.3'N 73°51.3'W	36.3	Middle Shelf	Upper Flank	2.65	*Placopecten magellanicus* *Placopecten magellanicus*
A-6	39°20.3'N 74°11.9'W	20.1	Inner Shelf	Crest	5.75	*Spisula solidissma* *C. virginica* Shell hash *Spisula solidissma* *Anomia simplex* *Echinarchnius parma*
A-7	39°20.1'N 74°11.5'W	21.3	Inner Shelf	Upper Flank	2.2	
5	39°08.3'N 74°09.9'W	27.4	Transitional	Crest	0.97	Shell hash of *Ensis directus* *Spisula soslidissma* *Anomia simplex* *Nassarius trivittatus*
6	39°08'N 74°09.4'W	34.7	Transitional	Trough	2.5	
7	39°07.8'N 74°09.1'W	39.3	Transitional	Flank	3.85	Shell hash *Spisula solidissma* *Ensis directus* *Nassarius trivittatus*
8	39°07.4'N 74°08.5'W	25.6	Transitional	Crest	3.98	
9	39°09.6'N 73°55.5'W	48.46	Middle Shelf	Trough	5.82	Shell fragments *Yoldia* *Spisula solidissma*
10	39°04.5'N 74°11.9'W	44.5	Middle Shelf	Trough	3.36	Fragments of *Spisula solidissima* Shell hash

verted to horizontal distances. For each bandwidth a 95% confidence interval was determined.

Cluster analysis.—Ridge azimuths and length/width ratios, taken from the closed contour map, were examined by cluster analysis in a method similar to Tiezzi *et al.* (1979). These variables are assumed to reflect in-trinsic differences resulting from processes of formation and modification. To ensure a statistically valid sample size, the study area for this portion of the work was expanded to include an area from the coast to the 45 m (25 fathom) contour, from Great Egg Valley northward to approximately 40°N (Fig. 1).

TABLE 2.—RELATIVE ABUNDANCE OF MICROFAUNA (VIBRACORES 6, 7, AND 9)[1]

Depth in centimeters from top of core	Benthic Foraminifera										Planktonic Foraminifera			Nannofossils			
	Eliphidium clavatum (Cushman)	Cibicides lobatulus (Walker & Jacob)	Bucella frigida (Cushman)	Pseudopolymorphina novangliae (Cushman)	Bolivina pacifica (Cushman & McCullock)	Nonionella labradrorica (Dawson)	Virgulina fusiforms/loeblichi (Williamson)	Islandiella islandica (Norvang)	Quinqueloculina artica (Cushman)	Quinqueloculina stalkeri (Loeblich & Tappan)	Globigerina bulloides	Globigernia pachyderma	Globorotalia crassaformis	Braarudosphaera bigelowi	Coccolithus pelagicus	Gephyrocepsa oceanica	Reworked Cretaceous species
Core 9																	
30-32	C													VR	VR		
60-63	C						C		VR								
144-146	C					R	VR					VR	VR				
224-226	A	R				R	VR				VR						
314-316	A	R				R		VR				VR					
394-396	A	R	VR	VR		R		R	VR		C	VR			VR	VR	R
Core 7														Barren			
260-262	R	VR				VR	VR		VR			VR					
262-264	R	VR				VR	VR		VR		VR	VR					
278-280	R					VR					VR	VR					
297-299	R	VR				VR	VR		VR			VR					
Core 6														Barren			
110-112	R																
120-122	R																
130-132	R																
243-245	R																
255-257	R																

[1]See Table 1 for vibracore position.   A = Abundant   C = Common   R = Rare   VR = Very Rare

RESULTS

Bathymetry

CLOSED CONTOUR MAP

The trend and characteristics of individual ridges are more apparent on the closed contour map (Fig. 3) than on a bathymetric map. The shoreface-connected and nearshore ridges, all in water depths less than 20 m, intersect the coast at 15°-30°. Seaward of these shoreface-connected ridges is a large complex of ridges that are approximately coast-parallel. These latter ridges, which are shown in stippled pattern in Figure 3,

are in water depths of 20-40 m. Seaward of these mid-shelf ridges is yet another ridge set, in water generally deeper than 40 m, which intersects the mid-shelf ridges at angles comparable to the angle of intersection of the nearshore ridges with the present coastline.

Of minor importance on the mid-shelf are a few arcuate ridges. These ridges, on the northern flank of Great Egg Valley, probably formed as large estuary mouth shoals (McKinney et al., 1974) and are thought to be anomalous when compared with the rest of the ridges in the study area. As a consequence, no future discussion will be directed toward these arcuate ridges.

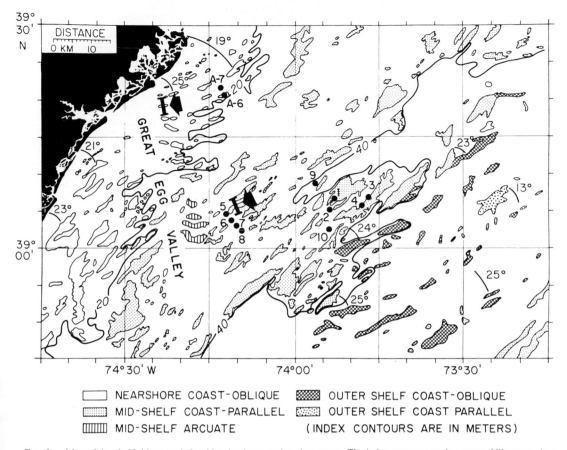

NEARSHORE COAST-OBLIQUE   OUTER SHELF COAST-OBLIQUE
MID-SHELF COAST-PARALLEL   OUTER SHELF COAST PARALLEL
MID-SHELF ARCUATE   (INDEX CONTOURS ARE IN METERS)

FIG. 3.—Map of the shelf ridges as defined by the deepest closed contours. The index contours are in meters. Vibracore sites are marked with a solid circle; two grab sample transects are shown by a solid line.

Unique to the outer portions of the mid-shelf and the inner portions of the outer shelf are a few closed contours with dual elongation directions. In these instances one azimuth corresponds with the stippled mid-shelf coast-parallel contours and the other with the more seaward, coast-oblique, cross-hatched contours of Figure 3. Examples of closed contours with dual elongation directions are found at 39°07'N, 73°40'W and 39°12'N, 73°30'W. The closed contours with dual azimuths plus the close proximity of both the single azimuth mid-shelf and outer shelf ridges results in the bimodality observed in areas B and E of Figure 4.

In addition to the contrast in azimuths, the ridge sets differ in general appearance. Based on closed contours, the mid-shelf coast-parallel ridges are generally more ragged in outline than are either the shoreward or seaward coast oblique ridges. The ragged appearance is especially apparent in the central portion of the mid-shelf ridge set.

Analysis of the closed contour data by cluster analysis supports the distributions of orientations and appearances recognized visually. When the length/width ratio is used in conjunction with the elongation azi-

muth for individual ridges, the mid-shelf coast-parallel ridges form a group distinct from the other ridge sets (Fig. 5). Of the 23 ridges from the mid-shelf, 19 were grouped by cluster analysis as a single population in terms of length/width ratio and azimuths. Similarly, 8 of the 12 nearshore coast-oblique ridges group together. But included in this nearshore grouping were two ridges from the outer shelf and one mid-shelf ridge. Thirteen ridges were visually identified as outer shelf coast-oblique in Figure 3. Nine of these outer shelf ridges, along with two nearshore coast-oblique ridges, were grouped together by cluster analysis. Interestingly, the mid-shelf coast-parallel and outer shelf coast-oblique ridges were the end members of the cluster analysis with the nearshore coast-oblique ridges being transitional. The nearshore coast-oblique ridges showed a strong bias toward the outer shelf coast-oblique grouping.

Cross-Shelf Profiles

Break-in-slope.—Certain topographic similarities are found in each of the shore-normal cross-sectional profiles (Fig. 6). For example, each profile shows two

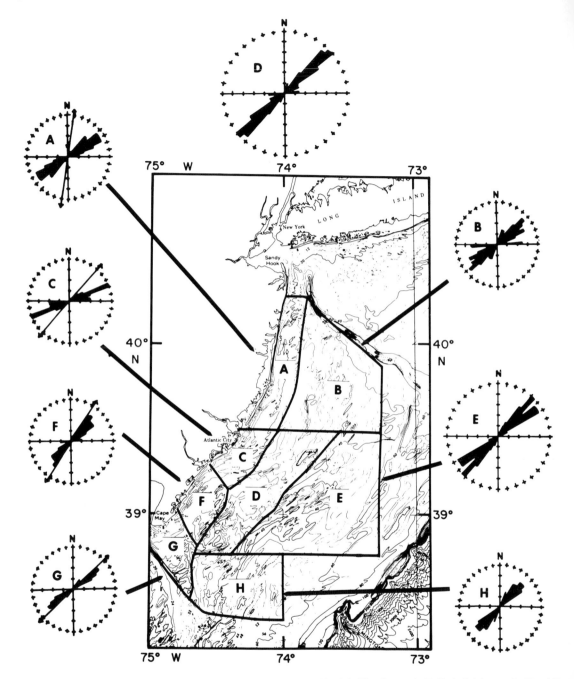

FIG. 4.—Rose diagrams of the ridge orientations in eight sectors (A-H) of the New Jersey shelf. Each division on the X and Y axes represents a value of two. The compass rose is divided into 10° units. The trend of the coastline for the nearshore sectors (A, C, F, G) is shown as a thin line. The study area discussed herein includes parts of sectors C, D and E.

breaks-in-slope. These breaks-in-slope divide the profiles into segments that closely correspond to the three ridge sets seen on the closed contour map. Additionally, the middle segment appears to possess a subtle break-in-slope, more obvious on some of the profiles than on others.

On all the profiles there is a difference in slope between the three major line segments. The seaward and

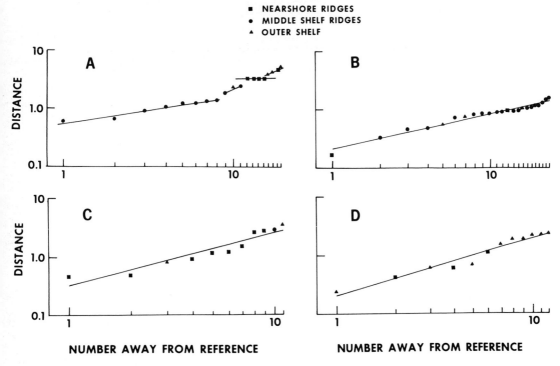

- ■ NEARSHORE RIDGES
- ● MIDDLE SHELF RIDGES
- ▲ OUTER SHELF

NUMBER AWAY FROM REFERENCE   NUMBER AWAY FROM REFERENCE

FIG. 5.—Data utilized in cluster analysis using the orientation and length/width ratios of closed contours on the New Jersey shelf. The analysis involves plotting euclidean distances from a reference value vs. the sample rank, from closest to farthest away from the reference point, on a log-log scale. The reference value used in plot A was an initial best-guess selection. The reference values for plots B-D were obtained from each of the three major line segments in panel A. A, Initial grouping into three populations. Due to the restriction in memory of the cluster analysis program (Tiezzi *et al.*, 1979), only a few of the ridges in each of the three shelf areas were used for this initial plot. Each population is identified with ridge type based on qualitative observations. N = 19. B, Grouping for mid-shelf coast-parallel ridges in panel A. N = 23. Grouping of all ridges with values statistically compatible to a reference point obtained utilizing least-square means and standard deviations. C, Grouping of ridges compatible with a reference point for the nearshore coast-oblique ridges of panel A. N = 11. D, Grouping of ridges compatible with a reference point for the outer shelf coast-oblique ridges of panel A. N = 12.

shoreward segments have comparable slopes that are greater than the slope of the middle line segment.

Spatial harmonics.—In each profile, a large spatial wavelength dominates throughout all segments except the nearshore. Superimposed on the dominant larger-scale ridges, especially in the middle line segment, are smaller, closely spaced ridges. Closely spaced ridges are observed in the very nearshore, however, the amplitude and spatial frequency of these nearshore ridges increase southward as evidenced by profiles A through F (Fig. 6).

To better evaluate the distribution of the various orders of ridge wavelength, spectral analysis was used. Fourier transform analysis of the spatial harmonics for the total line length of profile D was compared to similar analyses of the three line segments separated by breaks-in-slope. Each of these smaller line segments was then compared one to the other. Each line segment was found to be dominated by ridges with wavelengths in excess of 6 km (Fig. 7). However, the amplitude of the spectral energy for the topography represented by

the middle line segment (stippled pattern of Fig. 3) was twice that for the area represented by the other two line segments. The amplitude for the inshore and offshore segments is similar.

In addition to the larger ridges, each line segment represents areas possessing ridges with a horizontal wavelength of approximately 1,500 m. But as in the case of ridges with wavelengths in excess of 6 km, the amplitude of the spectral energy for the mid-shelf ridges is twice that for the inshore ridges and nearly an order of magnitude greater than the offshore ridges. The ridges in the nearshore segment possess a wavelength of approximately 1,500 m, with the middle portion decreasing to 1,475 m and the offshore segment being 1,350 m.

Figure 7 also indicates ridges with a smaller wavelength. This higher frequency approaches the limit of sampling resolution; thus, any inference to the meaning awaits additional closely gridded sampling.

In addition to the spectral analysis, the area represented by the three line segments can also be differenti-

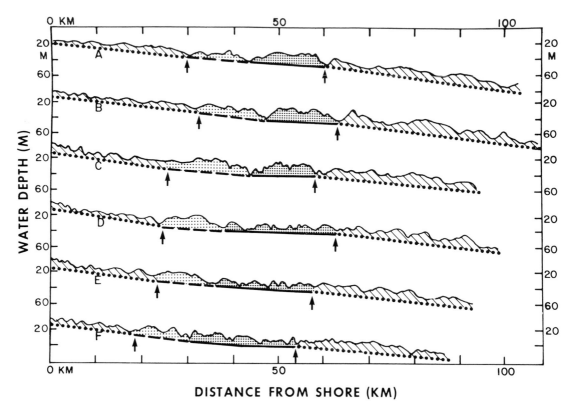

DISTANCE FROM SHORE (KM)

Fɪɢ. 6.—Cross-sectional view of profiles A through F of Figure 2. The nearshore and outer shelf coast-oblique ridges are shown in hatched pattern underlain by dots; the mid-shelf coast-parallel ridges are shown in stippled pattern as two segments underlain by solid and dashed lines. Two pronounced breaks-in-slope are observed on each line (arrows); steep seaward-facing escarpments are observed on most lines at the break-in slope. The lines or dots below the pattern are parallel to a line connecting the lowest lows, but are offset to allow placement of the stippled or dashed pattern.

ated by averaging the wavelength or spacing of all the ridges. All the ridges along profiles A through F indicate an average of 3.4 km wavelength for the nearshore ridges, 2.4 km wavelength for the mid-shelf coast parallel ridges, and 4.1 km wavelength for the outer shelf coast-oblique ridges (Fig. 8).

Seaward-facing escarpments.—Coincident with the seaward edge of the mid-shelf coast-parallel ridges is a steep seaward-facing escarpment (Fig. 6). The escarpment is especially pronounced on profiles A and C, and would be on B except for the anomalously high ridge immediately seaward of the middle line segment. The escarpment, however, is less pronounced or missing entirely from the three southern profiles.

A more shoreward, seaward facing escarpment is also found at the small break-in-slope within the middle line segment (stippled area of Fig. 6). This second escarpment is most pronounced on profiles A, B, D, and E.

Texture

Vibracore samples.—Two textural patterns are found in the cores. Vibracores from the nearshore ridges (A-6 and A-7; Fig. 9) consist of sand with a

coarse to medium-fine mean size with only a slight upward coarsening. The mean size of core A-7 coarsens upward from a medium-fine sand (2.4φ) to a medium sand (1.3φ) over a core length of 220 cm. Similarly, core A-6 coarsens upward, but only in the upper 240 cm; for over half of the 570 cm core the texture is a uniform medium sand (1.8φ).

The mid-shelf coast-parallel ridge cores are more pronounced in their upward coarsening. Their mean size is finer-grained, being generally between 2φ and 3φ (Fig. 9). Characterizing many of the mid-shelf cores is an upper portion, usually less than one meter thick, which is anomalously coarse-grained. In this upper portion, numerous coarsening and fining events occur. Such an example is core 7, where a sharp contact is found at 50 cm. Here, a coarse-grained, oxidized brown sand, which fines upward, unconformably overlies a massive, fine-grained, greenish-gray sand.

Surficial sands.—Like the texture of the sediment in the vibracores, the surficial sands are characterized by two different mean grain size distribution patterns. The distribution of grain size of the surficial sands on the

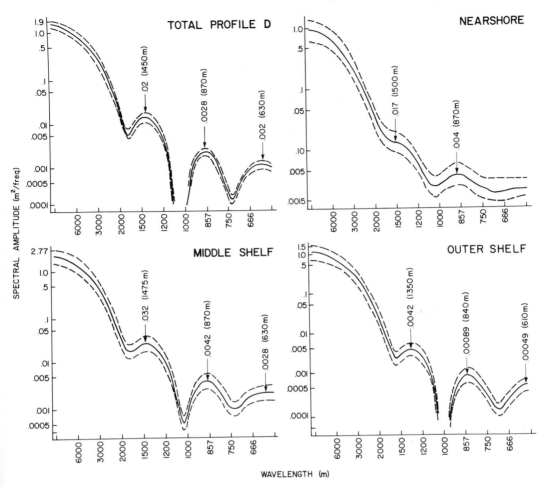

Fig. 7.—Spectral analysis of profile D (see Figure 2 for location). Wavelength spectra were determined by Fourier transform analysis for the total profile, the nearshore, middle shelf, and outer mid-shelf ridges of Figure 6. The vertical axis is the spectral amplitude of energy and the horizontal axis is the wavelength of the ridges in meters. The dashed line indicates an envelope of 95% confidence interval. Arrows mark peaks in the signal strength at various wavelengths.

nearshore ridges is asymmetrical: the sediment is coarsest on the lower shoreward flank and decreases in size across the crest to the lower seaward flank (Fig. 10). The trough samples are varied; their mean grain size ranges from quite coarse to fine sand size. In contrast, the middle shelf coast-parallel ridges are characterized by a more symmetrical size distribution. The sands on the upper shoreward flank are the coarsest with a fining toward both troughs.

Fauna.—Both the nearshore coast-oblique and mid-shelf coast-parallel ridges contain shell fragments of macrofauna typical of shallow marine water (Table 1). The only Holocene-aged material of a species restricted to a lagoon is *Crassostrea virginica* (Galtsoff, 1964; Emery and Merrill, 1979). This sample was collected in core A-6, which is from a nearshore coast-oblique ridge. The fragmented nature of the sampled *C. virginica* suggests it may have been exhumed from

an underlying lagoonal substrate and subsequently mixed with fragments of normal marine fauna. The presence of large fragments of *Spisula solidissma* implies deposition in 3-21 m of water for these nearshore coast-oblique ridges. Similarly, the presence of *S. solidissma* and *Placopecten magellanicus* implies a nearshore depostion for the mid-shelf coast-parallel ridges. Both of the pelecypods *S. solidissma* and *P. magellanicus* are common to shallow-water, open-marine environments.

A similar nearshore environment for the mid-shelf ridges is suggested by the microfauna. The microfauna found in two of the mid-shelf ridge cores (Table 2) are generally typical of sandy nearshore zones of normal marine salinity. Two species, *Elphidium clavatum* (Cushman) and *Quinqueloculina artica* (Cushman) have been reported in lagoons or tidal marshes, but only in areas where there was normal marine or hyper-

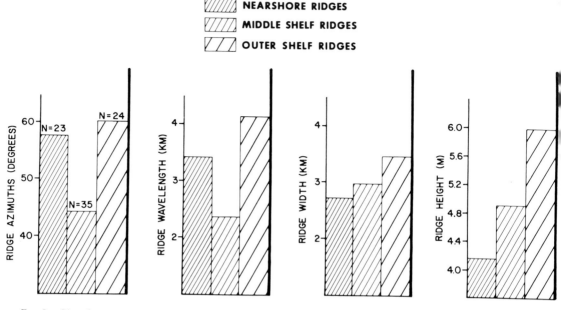

Fig. 8.—Plot of average ridge orientation (azimuth), wavelength (spacing), width, and height for the nearshore, middle, and outer shelf ridges. The values were obtained from the bathymetric profiles of Figure 6.

saline waters (Murray, 1973). Unfortunately, none of the microfauna provides depth restraints of adequate resolution to establish a maximum depth for depositional environment. The most common microfauna in Table 2, *Elphidium*, are living today on the more seaward portions of the mid-shelf ridge complex (Poag *et al.*, 1980). Poag *et al.*'s study did not extend to the shoreward portions of the mid-shelf ridges or the nearshore ridges, but based on Murray's (1973) work, *Elphidium* should be in these areas as well.

DISCUSSION

Ridge Variability

Nearshore coast-oblique ridges.—The nearshore coast-oblique ridges, including the shoreface-connected ridges, average 4.2 m in height, 2.7 km in width, and 3.4 km in wavelength (Fig. 8). These ridges, with an average orientation of 58°, intersect the coastline at an average of 25° with the standard deviations of 13° and 5.1°, respectively. Cores from these nearshore ridges indicate that the ridges internally consist of sand having a mean grain size ranging from medium-fine (2.4φ) to coarse-grained (0.8φ). The vertical sequence is characterized by uniform texture in the lower 300 cm of core A-6, but with upward-coarsening in the upper 260 cm of this core and in core A-7 (Fig. 9). Also common in the upper portions are numerous events of coarsening and fining. The surficial samples, however, are more varied. The surficial sands grade from coarse (0.4φ) on the shoreward flank to fine (2.6φ) on the seaward flank (Fig. 10).

Outer shelf coast-oblique ridges.—Unfortunately, the data for the outer shelf coast-oblique ridges (Fig. 3) are inadequate to permit total comparison to the other two ridge sets. However, where the data are available, the outer shelf coast-oblique ridges appear similar to the nearshore coast-oblique ridges. The average orientation for these ridges is within four degrees of the nearshore ridges. With a similar average width and wavelength, only the height is appreciably different between these two ridge sets (Fig. 8). The ridge's height systematically becomes greater as the water deepens toward the offshore.

Mid-shelf coast-parallel ridges.—The middle shelf ridges (stippled pattern in Fig. 3) have different characteristics compared to the nearshore coast-oblique ridges, and presumably, from the outer shelf coast-oblique ridges as well. The most obvious difference is in orientation. The mid-shelf ridges are approximately coast-parallel, thus their trend differs from the other two ridge sets by 15° to 30°. The ridge width and height are intermediate between the nearshore and the outer ridges (Fig. 8), implying that both ridge width and height increase with water depth. The wavelengths of the ridges, however, are apparently a result of factors other than increase in water depths. The middle shelf ridges have an average wavelength of 2.3 km, substantially less than that of the other two ridge sets. Spectral analysis of the spatial harmonic indicates a subtle trend of a first-order oscillation in the ridge field. For ridges with a wavelength in excess of 6 km, the amplitude of spectral energy is twice that of the other ridge sets (Fig. 7). A similar variance is found for the ridges of approximately 1.5 km spacing. This indicates that the larger-scale ridges are more common on the mid-shelf. Superimposed on the large-scale, mid-shelf ridges is a higher-frequency, second-order ridge system.

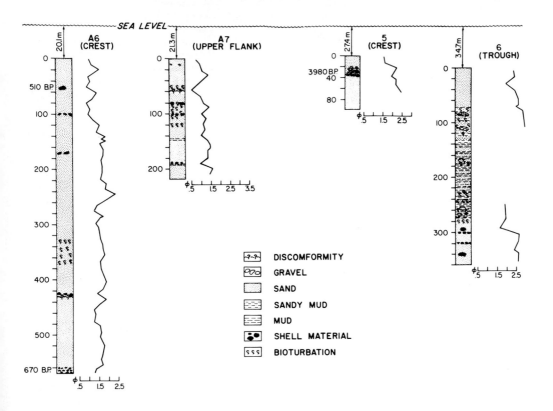

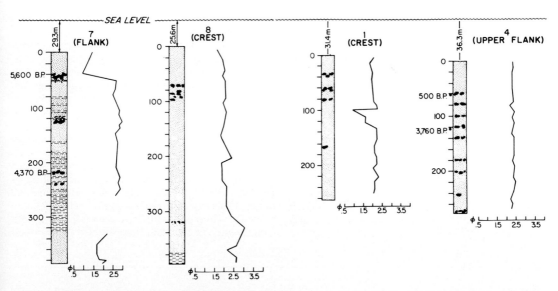

FIG. 9.—Lithic log and mean grain size of eight of the vibracores. Cores A-6 and A-7 are from a nearshore coast-oblique ridge; the other cores are from the mid-shelf coast parallel ridges. See Figure 1 for core site locations and Tables 1 and 2 for fauna content. Vertical scale is in centimeters. Mean grain sizes in quarter-phi units are plotted every 10 cm for the sand horizons. Information for cores 1 and 4 from Stubblefield *et al.*, 1975.

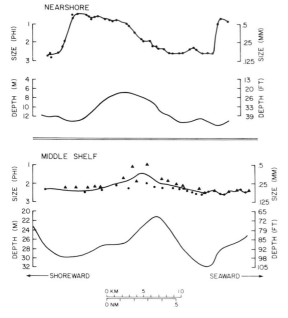

FIG. 10.—Plot of mean grain size distribution across a nearshore and a mid-shelf ridge (see Fig. 1 for transect locations). The bottom surface grab samples from the nearshore ridge were collected on a single transect; those from ther mid-shelf ridges were collected on two parallel, closely spaced transects. The sampling spacing averaged 75 m. Figure is from Stubblefield and Swift, 1981.

The mid-shelf coast-parallel ridges are also different in terms of morphology. These ridges, especially in the middle portions of the ridge set, appear ragged in outline, whereas the nearshore and outer shelf coast-oblique ridges are smoother. When the length/width ratio is compared to orientation, the mid-shelf ridges are readily separated by cluster analysis (Fig. 5). The other two ridge sets are distinguishable one from the other with cluster analysis, but less so than from the mid-shelf ridges.

Texturally, the mid-shelf coast-parallel ridges are different from the other ridges. Unlike the nearshore ridges, the surficial size distribution is approximately symmetrical with the coarsest sand being near the upper shoreward flank. Samples from the mid-shelf ridge cores indicate that these ridges are finer-grained, better sorted, tend toward more negative skewness, and may have a higher kurtosis.

In the vertical sequence, the sand coarsens uniformly upward to a surficial layer (Fig. 9). This upper portion may exceed a meter or so in thickness and is characterized by events of extreme coarsening and fining. This portion is interpreted as being a zone of mid-shelf reworking or aggradation. The recent activity is demonstrated by shell material, 0.7 m beneath the sediment surface, which is less than 500 years old (Stubblefield et al., 1975).

MULTIPLE GENESIS MODEL

Our data suggest that major differences exist between the ridges on the mid-shelf and those in the nearshore, and presumably some of those on the outer shelf. These differences indicate that either: (1) the ridges formed differently, or (2) after formation, they were differently modified. The bracketing of the mid-shelf ridges by the similar nearshore and outer shelf coast-oblique ridges suggests the former of these options. The suggestion that the ridges formed differently supports a model that the ridges formed during two stages of the last transgression, some as barriers and some as post-transgressive ridges. Such a model is in contrast to the bulk of previous investigations which have advocated that all the ridges formed in a single setting, either as barriers or as post-transgressive ridges, but not as both.

Bathymetry

Closed contour map.—The trend of the bathymetric features observed on the closed contour map is one line of evidence supporting a dual time of ridge formation. The differences between the trends of the mid-shelf coast-parallel ridges and the nearshore and outer shelf coast-oblique ridges suggest dual origins and possibly two times of formation for the ridges in the study area. Furthermore, the intersection angle between these three ridge sets and the coastline provides constraints for their time of formation. This intersection angle implies that if the nearshore ridges formed as shorelines during the Holocene transgression, the coast has undergone major realignment in the past few thousand years (Uchupi, 1968; Duane et al., 1972). By similar argument the difference in orientation between the nearshore coast-oblique and the mid-shelf coast-parallel ridges suggests that the mid-shelf coast-parallel ridges did not form as coast-oblique, shoreface-connected ridges. Otherwise, the paleo coastline, when the mid-shelf ridges formed, would have been oriented 15°-30° more westerly than today. However, as D. Swift (per-

FIG. 11.—Schematic representation for developing multiple-order wavelengths between ridges on the mid-shelf. During Sea Level Time 1 lagoonal deposits are deposited over barrier sands, behind the dunes, as part of a prograding barrier. At Sea Level Time 2, or during the early stages of marine transgression, the upper portions of the barrier and the lagoonal deposits are removed leaving an irregular erosional surface. As the water deepens through continual transgression, the erosional surface is modified. This modification is probably in the form of trough deepening and ridge aggradation as suggested by Stubblefield et al. (1975) and Stubblefield and Swift (1976), Previous bottom topography is shown as a dashed line.

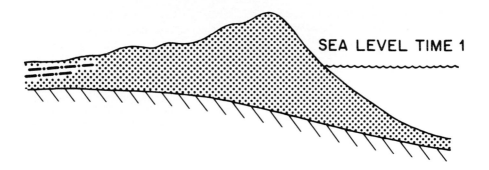

SEA LEVEL TIME 1

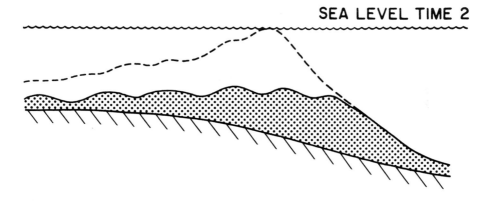

SEA LEVEL TIME 2

SEA LEVEL TIME 3

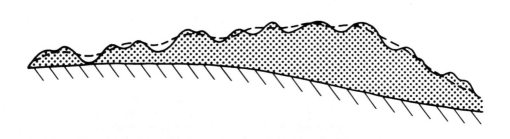

BOTTOM TOPOGRAPHY   BARRIER SANDS

PREVIOUS
BOTTOM TOPOGRAPHY   LAGOONAL MUDS

                                      PLEISTOCENE
                                      SUBSTRATE

sonal communication, 1980) points out, in places, the nearshore ridges are forming as coast-parallel features, rather than a 15°-30° oblique angle to the coast. Such rare examples are found off Cape May (New Jersey), Assateague (Maryland), and Virginia Beach (Virginia).

Also, Swift *et al.* (1972) and McKinney *et al.* (1974) suggest that certain mid-shelf ridges, specifically those in this study area, developed during earlier stages of the Holocene transgression. They note two scales of ridges, both having developed either totally or to a large measure in response to the hydraulic regime of the inner shelf. They suggest the larger or first-order ridge system (see Fig. 2a of McKinney *et al.*, 1974) perhaps inherited certain dimensions and spacings from an older topographic fabric, but was modified by the Holocene tidal regime while adjacent to a major estuary. They also suggest that the smaller-scale or second-order ridges were suggested to have formed in slightly deeper water subsequent to the first-order ridges (McKinney *et al.*, 1974).

The closed contour map (Fig. 9) differentiates between ridges based on orientation, however, it fails to differentiate between the suggested ridge orders as determined by dimensions and spacings. The two ridge trends seen in the closed contour map are not unlike the orientations of the first-order ridge systems of McKinney *et al.* (1974; see their Fig. 2a). The closed contour type of map should provide constraints for evaluating the models of post-transgressive ridge evolution as either coast-parallel shoreface-connected ridges (D. Swift, personal communication, 1980) or part of a shoal-retreat massif (Swift *et al.*, 1972; McKinney *et al.*, 1974).

Seaward of the mid-shelf coast-parallel ridges is the suite of coast-oblique ridges (Fig. 3). These seaward coast-oblique ridges are approximately parallel to the nearshore coast-oblique ridges and, as such, intersect the mid-shelf coast-oblique ridges at 15-30 degrees. If the mid-shelf ridges are abandoned shoreface-connected ridges, a mechanism must be invoked to change them from coast-oblique to coast-parallel in a very short lateral distance. Likewise, if all the mid-shelf ridges are parts of shoal retreat massifs, it is difficult to explain the two orientations.

Model for Ridge Formation

The orientation of the various ridges is a basis for a model for ridge formation. The intersection angle between the outer shelf coast-oblique ridges and the mid-shelf coast-parallel ridges is similar to the intersection angle betwen the near-shelf coast-oblique ridges and the coastline. This similarity prompts the suggestion that the outer shelf coast-oblique/mid-shelf coast-parallel ridge complex is an earlier Holocene analog to the modern nearshore ridge/coastline complex. This interpretation infers that the mid-shelf ridges are a part of a submerged barrier beach complex and that the outer shelf coast-oblique ridges were formed by processes similar to the modern shoreface-connected ridges.

Cross-Shelf Profiles

The cross-sectional profiles of the shelf ridges from the nearshore through the outer shelf ridges (Fig. 6) provide additional data regarding the ridge development. By connecting the topographic lows the profiles can be subdivided into four line segments. The two middle line subsegments are under the mid-shelf ridges; the break-in-slope between these two lines is much less than for the inner and outer breaks-in-slope. If the two middle line segments are grouped as one, which is reasonable, the profiles are characterized by three distinct line segments. This method divides the ridges in the study area into groups similar to those of the closed contour map (Fig. 3).

The portion of the shelf that is defined as mid-shelf coast-parallel ridges is characterized by at least one large-scale features. This is most apparent on the three northern profiles, A, B and C (Fig. 6), but is also noticeable on the three southern profiles. Superimposed on these large bathymetric features are the smaller-scale ridges. These large- and small-scale ridges are not unlike the first- and second-order ridge system of McKinney *et al.* (1974).

Shoreward of the mid-shelf, in the nearshore, the large-scale ridges are absent. Here the ridges are smaller, being comparable in size to the second-order ridges of the mid-shelf.

Seaward of the mid-shelf coast-parallel ridges, a consistent pattern of ridge distribution is more difficult to recognize. On the northern profiles, A and B, and on profile D, adjacent to the coast-parallel ridges, ridges of a size comparable to the nearshore ridges exist. These are bordered to seaward by ridges of a larger scale. These larger-scale ridges are not capped by the smaller-scale ridges as are the large-scale, coast-parallel mid-shelf ridges. Profile C very subtly shows a similar pattern, but the distinction between the smaller- and larger-scale ridges is less than that in profiles A, B and D. The two southernmost profiles, E and F, show the larger-scale ridges.

The lack of a definitive distinction between the large- and small-scale ridges in the offshore portion of the study area is not surprising when the closed contour map (Fig. 3) is critically examined. Based solely on ridge orientation, ridges with either a coast-parallel or a coast-oblique orientation are found on the outer shelf. As is observed between the nearshore and mid-shelf ridges, these outer shelf coast-oblique and coast-parallel ridges are in zones aligned approximately parallel to the shoreline. The distribution of both coast-oblique and coast-parallel ridges on the outer shelf suggests a series of small paleobarriers with their related shoreface-connected ridges which were overstepped by the Holocene transgression. The imprint of this relationship between the ridges with these two orientations has not been completely obscured by the subsequent nearshore and mid-shelf currents.

Spectral analysis.—A quantitative examination of the ridge spacings does not exclude any model for ridge formation on the New Jersey shelf. However, the tendency for a wavelength of 6 km to be greater in the mid-shelf coast-parallel ridges than in the nearshore coast-oblique ridges (2.77 versus 1.0 m²/freq.) is striking. The scaling of these large-scale ridges is comparable to the wavelength of the subaerial Pleistocene barriers on the South Carolina coastal plain (Colquhoun, 1969). As such, the large-scale wavelength on the mid-shelf is most compatible with a model of submerged barriers.

Also more common to the mid-shelf coast-parallel ridges are ridges with a spacing of approximately 1.5 km (Figure 7). The shorter-wavelength ridges, seen as the superimposed ridges in Figure 6, are easily explained by erosion of the upper portions of the barrier during transgression followed by later aggradation on the remnant barrier (Fig. 11). The ragged outline of these mid-shelf ridges in the closed contour map also suggests erosion at some stage of ridge development.

Elevation differences in the mid-shelf ridges.—The cross-sectional profiles of the mid-shelf ridges give an impression of a submerged barrier system (Swift *et al.*, 1972). But, Swift *et al.* argue against the barrier concept principally on the relative heights of adjacent large-scale ridges. They invoke Gilbert's (1880) model of a staircase sequence of barrier development during a marine transgression. In Gilbert's model, the crest of each barrier should be at the altitude of the toe of the adjacent landward barrier. This vertical relationship is valid only if there has been no erosion or the amount of erosion was equal throughout the barrier complex. Textural evidence, as will be discussed later, suggests that erosion has removed the upper and probably most of the middle shoreface of the mid-shelf shore-parallel ridges. Submersible dives (McKinney *et al.*, 1974) and seismic profiles (Stubblefield and Swift, 1976) indicate outcropping of late Pleistocene or early Holocene mud units on the lower flanks of the deeper mid-shelf troughs. Thus the erosion is subsequent to the deposition of the sand over the mud material. Some measure of time for this erosion is possible from radiocarbon dates of fauna obtained by Stubblefield *et al.* (1975) These ages indicate that as much as 0.6 m of aggradation has occurred in the past 500 years B.P. A piece of coal, probably from a steamship, found 0.3 m beneath the surface of core 7 also indicates recent ridge activity. Stubblefield *et al.* (1975) suggest that the source of this recently reworked sand is probably from the trough, suggesting that some erosion of the troughs is still occurring, albeit probably at a much reduced scale than what occurred shortly after submergence. This trough erosion and the adjacent ridge aggradation would account for the elevation differences betwen the "toe" of the barrier and the adjacent seaward barrier system. Thus Gilbert's (1880) stairstep criterion is not violated.

Perhaps a more meaningful measure is the difference in elevation between the adjacent large-scale mid-shelf ridges. On profiles A-F (Fig 6) the shoreward of the two large-scale ridges (Stippled patterns in figure) varies in elevation from 1 to 9 m higher than the seaward ridge. Even the lesser of these elevation differences is significantly greater than expected when considering only the regional slope of the shelf. If a value of 0°3' is used for the shelf's declivity (Uchupi, 1968), the maximum elevation difference from the shelf's gradient, over the horizontal distances between the large-scale mid-shelf ridges, is 0.07 m. A comparison between the shelf gradient and the ridge elevation differences is schematically shown in Fig. 12. The 1 to 9 m difference in height is too great to be associated with the shelf's gradient, but it is compatible with two separate but adjacent paleobarriers.

Texture

Textural variations in the ridges also offer some indication as to ridge genesis and hence the site of ridge formation relative to the marine transgression. The mid-shelf coast-parallel ridges are characterized by: (1) an upward coarsening, (2) subtle grain-size coarsening and fining trends, and (3) zones of thinly interlayered mud and fine sand. These textural trends are contrasted to ridges interpreted as post-transgressive. In the latter ridges, the vertical grain size gradient is more diverse, the coarsening and fining events are frequent and more pronounced, and the interlayered mud and fine sand horizons are rare.

Vertical grain size gradients.—Houbolt (1968) suggested that the modern tidally formed ridges of the North Sea possess a uniform grain size gradient throughout the vertical sequence. A more diverse grain size gradient is found in other post-transgressive

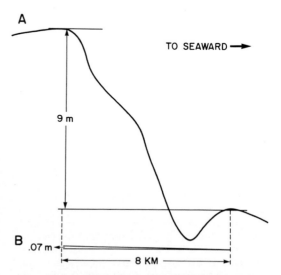

FIG. 12.—A, Line drawing of the maximum height differences between two adjacent mid-shelf ridges on profile D (Fig. 6). B, Elevation difference due to the shelf's slope over a horizontal distance equal to the horizontal distance between ridges in part A. The vertical scale is the same in A and B.

ridges. For example, in the cores from the near-shore coast-oblique ridges in this study, both uniform and upward-coarsening textures are found (Fig. 9). The lower 330 cm of core A-6 have a uniform mean size in the vertical sequence, whereas the upper 240 cm of this core and the entire core A-7 have a systematic upward coarsening. In a similar study of 21 cores from Assateague, Maryland, Field (1976) describes approximately even distribution between cores of upward-coarsening, upward-fining, and uniform vertical grain size.

The upward coarsening and fining observed in this and Field's (1976) studies may be a function of the ridges' mobility. As noted initially by Moody (1964) and later by, among others, Swift and Field (1981) and Stubblefield and Swift (1981), the grain size is usually out of phase with topography on the nearshore ridges. The coarsest sand is on the lower shoreward flank, with a general fining of grain size over the crest to the seaward flank. Due to the grain size distribution, the shoreward flank is interpreted as the up-current side. With an accreting but otherwise stationary ridge, the mean grain size is expected to be basically uniform in vertical mean grain size gradient with discrete coarsening and fining occurrences. However, with ridge migration, the size distribution may either fine or coarsen upward, depending on whether or not the bathymetry and grain size are in phase (Figueiredo et al., 1982). Such ridge migration has been documented for the nearshore ridges by Moody (1964), Duane et al. (1972), and Field (1976).

Occurrences of coarsening and fining.—Both the mid-shelf coast-parallel and the nearshore coast-oblique ridges have zones of coarsening and fining. The differences betwen ridge sets is in the frequency and magnitude of these occurrences. In the nearshore ridges these occurrences, which are probably related to discrete storm events, are numerous and have a mean size range of 0.4 to 1.1 phi (Fig. 9). The mid-shelf ridges appear to have two genetic populations. The upper one meter or so contains storm events similar to the nearshore ridges. Beneath this upper zone the variations (storm events) are much fewer and more subdued. The deflection range of this lower zone is only 0.1 to 0.2 phi (Fig. 9). From ages obtained from radiocarbon dates (Stubblefield et al., 1975), the upper portion of the mid-shelf ridges is interpreted as being recently aggraded, some of which has been quite recent. In contrast, the grain size gradient of the lower portion is believed to be a result of initial formation.

Interbedded mud and fine sand .—The common occurence of zones of thinly interbedded mud and fine sand in the mid-shelf ridges, and the virtual absence of such zones in the nearshore ridges, provide more contrast between the ridge sets. Similar mud horizons are found in other mid-shelf ridge cores (R. Tillman and J. Rine, Cities Service Oil Co., unpublished data).

Any model for the genesis of the ridges must consider the mud within the mid-shelf ridges. A similar interlayering of mud and fine sand was observed by Swift et al. (1978) in the Platt Shoal ridges on the North Carolina shelf. They only observed this interlayering in an intermediate horizon that overlaid a Pleistocene substrate and which was overlain by medium to coarse, well sorted sands. These two horizons prompted Swift et al. to propose that these ridges formed in stages, first as part of a lunate estuary mouth shoal as described by Howard and Reineck (1972) and Oertel and Howard (1972), and later as part of a shoal retreat massif (Swift et al., 1972).

But, if as suggested by the previously presented data the mid-shelf coast-parallel ridges are remnants of a former coastline, the muds and fine sands may have been deposited as discrete events in either the backbarrier or the shoreface. Of the two environments, the backbarrier or lagoon is perhaps the easiest to envision. Swift et al. (1973) proposed a model involving lagoonal sediments for the New Jersey mid-shelf ridges; they suggested a thick accumulation of lagoonal deposits. In this model the sands were considered to have been introduced over lagoonal deposits as washover during storms. Between the washovers, the lagoonal silts and clays would have hosted a variety of benthic organisms that reworked the sediment. The sedimentary structures of the fine sediment, not visible in the core samples, may have been destroyed by burrowing of organisms or an artifact of coring operations.

The presence and/or lack of certain biogenic components within mud layers sampled from mid-shelf ridges indicate to the authors that the layers were deposited on an open shoreface. Evidence that is supportive of this environment is twofold: (1) scarcity of rooting or plant structures in the mud layers, and (2) abundance of open-marine macrofauna and microfauna as compared to the absence of fauna known to be restricted to lagoons. First discussed is the lack of rooting and plant material. The cores examined for this study, or those from the same area examined by Cities Service Oil Company (R. Tillman and J. Rine, personal communication), contain no evidence of rooting, and only once were plant fragments found. The plant material was in a single, very thin mud lamina within thick deposits of sand. A similar lamina of mud with plant fragments was found in a core from the nearshore coast-oblique ridges. Obviously negative evidence is tenuous, but among the 11 cores examined in these two studies, rooting and/or plant material should have been more in evidence if a paleolagoon were cored, especially so since Kraft and John (1979) found abundant plant structures in their cores through a lagoonal sequence in a study of the modern Delaware barrier system.

The second piece of evidence for a foreshore rather than a lagoonal environment of the muds is paleontological. The mid-shelf ridge cores are totally lacking in macrofauna restricted to a lagoon, e.g., *C. virginica*. Instead, shell materials of *S. solidissma* and *P. magellanicus*, both of which are open-marine species, occur throughout the cores. Complementary evidence is provided by the microfauna. The salinity range of the microfauna present in the cores (Table 2) restricts the depositional environment to open marine or to lagoons with either normal marine salinity or hypersalinity (providing there was no transport of the tests outside their environment of origin).

If, as suspected, the depositional environment was the shoreface, the availability and sedimentation of clay and silts must be considered. Modern corollaries for interbedded deposition on the lower shoreface are found in the Galveston Island barrier in water depths of 8 to 10 m (Bernard *et al.*, 1962) and on Sapelo Island in water depths of 2 to 5 m (Howard and Reineck, 1972). But under present conditions, mud is not accumulating on the shoreface of the New Jersey barriers (D. J. P. Swift, personal communication). This is not to say, however, that during late Pleistocene or early Holocene, fine sediments were not being introduced to the nearshore in quantities comparable to the present Georgia and Texas coasts. During this time the glacial deposits were conceivably still rich in fine sediments that were available for winnowing, transport, and eventual incorporation into the nearshore sediments. The actual input of the fine sediment into the nearshore may be from estuary bypassing during stratified conditions (Meade, 1972a), reworking of shelf sediments (Hathaway, 1972; Meade, 1972b; Pevear, 1972) or erosion of Pleistocene outcrops (Pevear, 1972). An example of the former method is observed in the modern Delaware River estuary. The fine sediments move seaward through the central portion of the estuary (Oostdam and Jordan, 1972) and are transported northward along the New Jersey coast (as observed in satellite imagery).

Deposition of fine sediment in the nearshore is not without controversy. Meade (1972b) questions such deposition, contending instead that the suspended material will be eventually transported to the backbarrier for deposition. Doyle *el al.* (1968), however, argue for nearshore deposition at least along the coasts of southeastern U.S. Oertel and Howard (1972) cite evidence for similar nearshore deposition on the inlet-associated shoals of Georgia. From a theoretical approach, McCave (1972) proposed a model for deposition of fine sediment where a concentration of fine particles in the coastal water serves to dampen the wave action, permitting deposition. They established a relationship between sediment concentration and the "activity" of water movement, a relationship that determined the site of deposition.

Also compatible with the suggestion for deposition on the shoreface is the previously discussed upward coarsening of the mean grain size within the mid-shelf ridges. Within a prograding barrier sequence the mean grain size of the sediment gradationally coarsens upward (Davies *et al.*, 1971; Reineck and Singh, 1971; Bernard *et al* .,1962; Fig. 9). This coarsening-upward sequence is considered to reflect a passage through the shoreface into the foreshore and aeolian dune environment (Elliott, 1978).

Modification of mid-shelf ridges..—The absence of a coarser grain size population within the medial and lower portions of the mid-shelf ridges implies that possibly only the lower portions of the shoreface were preserved during the marine transgression. Removal of portions of the barriers during transgression is also suggested by the difference in elevation between the modern or preserved barrier systems and the ridges on the middle and outer shelf. Linear sand bodies interpreted as barrier systems vary from slightly under 12 m in height for the Galveston Island barrier (Bernard *et al.*, 1962) to 17 m for the Cape Henlopen barrier spit (Kraft *et al.*, 1978), and in excess of 12 m for the Lower Cretaceous Muddy Sandstone (Davies *et al.*, 1971). When these values are compared to the 4-6 m maximum and 2 m averge difference in height between the mid-shelf coast-parallel ridge and the adjacent outer shelf coast-oblique ridges (Fig. 6), the conclusion is reached that sediment has been removed from the original barrier. If the thickness of the modern Galveston Island barrier is used for comparison, the 2-6 m difference in height indicates that much of the original barrier has been eroded, leaving only the lower shoreface and lowermost portions of the middle shoreface.

Classification

If the interpretation of two modes of origin is correct, the ridges, as represented by the stippled pattern of Fig. 3, are the remains of a barrier system that maintained partial geometric integrity during marine transgression. The outer shelf coast-oblique ridges were formed as shoreface-connected ridges similar to the present nearshore ridges. Adoption of this model encourages a classification of the ridges which reflects a generic classification system rather than one which utilizes only the location and trend of the ridges. The shoreface-connected and nearshore coast-oblique ridges are actively responding to the nearshore hydraulic regime, thus a classification of *active shoreface-connected/nearshore ridge* is appropriate. The mid-shelf coast-parallel ridges are the lower portions of a relict barrier system that partially survived transgression, prompting a classification of *degraded barriers*. The outer shelf coast-oblique ridges, while being formed as shoreface-connected ridges, are no longer a part of the coastline system; as such, they may be considered as relict shoreface-connected nearshore ridges. The term "relict" is used in the sense that these ridges are no longer a part of a nearshore system, and is not intended to imply that these ridges are completely inactive. For simplicity, these are called *relict nearshore ridges*.

A schematic representation of the development of these three ridge sets appears as Fig. 13. The nearshore ridges form as shoreface-connected ridges at a 15°-30° angle of incidence to the coastline. The relict nearshore ridges retain this angle of intersection to the degraded barrier.

SUMMARY

Most of the models addressing the origin of the sand ridges on the storm-dominated continental shelves of the U.S. Middle Atlantic Bight suggest that the ridges formed: (1) previous to the last marine transgression, (2) as barriers during passage of the shoreline, or (3) after passage of the shoreline as nearshore ridges. Few workers have suggested that these sand ridges may have formed during more than one of these times. Bathymetric and geological data from the New Jersey shelf indicate that some of the sand ridges formed as barriers, whereas other ridges formed on the nearshore as post-transgressive features.

Close contour mapping on which the highs are delimited reveals a variance in ridge orientation. The nearshore and shoreface-connected ridges intersect the coast at 15°-30°. This intersection angle, characteristic of many nearshore ridges on the world's shelves, was a critical point in Duane et al.'s (1972) and Swift et al.'s (1972) argument against the ridges being relict coastlines. However, seaward of the nearshore ridges is a complex of mid-shelf ridges that is approximately coast-parallel. Farther seaward there is yet another ridge set that intersects the mid-shelf ridges at 15°-30°. When the orientation and length/width ratios of the ridges are examined statistically by means of cluster analysis, three distinct groupings develop. The coast-parallel mid-shelf ridges are grouped as one category, the nearshore ridges as a second, and the seaward ridges that intersect the mid-shelf coast-parallel ridges as the third.

Other differences are found between the ridge groups. Examination of a series of cross-shelf profiles suggests that when the lowest lows are connected, three straight line segments emerge. The middle line segment may be further subdivided into two line segments, but with a more subtle break-in-slope than is found between the major divisions. The three major line segments independently separate the shelf ridges in a manner similar to the closed contour map. Spectral analysis of the profiles indicates that the portion representing the mid-shelf coast-parallel ridges is characterized by a dominant first-order wavelength of 6 km. Superimposed are smaller-scale wavelengths of 1.5 km or less. The nearshore and outer shelf coast-oblique ridges have an indication for similar ridge spacings, but to a lesser degree.

Textural variations are also found between the ridge sets. Vibracores of the nearshore coast-oblique ridges reveal medium to medium-coarse sand with numerous records of high-energy events. The vertical mean grain size profile is either of uniform texture or upward-coarsening. In a similar study Field (1976) observed some upward fining in addition to the other two vertical gradients. The surficial grain size distribution over the nearshore ridges is asymmetrical with the coarsest fraction near the base of the shoreward flank (Swift and Field, 1981; Stubblefield and Swift, 1981). These textural trends are contrasted to the mid-shelf coast-parallel ridges.

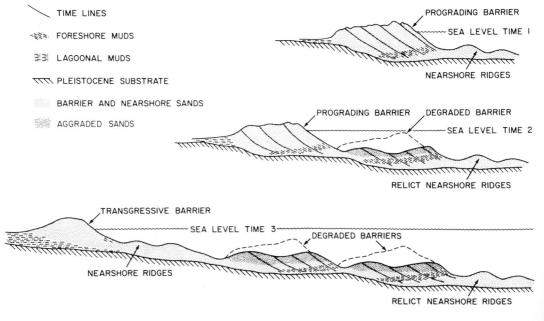

FIG. 13.—Schematic showing the development of the New Jersey shelf ridges. The figure is not drawn to scale; the barriers are somewhat enlarged relative to spatial dimensions in order to emphasize various features.

Vibracores through the mid-shelf ridges reveal two distinct depositional units. The upper one meter or so is similar to the nearshore ridges, with a medium to medium-coarse grain size and numerous occurrences of coarsening and fining (high-energy events). Radiocarbon dating of fauna and a piece of coal suggest that this upper zone was aggraded subsequent to transgression. Below this upper zone, however, the mid-shelf ridges are characterized by: (1) a fine to medium-fine grain size, (2) upward coarsening, (3) fewer and less pronounced occurrences of coarsening and fining, and (4) zones of interlayered clays and fine sand. The surficial grain size distribution of the mid-shelf ridges is somewhat symmetrical, with the coarsest fraction near the upper shoreward flank (Stubblefield and Swift, 1981).

The model that emerges from the data is that the outer shelf coast-oblique ridges formed adjacent to a barrier-beach complex. With a rise in sea level, the barriers were overstepped and partially degraded, preserving only the lower parts of the shoreface. At a later period, sea level slowed relative to sediment supply, and a new set of nearshore ridges developed. This model results in a cross-shelf sequence of topographic features which, when proceeding seaward from the coast, consist of: (1) modern barriers, (2) modern shoreface-connected and nearshore ridges, (3) degraded barriers, and (4) relict nearshore ridges. The degraded barriers continue to be modified in the form of aggradation by the mid-shelf hydraulic regime. Textural variations, radiocarbon dates, and a piece of coal indicate some ridge aggradation during historical times.

ACKNOWLEDGEMENTS

The authors are grateful to B. A. McGregor, H. J. Knebel, R. R. Berg, and J. W. Kofoed for constructive criticism offered to this study. Special appreciation is extended to D. J. P. Swift who, while not accepting all the tenets of our model, nevertheless freely provided many useful suggestions to improve both the model and the paper. The data presented in this paper were collected by NOAA's Marine Ecosystems Analysis Program as part of the New York Bight Project. Fauna identifications were by Don Moore, Rosenstiel School of Marine and Atmospheric Science, University of Miami, Florida, R. J. Stanton, Department of Geology, Texas A&M University, and Ming-Jung Jiang, Department of Oceanography, Texas A&M University.

REFERENCES

AGASSIZ, A., 1894, A reconnaissance of the Bahamas and of the elevated reefs of Cuba: Mus. Compar. Zoology Bull., v. 26, p. 1–203.

BELDERSON. R. H., N. H. KENYON, AND A. H. STRIDE, 1971, Holocene sediments on the continental shelf west of the British Isles, in F. M. Delaney (ed.), ICSU/SCOR Working Party 31 Symposium, Cambridge, 1970: Geology of East Atlantic Continental Margin, Inst. Geol. Science Rep. 70/14, p. 157–170.

BERNARD, H. A., R. J. LEBLANC, AND C. F. MAJOR, JR., 1982, Recent and Pleistocene geology of southeast Texas: Geology Gulf Coast and Central Texas and Guidebook of Excursion, Houston Geol. Soc., p. 175–225.

CASTON, V. N. D., 1972, Linear sand banks in the southern North Sea: Sedimentology, v. 18, p. 63–78.

COLQUHOUN, D. J., 1969, Geomorphology of the lower coastal plain of South Carolina: Div. Geol. State Development Board, Columbia, South Carolina, 36 p.

DAVIES, D. K., F. G. ETHRIDGE, AND R. R. BERG, 1971, Recognition of barrier environments: Am. Assoc. Petroleum Geologists Bull., v. 55, p. 550–555.

DOYLE, L. J., W. J. CLEARY, AND O. H. PILKEY, 1968, Mica: its use in determining shelf-depositional regimes: Marine Geol., v. 6, p. 381–389.

DUANE, D. B., M. E. FIELD, E. P. MEISBURGER, D. J. P. SWIFT, AND S. J. WILLIAMS, 1972, Linear shoals on the Atlantic inner shelf, Florida to Long Island, in D. J. P. Swift, D. B. Duane, and O. H. Pilkey (eds.), Shelf Sediment Transport: Process and Pattern: Dowden, Hutchinson, and Ross, Stroudsburg, Pennsylvania, p. 447–449.

ELLIOTT, T., 1978, Clastic shorelines, in H. G. Reading (ed.), Sedimentary Environments and Facies: Elsevier, New York, p. 143–177.

EMERY, K. O., 1967, Estuaries and lagoons in relation to continental shelves, in G. F. Lauff (ed.), Estuaries: Am. Assoc. Advancement Science, Washington, D. C., p. 9–14.

_____, AND A. S. MERRILL, 1979, Relict oysters on the United States Atlantic continental shelf: A reconsideration of their usefulnes in understanding late Quaternary sea level history: Discussion and reply: Geol. Soc. America Bull., v. 90, p. 689–694.

FIELD, M. E., 1976, Quaternary evolution and sedimentary record of a coastal plain shelf: Central Delmarva peninsula, Mid-Atlantic Bight, U.S.A.: Unpub. PhD Dissertation, George Washington Univ., Washington, D. C., 199 p.

FIGUEIREDO, A. G., J. E. SANDERS, AND D. J. P. SWIFT, 1982, Storm-graded layers on inner continental shelves: Examples from southern Brazil and the Atlantic coast of central United States: Sedimentary Geol., v. 31, p. 171–190.

GALTSOFF, P. S., 1964, The American oyster, *Crassostrea virginica* Gmelin: U. S. Dep. Interior Fish and Wildlife, Fisher Bull., v. 64, 480 p.

GILBERT, G. K., 1880, Contributions to the history of Lake Bonneville: U. S. Geol. Survey, 2nd Ann. Rep., p. 169–200.

HATHAWAY, J. C., 1972, Regional clay mineral facies in estuaries and continental margin of the United States east coast: Geol. Soc. America Mem. 133, p. 293–316.

HOUBOLT, J. J. H. C., 1968, Recent sediments in the southern bight of the North Sea: Geol. en Mijnbouw, v. 47, p. 245–273.

HOWARD, J. D., AND H. E. REINECK, 1972, Physical and biogenic sedimentary structures of the nearshore shelf: Senckenbergiana Marit., v. 4, 217–223.

JORDAN, G. F., 1962, Large submarine sand waves: Science, v. 136, p. 839–848.

KRAFT, J. C., 1971, Sedimentary facies patterns and geologic history of a Holocene transgression: Geol. Soc. America Bull., v. 82, p. 2131–2158.

_____,E. A. ALLEN, AND E. MAURMEYER, 1978, The geological and paleogeomorphological evolution of an estuarine and coastal barrier system [abs.]: Geol. Soc. America, Abstracts with Programs, v. 10, p. 72.

_____,AND C.J. JOHN, 1979, Lateral and vertical facies relations of transgressive barrier: Am Assoc. Petroleum Geologists Bull., v. 63, p. 2145–2163.

McCAVE, I. N., 1972, Transport and escape of fine-grained sediment from shelf areas, in D. J. P. Swift, D. B. Duane, and O. H. Pilkey (eds.), Shelf Sediment Transport: Process and Pattern: Dowden, Hutchinson, and Ross, Stroudsburg, Pennsylvania, p. 225–248.

McCLENNEN, C. E., 1973, Nature and origin of the New Jersey continental shelf topographic ridges and depressions: Unpub. PhD Dissertation, Univ. Rhode Island, Kingston, Rhode Island, 94 p.

_____, AND R. L. McMASTERS, 1971 Probable Holocene transgressive effects on the geomorphic features of the continental shelf off New Jersey, United States: Maritime Sed., v. 7, p. 69–72.

McKINNEY, T. F., AND G. M. FRIEDMAN, 1970 Continental shelf sediments of Long Island, New York. Jour. Sed. Petrology, v. 40, p. 213–248.

_____,W. L. STUBBLEFIELD, AND D. J. P. SWIFT, 1974, Large scale current lineations on the Great Egg Shoal retreat massif, New Jersey shelf: Investigations by sidescan sonar: Marine Geol., v. 17, p. 79–102.

McMASTER, R. L., AND L. E. GARRISON, 1967, A submerged Holocene shoreline near Block Island, Rhode Island: Jour. Geol., v. 75, 335–339.

MEADE, R. H., 1972a, Transport and deposition of sediments in estuaries, in B. W. Nelson (ed.), Environmental Framework of Coastal Plain Estuaries: Geol. Soc. America Memoir 133, p. 91–120.

_____, 1972b, Sources and sinks of suspended matter on the continental shelves, in D. J. P. Swift, D. B. Duane, and O. H. Pilkey (eds.), Shelf Sediment Transport: Process and Pattern: Dowden, Hutchinson, and Ross, Stroudsburg, Pennsylvania, p. 249–262.

MOODY, D. W., 1964, Coastal morphology and processes in relation to the development of submarine sand ridges off Bethany Beach, Delaware: Unpub. PhD Dissertation, Johns Hopkins Univ., Baltimore, Maryland, 167 p.

MURRAY, J. W., 1973, Distribution and Ecology of Living Benthic Foraminiferids: Heinemann Educational Books, London, 251 p.

NELSEN, T. A., 1976, An Automated rapid sediment analyzer (ARSA): Sedimentology, v. 23, p. 867–872.

OERTEL, G. F., AND J. D. HOWARD, 1972, Water circulation and sedimentation at estuary entrances on the Georgia coast, in D. J. P. Swift, D. B. Duane, and O. H. Pilkey (eds.), Shelf Sediment Transport: Process and Pattern: Dowden, Hutchinson, and Ross, Stroudsburg, Pennsylvania, p. 411–427.

OFF, T., 1963, Rhythmic linear sand bodies caused by tidal currents: Am. Assoc. Petroleum Geologists Bull., v. 47, p. 324–341.

OOSTDAM, B. L., AND R. R. JORDAN, 1972, Suspended sediment transport in Delaware Bay, in B. W. Nelson (ed.), Environmental Framework of Coastal Plain Estuaries: Geol. Soc. America Memoir 133, p. 143–149.

PEVEAR, D. R., 1972, Source of recent nearshore marine clays, southeastern United States, in B. W. Nelson (ed.), Environmental Framework of Coastal Plain Estuaries: Geol. Soc. America Memoir 133, p. 317–335.

POAG, C. W., H. J. KNEBEL, AND R. TODD, 1980, Distribution of modern benthic foraminifers on the New Jersey outer continental shelf: Marine Micropaleontology, v. 5, p. 43–69.

REDFIELD, A. C., 1956, The influences of the continental shelf on the tides of the Atlantic Coast of the United States: Jour. Marine Res., v. 17, p. 432–448.

REINICK, H. E., AND I. B. SINGH, 1971, Der Gulf von Gaeta (Tyrrhenisches Meer) III. Die Gefüge von vorstrandund Schelfsedimenten: Senckenbergiana Marit., v. 3, p. 135–183.

RIGGS, S. R., AND D. H. FREAS, 1968, Submerged shoreline features on the shelf near Cape Fear, North Carolina [abs.]: Geol. Soc. America Spec. Paper 115, p. 496–497.

SANDERS, J. E., AND N. KUMAR, 1975a, Evidence of shoreface retreat and in-place "drowning" during Holocene submergence of barriers, shelf off Fire Island, New York: Geol. Soc. American Bull., v. 86, p. 65–76.

_____AND_____, 1975b, Holocene shoestring sand on inner continental shelf off Long Island, New York: Am. Asoc. Petroleum Geologists Bull., v. 59, p. 997–1009.

SHEPARD, F. P., 1973, Submarine Geology, 3rd Ed.: Harper and Row, New York, 517 p.

SMITH, J. D., 1969, Geomorphology of a sand ridge: Jour. Geol., v. 72, p. 39–55.

_____, 1970, Stability of a sand bed subjected to a shear flow of low Froude number: Jour. Geophys. Res., v. 75, p. 5928–5940.

STEARNS, F., 1967, Bathymetric maps of the New York Bight, Atlantic continental shelf of the United States, scale 1:125,000: National Ocean Survey, National Oceanic and Atmospheric Administration, Rockville, Maryland.

STEWART, H. B., JR., AND G. F. JORDAN, 1964, Underwater sand ridges on Georges Shoal, in R. L. Miller (ed.). Papers in Marine Geology, F. P. Shepard Commemorative Volume: McMillan and Co., New York, p. 102–114.

STUBBLEFIELD, W. L., J. W. LAVELLE, D. J. P. SWIFT, AND T. F. McKINNEY, 1975, Sediment response to the present hydraulic regime on the central New Jersey shelf: Jour. Sed. Petrology, v. 45, p. 337–358.

_____,AND D. J. P. SWIFT, 1976, Ridge development as revealed by subbottom profiles on the central New Jersey shelf: Marine Geol., v. 20, p. 315–334.

_____AND_____, 1981, Grain size variations across sand ridges, New Jersey continental shelf, U.S.A.: Geo-Marine Letters, v. 1, p. 45–48.

WIFT, D. J. P., J. W. KOFOED, F. P. SAULSBURY, AND P. SEARS, 1972, Holocene evolution of the shelf surface, central and southern Atlantic shelf of North America, *in* D. J. P. Swift, D. B. Duane, and O. H. Pilkey (eds.), Shelf Sediment Transport: Process and Pattern: Dowden, Hutchinson, and Ross, Stroudsburg, Pennsylvania, p. 499–575.

_____,AND M. E. FIELD, 1981, Evaluation of a classic sand ridge field: Maryland sector, North American inner shelf: Sedimentology, v. 28, p. 461–482.

_____,AND J. C. LUDWICK, 1976, Substrate response to hydraulic processes: Grain-frequency distributions and bedforms,*in* D. J. Stanley and D. J. P. Swift (eds.), Marine Sediment Transport and Environmental Management: John Wiley and Sons, New York, p. 159–196.

_____,T. A. NELSEN, J. McHONE, B. HOLLIDAY, H. PALMER, AND G. SHIDELER, 1977, Holocene evolution of the inner shelf of southern Virginia: Jour. Sed. Petrology, v. 47, p. 1454–1474.

_____, P. C. SEARS, B. BOHLKE, AND R. HUNT, 1978, Evolution of a shoal retreat massif, North Carolina shelf: Inferences from areal geology: Marine Geol., v. 27, p. 19–42.

'IEZZI, L. J., H. D. TOLLEY, AND R. B. SCOTT, 1979, A new clustering analysis method for geologic data: Application to MOR basalts: EOS, v. 60, p. 971.

JCHUPI, E., 1968, Atlantic continental shelf and slope of the United States—Physiography: U. S. Geol. Survey Prof. Paper 529C, 30 p.

_____, 1970, Atlantic continental shelf and slope of the United States—Shallow structure: U. S. Geol. Survey Prof. Paper 529I, 44 p.

'AN VEEN, J., 1935, Sand waves in the North Sea: Hydrogr. Rev., v. 12, p. 21–28.

RECOGNITION OF TRANSGRESSIVE AND POST-TRANSGRESSIVE SAND RIDGES ON THE NEW JERSEY CONTINENTAL SHELF: DISCUSSION

DONALD J. P. SWIFT[1], THOMAS F. MCKINNEY, AND LLOYD STAHL
Atlantic Oceanographic and Meteorological Laboratories, NOAA, Virginia Key, Miami, Florida 33149;
Dames and Moore, Inc., 6 Commerce Drive, Cranford, New Jersey 07016;
and Dames and Moore Inc., 301 W. Camino Gardens Blvd., Boca Raton, Florida 33432

ABSTRACT

It has been proposed by Stubblefield and colleagues (this volume) that the sand ridges of the central new Jersey shelf contain a basal muddy sand stratum deposited as a lower shoreface facies during a period of coastal progradation. They conclude that the ridge morphology above the mid-shelf scarp is in part a relict strand plain.

The coastal progradation hypothesis for the origin of the lower muddy sand facies is a reasonable one, but to date there is not enough evidence to discriminate among facies. Examination of the inner shelf surface above the mid-shelf scarp reveals topographic, stratigraphic, and grain-size patterns that may be interpreted as being in conflict with the relict strand plain model. We conclude that the ridge topography on the surfaces above the scarp on the New Jersey Shelf is a response to storm flows subsequent to transgression.

NATURE OF THE PROBLEM

Our colleagues have been able to shed light on the shallow stratigraphy of the New Jersey Shelf by the analysis of data made available to them during the course of NOAA studies in the New York Bight. However, we are not convinced that their conclusions uniquely fit the data; we wish to consider other possibilities.

In order to better understand the nature of the problem, we would like to briefly review the results of the New York Bight Project. In particular, we wish to summarize the geomorphology of the region which we view as critical to the interpretation of the region's Holocene history. The area in question lies in a region where first-order topographic elements trend normal to shore, across the New Jersey inner and central shelf. The most important of these is the Great Egg Shelf Valley (Swift *et al.*, 1972; McKinney *et al.*, 1974; Fig. 1, this paper). To the northeast of the shelf valley lies a shore-normal zone corrugated by (approximately) shore-parallel sand ridges. Individual ridges lie athwart the zone, like the teeth of a comb. We have argued that the shelf valley is the retreat path of the ancestral Great Egg Estuary Mouth during Holocene rise of sea level and have inferred that the corrugated zone correspondingly marks the retreat of a depositional center on the up-drift side of the estuary mouth (McKinney *et al.*, 1974).

The corrugated zone is designated as a shoal retreat massif; the term "massif" serving in this case as shorthand for a large-scale shelf high, itself composed of smaller topographic highs. A comparison of the New York Bight with the adjacent Delaware Shelf Valley (Swift, 1973) is useful. The Delaware Shelf Valley is still connected with the estuary mouth that formed it, and a uniformitarian analysis is thus possible. As the valley is traced landward, it ends in a tidal scour trench in the estuary mouth. Similarly, the low, corrugated ridge beside it can be traced landward into a depositional zone for littoral drift at the southern end of the New Jersey coastal compartment. The arguments are mainly geomorphologic in nature, but have been in large measure confirmed by the stratigraphic studies of Sheridan *et al.* (1974). These workers have shown that the massif is a planoconvex thickening in a Holocene sand sheet that disconformably overlies a unit of diverse late Pleistocene facies, and that the seafloor valley is offset from the buried river by several kilometers.

The comb-like array of sand ridges that lies athwart the Great Egg Massif, we believe, is a secondary phenomenon, a consequence of post-transgressive erosion and deposition by cross-massif storm flows (McKinney *et al.*, 1974).

A second morphologic element in the New York Bight area of major importance is a discontinous scarp with up to 8 m of relief. It can be traced intermittently from Cape Hatteras to the Long Island shelf. It has been designated the mid-shelf shore (Swift *et al.*, 1972), the Block Island Shore (Emery and Uchupi, 1972), and the Atlantis Shore (Dillon and Oldale, 1978). We have inferred that it is a relict lower shoreface, marking a period of near stillstand during the post-glacial rise in sea level.

The Great Egg Shelf Valley complex near its intersection with the mid-shelf shoreline was selected for further NOAA study because of the diversity of well-defined, large-scale morphologic elements in this area (scarp, shelf valley, massif, and the well-developed sand ridge topography characteristic of the massif). We

[1]Present Address: ARCO Research, Box 2819, Dallas, Texas 75221

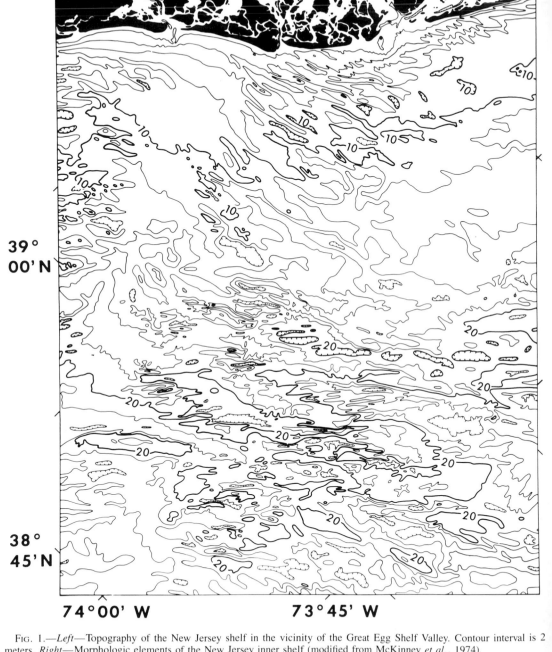

FIG. 1.—*Left*—Topography of the New Jersey shelf in the vicinity of the Great Egg Shelf Valley. Contour interval is 2 meters. *Right*—Morphologic elements of the New Jersey inner shelf (modified from McKinney *et al.*, 1974).

felt that if we could understand the morpho-strat-igraphic relationships here, we would similarly under-stand them in most parts of the Middle Atlantic Bight.

Stubblefield and others have presented data of major

significance from their portion of the NOAA study; they find that the basal Holocene beds of the study area are distinctly more muddy than the upper strata. They present a scenario to account for this twofold Holocene

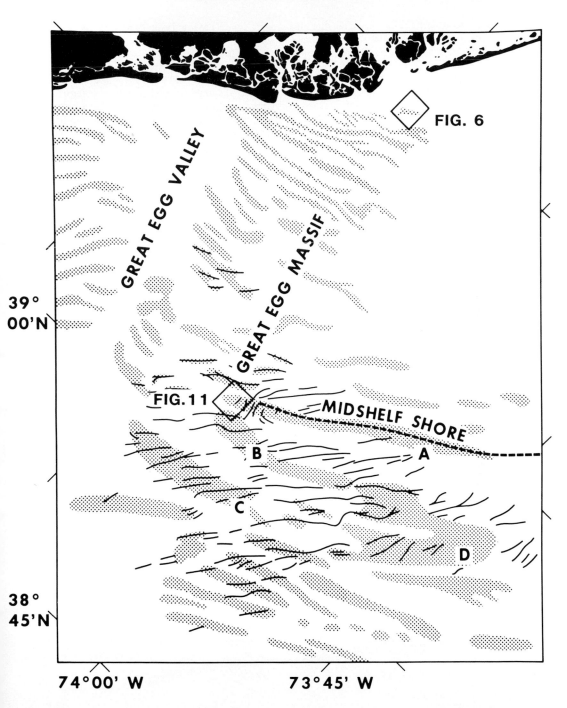

stratigraphy that is basically plausible, but leads them to corollary conclusions that are unacceptable. They believe that the muddy facies is related primarily to processes of deposition which were active at the time that a mid-shelf shoreline occupied a seaward portion of the present shelf. They argue that the scarp is the result of deposition during a period not merely a still-stand, but of progradation, and that the lower muddy sands are lower shoreface beds which were subsequently overrun by clean upper shoreface beds during the progradation. Stubblefield and his colleagues (this volume) also note that the topographic grain on the terrace surface immediately landward of the scarp is more nearly shore-parallel than it is either further seaward or further landward. They conclude that this shore-parallel trend is in fact a strand plain topography whose shoreline portions are still recognizable despite transgression.

The two major points of contention are: (1) the origin of the muddy facies, and (2) the origin of the ridge topography on the terrace above the scarp.

MUDDY EVENTS ON THE INNER SHELF

We believe that the scenario in which the muddy facies is generated by shoreface progradation (Stubblefield, 1981) is a reasonable one, but note that several other models may be equally applicable. One is a stillstand model that we have presented earlier (Swift *et al.*, 1973; Fig. 2, this paper), in which we envisioned that near-stillstand conditions lead to shoreline stabilization and thickened lagoonal deposits under the mid-shelf scarp. In this model the lower muddy facies appear in the appropriate stratigraphic position for lagoonal deposits. The existing scarp is seen as remnant of the lower shoreface, the upper shoreface, beach

and dunes having been destroyed as transgression resumed.

This interpretation is in some respects more compatible with the data than is Figure 11 (Fig. 8, this paper) of Stubblefield and colleagues. The cores which contain the muddy sand and mud interpreted to be shoreface deposits are present only in cores 6 and 7 and the lower 20 cm of core 8. When this core stratigraphy is projected onto the closest regional profile, profile A of Stubblefield and others, the muddy deposits occur too high on the profile to support the proposed origin. In the schematic model, the muddy shoreface deposits represent the lowermost facies in the profile in Figure 3. The projected "stratigraphic position" of the muddy sediments is instead more consistent with an inferred back-barrier origin.

Our more recent studies suggest yet a third model, in which the muddy facies is due not to the proximity of the shoreline, as in the stillstand and stillstand-regression models, but to the proximity of the Great Egg estuary (Swift *et al.*, 1978; Fig. 4, this paper). We have

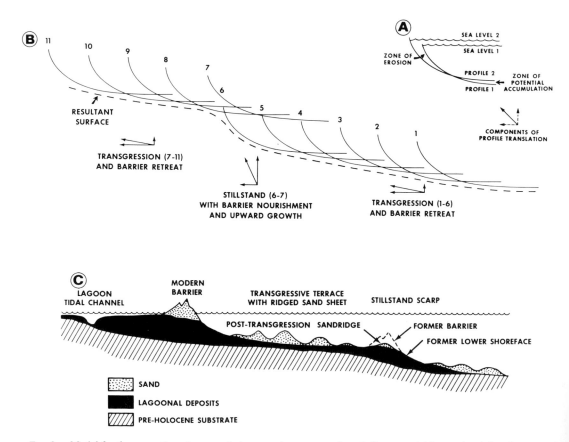

FIG. 2.—Model for the generation of a scarp during a marine transgression. *A*, BRUUN model for erosional shoreface retreat. *B*, Generation of a scarp by erosional shoreface retreat. *C*, Stratigraphic response model corresponding to B. Ridges on terrace have been formed by post-transgressive processes, including cap on truncated stillstand barrier (modified from Swift *et al.*, 1973).

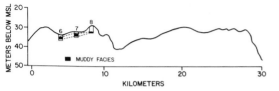

FIG. 3.—Projection of cores 6, 7, and 8 of Stubblefield's Figure 9 onto profile A of Stubblefield *et al.* (this volume; Fig. 9 of this paper).

found a very similar Lower Holocene muddy facies in the Albemarle Massif, adjacent to the Albemarle Shelf Valley on the northern North Carolina shelf, between the modern shoreline and the mid-shelf shore. A companion shoal, the Platt Massif, lies south of the shelf valley. These two massifs, their intervening shelf valley, and their muddy sand facies resemble the arcuate shoal systems and associated muddy sands that lie in front of estuary mouths on the inner shelf floor of the Georgia–South Carolina coast (Howard and Reineck,

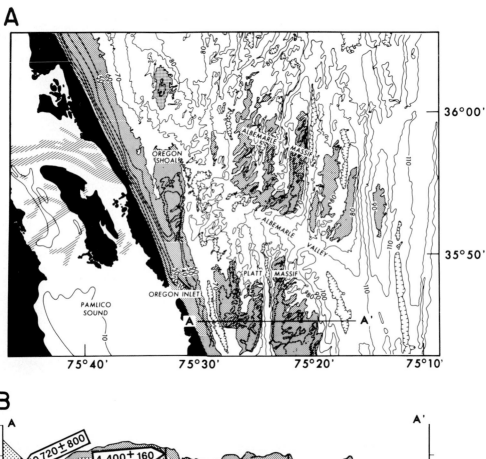

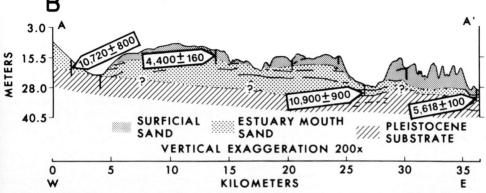

FIG. 4.—Albermarle Shelf Valley complex showing lower muddy sand facies deposited during estuarine stage (modified from Swift *et al.*, 1978). Diagonal ruling in lagoon west of Oregon Shoal indicates buried channel revealed by seismic profile.

1972). We have argued that the Platt Massif–Albermarle Massif system was initiated as an estuary mouth shoal during the stillstand associated with the mid-shelf shoreline. We have suggested that since the closure of the estuary by the modern Currituck Spit, the massifs have undergone dissection and reshaping as inner shelf sand ridge systems, with attendant reworking of the upper clean sand facies that presently overlies the muddy sands (Fig. 5). This hypothesis is certainly transferable to the Great Egg Massif. The latter, like the Platt-Albermarle system, presumably underwent submarine dissection after the estuarine tidal discharge was greatly reduced, either from piracy of the ancestral Schuykill River by the Delaware (Swift *et al.*, 1980) or by diversion of the Great Egg River toward Cape May (Kelly, 1981). The coast-oblique inner shelf ridges of Stubblefield and others (1981) must date from this period, and are closely analogous to the coast-oblique ridge of the Albermarle and Platt Massifs.

Based on data presently available, it is difficult to distinguish among the three proposed models. Is the lower muddy sand facies a lagoonal facies, a lower

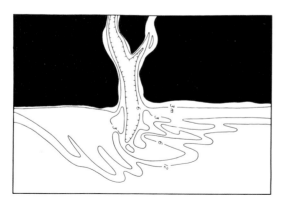

Fig. 5.—Conceptual model for transformation of arcuate estuary mouth shoal (Top) into shelf valley complex (Bottom) during marine transgression (Swift *et al.*, 1978).

shoreface facies, or an estuary mouth facies? The argument of Stubblefield and colleagues concerning the absence of roots in the muddy facies seems a weak one. If the muddy facies were a lagoonal deposit, the back-barrier marshes could have at least in part been destroyed by transgression. The modern mainland marshes constitute a horizon less than a meter thick, and are laterally discontinous. The presence of marsh deposits would help to establish the muddy facies as lagoonal, but their absence is less useful. Faunal analysis may provide the answer, but the faunal analysis by Stubblefield and colleagues of cores penetrating the muddy facies seems rather inconclusive.

In other studies, detailed grain size and analysis of sedimentary structures have been used with some success. Sanders and Kumar (1975) present a detailed description of the modern shoreface deposits of Fire Island, New York, and identified similar shoreface deposits on the inner Long Island shelf, inferred to represent the seaward portion of a former "drowned barrier." The landward portion of this feature was deduced from cores interpreted as a back-barrier sequence, thus bracketing a linear zone on the Long Island shelf where a "drowned barrier" had existed. Sands interpreted to be shoreface deposits are characterized by distinctive thinly laminated fine sands with gravel couplets interpreted to be storm deposits. The fine sands are distinguished by the absence of structures such as heavy mineral layers, ripple cross-lamination, high angle cross stratification and bioturbation. Sanders and Kumar (1975) also report interbedded layers of mud up to 2 to 3 mm in thickness in the fine, thinly laminated sands of the shoreface. In the absence of a comparable analysis, there is insufficient evidence for the interpretation that the muddy sediments encountered in the cores by Stubblefield and others represent shoreface deposits.

DEGRADED BARRIERS?

"Degraded barriers" are a concept with which we have difficulty. The "drowned barrier" of Sanders and Kumar (1975) was inferred to have been truncated by erosional shoreface retreat, and remnants of the purported barrier remain in the shallow stratigraphy of the shelf, but its shoreface expression has been destroyed. On the New Jersey Shelf, the Mid-Shelf shore is in fact a primary expression of a former shoreline (Swift *et al.*, 1972, 1973). However, the scarp does not represent only the lower shoreface, but also a post-transgressive marine sand ridge, molded perhaps by flow convergence on the crest of the scarp. In any case, the entire surficial sand sheet is interpreted as a "degraded barrier" (is "reworked barrier" a less morally-colored term?) in the sense that it includes the debris sheet produced by erosional shoreface retreat of the barrier face. This is seen most clearly in one part of our study area where the site study for the Atlantic Generating Station (Stahl *et al.*, 1974) has resulted in very closely spaced drill cores and a dense seismic profile net. This exceptional stratigraphic control has allowed us to re-solve the Holocene facies in considerable detail. We were able to identify a back-barrier sand facies between lagoonal muds and the overlying transgressive sand sheet (Stahl, *et al.*, 1974; Fig. 6, this paper). In this area the beach and dune facies are missing and are represented by a disconformity. We conclude that in this area the transgressive sand sheet is a destroyed barrier, smeared out by erosional shoreface retreat. No "barrier" topography has been preserved. Compare Figure 6 with the model presented in Figure 7.

Stubblefield *et al.* (this volume) wish to set forth a more radical interpretation, namely, that a belt of mid-shelf sand ridge topography 30 km wide in fact pre-serves the original fabric of a prograding strand plain (beach ridge plain) of the sort described by Curray *et al.* (1969). We find this interpretation radical because it is contrary to the generalization, first set forth in Swift *et al.* (1972), that during the Holocene transgression of the Middle Atlantic Bight erosional shoreface retreat has erased the subaerial topography of the coastal plain and has replaced it with a marine topography of storm-

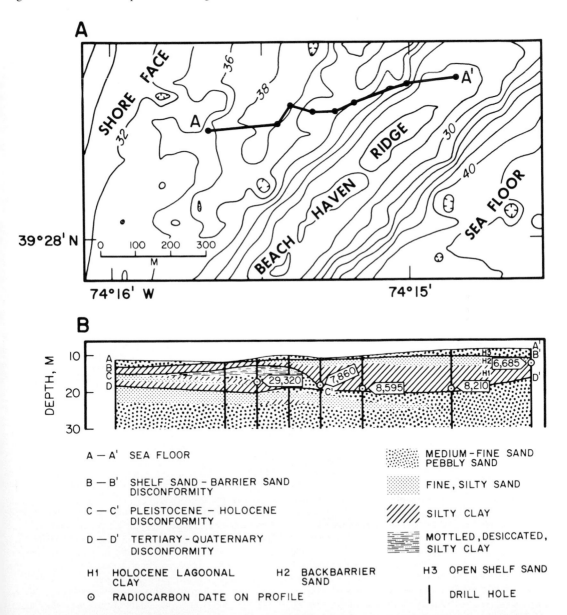

FIG. 6.—Stratigraphy of Beach Haven Ridge (based on Stahl *et al.*, 1974, Fig. 3). Note preservation of back-barrier sand (H2) between lagoonal mud (H1) and transgressive shelf sand (H3). See Figure 1 for location.

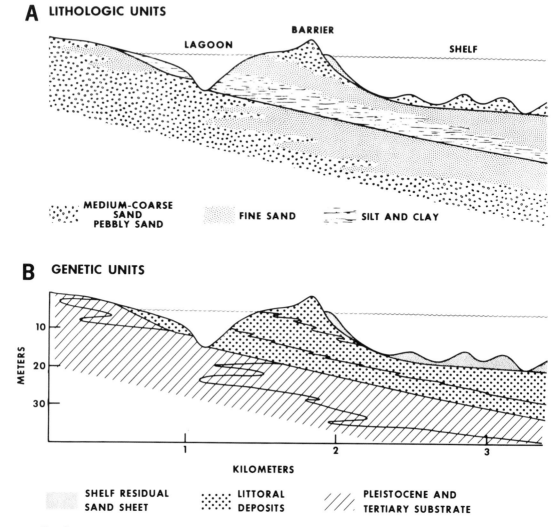

A   LITHOLOGIC UNITS

BARRIER

LAGOON

SHELF

····· MEDIUM-COARSE SAND PEBBLY SAND   FINE SAND   SILT AND CLAY

B   GENETIC UNITS

METERS

10

20

30

KILOMETERS

SHELF RESIDUAL SAND SHEET   LITTORAL DEPOSITS   PLEISTOCENE AND TERTIARY SUBSTRATE

FIG. 7.—A schematic model for the evolution of transgressive inner shelf stratigraphy at Beach Haven Ridge.

built sand ridges formed at the foot of the retreating shoreface. Variations in these processes over time have resulted in terraces with ridged shorefaces separated by scarps (Fig. 2), but the present terrace morphology is interpreted as post-transgressional in origin. However, generalizations require constant re-evaluation as new evidence becomes available, and we wish to reconsider this one: Do the ridges of the central New Jersey demonstrate the characteristics of a prograding beach ridge plain or the characteristics of an inner shelf ridge field?

Stubblefield and colleagues argue for two separate episodes of progradation, still discernible in seafloor morphology. Any model of transgressive shelf sedimentation that purports to result in multiple barrier morphology seaward of the shoreface must come to terms with the stairstep criterion of Gilbert (1880). Gilbert noted that if successive barriers are created and

overstepped during a transgression, the crest of one should extend no higher than the altitude of the beach of the next landward barrier. The rule says in effect, "thou shalt not build thy barrier in someone else's lagoon," since, as a consequence of limited wave fetch, wave power would not be adequate there to accomplish the construction.

Stubblefield and others argue that the inner of their two large-scale ridges has a crest "two to six meters" higher than that of its seaward companion, a difference that they argue is "significantly greater" than the regional slope. This argument appears to apply to the idealized drawing, Figure 8 (this paper), but is beyond resolution in Figure 9 (this paper) where, at least to the eye, the crests are concordant. In the text, they argue that the stairstep criterion once held, but that the evidence has since been destroyed by subsequent sub-

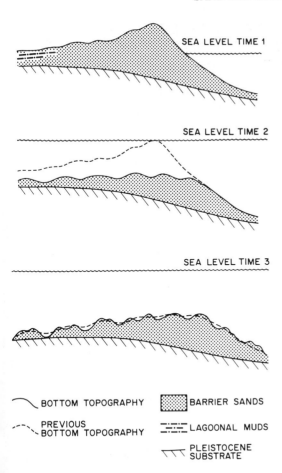

BOTTOM TOPOGRAPHY

PREVIOUS BOTTOM TOPOGRAPHY

BARRIER SANDS

LAGOONAL MUDS

PLEISTOCENE SUBSTRATE

FIG. 8.—Schematic diagram showing proposed development of the New Jersey shelf ridges (Stubblefield *et al.*, this volume).

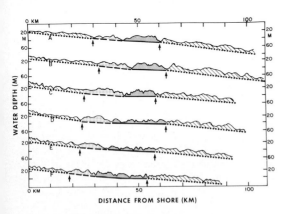

FIG. 9.—Cross-sectional views of profiles A through F from Stubblefield *et al.* (this volume). Facies interpreted by them are: (1) nearshore and outer shelf coast-oblique ridges—hachured pattern underlain by dots, (2) mid-shelf coast-parallel ridges—stippled pattern; two segments underlain by solid and dashed lines.

marine erosion. Figure 8 attempts to justify this: the first barrier is created, then largely demolished, so that its crest will properly accord with a second barrier that in turn is created and demolished, to result in the present configuration. At time one in Figure 8, the first barrier has lagoonal muds accumulating landward of it. The lagoonal muds should be overridden at time two by the lower foreshore beds of the prograding second barrier, but the lagoonal muds have been omitted there.

Beyond the stairstep criterion, there are several geometric problems to be considered if the terrace is to be considered a relict strain plain. One such problem is the extent to which the mid-shelf ridges are truly "shore-parallel." In fact, the mid-shelf ridges converge with the present shoreline at angles ranging from 5° to 15°, depending on how one determines the mean coastline trend and the mean ridge trend. Figure 3 of Stubblefield *et al.* (1981) maximizes the apparent shore-parallel nature by breaking B and C ridges of our Figure 1 into *en enchelon* shore-parallel segments. The big central ridges tend to have *en echelon* peaks, we believe, because a secondary, small-scale pattern has been superimposed on them subsequent to their formation. Even the secondary ridges tend to form small, northward opening angles with the regional trend of the contours. This relationship is often concealed by the fact that for long distances, contours do parallel to mean trend of the coast (Fig. 10). The contours, however, describe a series of zig-zags as they pass down the shelf. The coast-parallel sections of contours occur on the central portions of the "Z"s thus defined, while the kinks define ridge crests that are oblique with respect to the coast. Furthermore, the "Z"s are always normal, or left-handed "Z"s, almost never right-handed "Z"s (except along the northeast sides of shelf valleys). In other words, ridge crests, even in the mid-shelf ridges, make a northward opening angle with the coast, and this behavior is not explainable in terms of beach ridges. The main exception to this rule occurs on the seaward margins of C and D ridges (Fig. 1), where they are crenulated by the small-scale ridges.

More detailed study of mid-shelf ridges also indicates that if they were once beach ridges, they have suffered a sea change. We have noted (Swift *et al.* 1972, Fig. 2; 1983, Fig. 3) a tendency of shoreface and nearshore ridges to divide into 2 or 3 subridges at their northeastern ends and peak at their southwestern ends. While beach ridges do bifurcate, the pattern shows little resemblance to that seen in Figure 11. Beach ridge spacing is generally on the order of 50 m (Curray *et al.* 1969); there is a much coarser fabric here.

Finally, our studies of the Maryland inner shelf have led us to conclude that nearly coast-parallel ridges would be expected in the vicinity of a stillstand scarp. On this coast, ridges are forming on a shoreface undergoing erosional shoreface retreat, in response to southwesterly stream flows (Swift and Field, 1981). We have argued that the ridge angle may be determined by the relationship between the retreat rate of the coastline

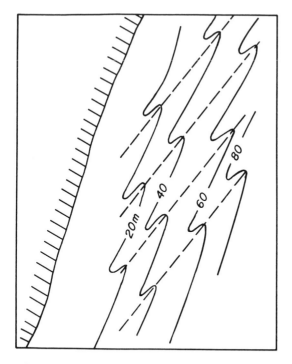

FIG. 10.—Schematic model of contours on a shelf with sand ridges inclined at a small angle to the coast.

and the down-coast rate of migration of a generating zone confined to water less than 10 m deep. If these two rates are subequal, a ridge forming a 45° angle with the coast is left as a trail by a generating zone subject to these two simultaneous motions (Fig. 12).

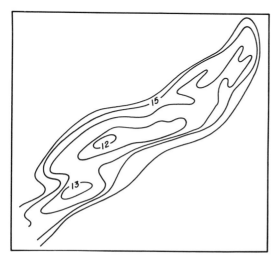

FIG. 11.—Peaked, bifurcated ridge from the central New Jersey shelf. See Figure 1 for location. This pattern is characteristic of nearshore ridges.

However, as stillstand is approached, the retreat rate is slowed, and ridges are formed that are more nearly parallel to the coast. Thus, a stillstand scarp should be associated with submarine sand ridges that are more nearly parallel to it than those further away. On the Maryland coast, the ridge angle decreases steadily to the south, from almost 45° near the Delaware border to 10° or less at Chincoteague. We infer that as one proceeds south on the Maryland coast, the ridge migration rate increases with respect to the coastal retreat rate.

We do not wish to place excessive emphasis on the shoreface-detachment mode of ridge formation, however, since there is evidence to indicate that several other processes have contributed to the mid-shelf ridges. In addition to the shoreface detachment mode described above, ridges may have formed by interaction of the massif with the estuary tide. We have argued that the arcuate ridges inherited orientations from a period when the Great Egg Shelf Valley was still subject to strong tidal currents, creating tidal streamlines across the Massif that curved into the estuary mouth (McKinney et al., 1974). We include in this category not only the arcuate ridges dismissed by Stubblefield and colleagues as anomalous, but also the greater outer ridges (A-D, Fig. 1b). Finally, storm flows on the central shelf appear to have imprinted the early, large-scale ridges with a fine-grained ridge pattern (McKinney et al., 1974; Fig. 1, this paper). These ridge-forming processes are successive in time, hence Stubblefield and others are incorrect when they state that theirs is the first model to suggest multiple successive modes of ridge formation.

In order to resolve the issue of the respective origins of inner shelf and mid-shelf ridges, we have recently collaborated with Stubblefield in order to compare grain-size distribution across a mid-shelf ridge with that across an inner shelf ridge of uncontested origin (Stubblefield and Swift, 1981). Stubblefield et al. (this volume) present the results as their Figure 10. Unlike Stubblefield and others, we do not view the results as indicative of markedly different genesis. For both ridges, the coarsest sand occurs on the landward flank, facing southwestward-trending storm currents (compare with data in Swift and Field, 1981). If either of these were barriers, the high-energy coarse-grained side would be expected on the seaward side. The mid-shelf ridge grain-size maximum is higher on the landward flank than it is on the nearshore ridge, but following Taylor and Dyer (1977), we attribute this to the greater asymmetry of the ridge: with an up-current slope longer and gentler than the down-current slope, the maximum shear stress should be experienced closer to the ridge crest.

CONCLUSIONS

The sand ridges of the central New Jersey shelf are interpreted as "degraded barriers" by Stubblefield et al. (this volume). We suggest that preferable model is one that explains the morphology and shallow strat-

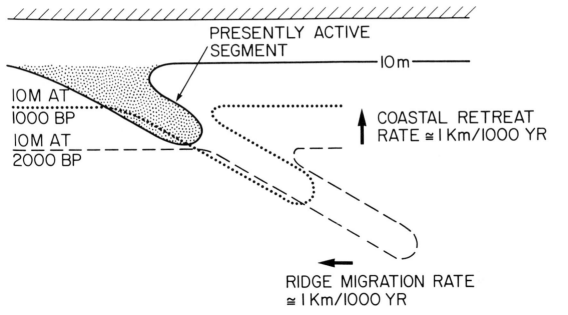

FIG. 12..—Model of sand ridge formation by the down-coast migration of a ridge-generating zone on a retreating shoreface (Swift and Field, 1981).

igraphy of this region by means of a period of coastal progradation associated with the formation of the mid-shelf shoreline. However, we see little evidence that allows us to discriminate between this model and alternative models that explain the lower muddy sands of the Holocene section as lagoonal deposits (Swift *et al.*, 1973), or as strata deposited during an earlier estuary mouth environment (Swift *et al.*, 1978). Furthermore, these deposits appear to be too high in the cored strata to be consistent with the "degraded barrier" model of origin.

The contention that the terrace on the New Jersey shelf above the mid-shelf scarp bears the relict pattern of two separate episodes of coastal progradation causes us difficulty. Careful examination reveals topographic patterns and topography-grain size relationships that characterize the ridge fields of post-transgressive origin elsewhere on the shelf, and no characteristics that are unique to beach ridges or barriers. Stubblefield and others invoke massive erosion and redeposition within this terrace in order to explain why their barriers fail to meet the stairstep criterion of Gilbert (1880). We conclude that the terrace certainly "hath suffered a sea change into something rich and strange" and that its surface has been restructured by post-transgressive processes, resulting in a sand ridge topography similar to that which we see elsewhere on the North American Atlantic shelf.

REFERENCES

CURRAY, J.R., F.J. EMMEL, AND P.J.S. CRAMPTON, 1969, Lagunas costeras, un simposio, *in* Memo. Simp. Int. Lagunas Costeras: UNAM-UNESCO, Nov. 28-30, 1967, Mexico, p. 63-100.

DILLON, W.P., AND R.M. OLDALE, 1978, Late glacial sea-level curve: Reinterpretation based on glacio-eustatic influence: Geology, v. 6, p. 56-60.

EMERY, K.O., AND E. UCHUPI, 1972, Western North Atlantic Ocean: Topography, rocks, structure, water, life, and sediments: Am. Assoc. Petroleum Geologists Memoir 17, 253 p.

GILBERT, G.K., 1880, Contributions to the history of Lake Bonneville: U.S. Geol. Survey, 2nd Ann. Rep., p. 169-200.

HOWARD, J.D., AND H.E. REINECK, 1972, Physical and biogenic sedimentary structures of the nearshore shelf: Senckenbergiana Marit., v. 4, p. 81-124.

KELLEY, J.T., 1981, Comment on Quaternary rivers on the New Jersey Shelf: Relation of seafloor to buried valleys: Geology, v. 9, p. 98.

MCKINNEY, T.G., W.L. STUBBLEFIELD, AND D.J.P. SWIFT, 1974, Large-scale current lineations, central New Jersey shelf: Investigations by side scan sonar: Marine Geol., v. 17, p. 79-102.

SANDERS, J.E., AND N. KUMAR, 1975, Evidence of shoreface retreat and in place "drowning" during Holocene submergence of barriers, shelf off Fire Island, New York: Geol. Soc. America Bull., v. 86, p. 65-76.

SHERIDAN, R.E., C.E. DILL, AND J.C. KRAFT, 1974, Holocene sedimentary environment of the Atlantic inner shelf off Delaware: Geol. Soc. America Bull., v. 85, p. 1319-1328.

STAHL, J., J. KOCZAN, AND D.J.P. SWIFT, 1974, Anatomy of a shoreface-connected sand ridge on the New Jersey shelf: Implications for the genesis of the shelf surficial sand sheet: Geology, v. 2, p. 117-120.

STUBBLEFIELD, W.L., AND D.J.P. SWIFT, 1981, Grain-size variations across sand ridges, New Jersey continental shelf, U.S.A.: Geo-Marine Letters, v. 1, p. 45-48.

SWIFT, D.J.P., 1973, Delaware Shelf Valley: Estuary retreat path, not drowned river valley: Geol. Soc. America Bull., v. 84, p. 2743-2748.

_____,D.B. DUANE, AND T.F. MCKINNEY, 1973, Ridge and scale topography of the Middle Atlantic Bight, North America: Secular response to the Holocene hydraulic regime: Marine Geol. v. 15, p. 227-247.

_____,AND M. FIELD, 1981, Evolution of a classic sand ridge field: Maryland sector, North American inner shelf: Sedimentology, v. 28, p. 461-482.

_____,J.W. KOFOED, F.P. SAULSBURY, AND P. SEARS, 1972, Holocene evolution of the shelf surface, south and central Atlantic shelf of North America, *in* D.J.P. Swift, D.B. Duane, and O.H. Pilkey (eds.), Shelf Sediment Transport: Process and Pattern: Dowden, Hutchinson, and Ross, Stroudsburg, Pennsylvania, p. 499-574.

_____,R. MOIR, AND G.L. FREELAND, 1980, Quaternary rivers on the New Jersey shelf: Relation of seafloor to buried valleys: Geology, v. 8, p. 276-280.

_____,P.C. SEARS, B. BOHLKE, AND R. HUNT, 1978, Evolution of a shoal retreat massif, North Carolina shelf: Inferences from areal geology: Marine Geol., v. 27, p. 19-42.

TAYLOR, P.A., AND K.R. DYER, 1977, Theoretical models of flow near the bed and their implications for sediment transport, *in* E.D. Goldberg (ed.), The Sea; Ideas and observations on progress in the study of the seas, v. 6, Marine modeling: John Wiley and Sons, New York, p. 579-601.

RECOGNITION OF TRANSGRESSIVE AND POST-TRANSGRESSIVE SAND RIDGES ON THE NEW JERSEY CONTINENTAL SHELF: REPLY

W.L. STUBBLEFIELD, D.W. MCGRAIL, AND D.G. KERSEY

National Oceanic and Atmospheric Administration, National Sea Grant Program, 6010 Executive Boulevard, Rockville, Maryland 20852; Department of Geological Oceanography, Texas A&M University, College Station, Texas 77843; and Reservoirs Inc., Houston, Texas 77043

ABSTRACT

Arguments proposed by Swift *et al.* (this volume) suggest that Stubblefield *et al.* (this volume) incorrectly infer that the New Jersey mid-shelf sand ridges represent a relict strandline. We suggest that the ridges are a product of three processes: barrier progradation, barrier degradation during marine transgression, and ridge aggradation by mid-shelf currents. Fauna and grain-size data argue strongly that the muds encased within the ridges were originally deposited as part of the open shoreface and not in either lagoonal or estuarine environments. Our topograhic and stratigraphic patterns are in conflict with a relict strand plain model. These patterns support a model for multiple-stage evolution.

INTRODUCTION

We thank Swift *et al.* (this volume) for their discussion of our paper (Stubblefield *et al.*, herein) in which they focused their differences with our paper toward two main areas: muddy events on the inner shelf and the concept of "degraded barriers". We will refer to new data from the same general area which strongly supports our interpretation of various aspects of our model. These data are from a cooperative cruise involving NOAA and a consortium of oil companies headed by Cities Service. In this reply we will refer to these data as "unpublished data".

MUDDY EVENTS ON THE INNER SHELF

Both our paper and that of Swift *et al.* (this volume) discuss three possibilities for the mud lenses within the New Jersey inner shelf ridges. We wish to emphasize that the ridges on which we have a difference of opinion are on the mid-shelf. In our paper we differentiated between the nearshore and mid-shelf "provinces" of the inner shelf. The mid-shelf ridges, in water depths of 20-40 m on the New Jersey shelf, were differentiated from the nearshore ridges on the basis of bathymetry (Stubblefield *et al.*, this volume).

One of the models suggested purports that the muds on the present mid-shelf are part of a thick lagoonal sequence. Swift *et al.* project the mud horizons in several of our cores into profile A (see Fig. 3, Swift *et al.*, this volume). However, the cores should be properly projected on profile F instead of profile A (see Stubblefield *et al.*, this volume, Figs. 1 and 2). When projected on profile F, the muds are substantially lower in the stratigraphic sequence (Fig. 1, this paper). This lower stratigraphic position of the muds may be used as evidence to support a multiple stage model for deposition on the lower shoreface.

In a ridge in comparable water depths and adjacent to the ridge in our Figure 1, similar muddy lenses are found in approximately the same stratigraphic position (unpublished data). In Figure 1 the uppermost muddy horizon is 4.75 m below the ridge crest, whereas the first observable mud is 4.5 m beneath the crest in the unpublished data. The microfauna in the muddy horizons of the cores contain the foraminifera *Elphideum clavatum*, *Ammonia beccarii*, and *Nonion* sp. (Jack Etter, personal communication, 1981). All of these species are common in the shallow open-marine waters of the New Jersey shelf (Culver and Buzas, 1980). No species which are restricted to marshy coastal environments (Culver and Buzas, 1980) were found. We cannot say with absolute certainty that the microfauna in these cores were not found in the late Pleistocene or early Holocene New Jersey lagoons. In fact, little is known of the microfauna distribution in modern New Jersey lagoons. However, comparison with the living microfauna in adjacent coastal areas indicates that those found in these cores were not living in a lagoon. In Long Island Sound the only microfauna common to our microfauna is *E. clavatum*, in abundance, and trace amounts of *Nonion* sp. (Buzas, 1965). Slightly farther north in Buzzard's Bay and Vineyard Sound, Massachusetts, with the exception of *E. clavatum* and *A. breccarii*, all the identified species are either absent or very rare (Murray, 1968, 1969). All of these areas are significantly less restricted than the typical lagoons, implying that the absent or very rare species would be even rarer in the lower saline lagoons.

The identified microfauna, plus the absence of any microfauna restricted to a marshy environment, strongly indicate that the muds were deposited on the open shoreface. As such, the species from these muddy horizons complement those found by us in the sand fraction adjacent to the muds (Stubblefield, *et al.*, this volume, Table 2) and support our model of a prograding barrier.

Swift *et al.* (this volume) discuss the possibility that the muddy facies was deposited in an estuary. Our model of deposition of the muddy facies on the shoreface of a prograding barrier is rejected by Swift *et al.* (this volume) as being "not well founded." Yet the

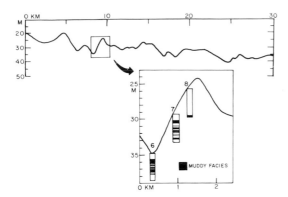

FIG. 1.—Projection of cores 6, 7, and 8 onto profile of Figure 2 of Stubblefield et al. (this volume). The O km position in this figure equates to approximately the 23 km position in Figure 2.

model of Swift et al. (1978) for the North Carolina shelf ridges is based on a similar data set as ours where they examine, among other parameters, the grain-size gradient of a series of vibracores. They base their interpretation of an estuarine depositional environment on a grain-size discontinuity and four radiocarbon dates (Swift et al., 1978, Fig. 8b; Fig. 2, this paper). In the three cores which contained both units, this discontinuity varies from 0.5φ size (cores P1 and P3) to in excess of 1φ (core P6). With one exception, a comparable grain-size discontinuity is not found in the New Jersey mid-shelf ridge cores (compare Stubblefield et al., this volume, Fig. 9 to Swift et al., 1978, Fig. 8b; Fig. 2, this paper). Our only core resembling a discontinuity of the magnitude described by Swift et

al. (1978) is in core 7. This particular event was intepreted by Stubblefield et al. (1975) and Stubblefield and Swift (1976) as indicative of a storm event on the midshelf. This interpretation was based on radiocarbon dates and a piece of coal, presumably from a steamship, which indicate that these near-surface grain-size discontinuities occurred during historical times. We view these discontinuities as events associated with ridge modification and not ridge genesis. Thus, our grain-size data do not support an interpretation of an estuarine deposit overlain by shelf sands on the New Jersey shelf because, with one exception, there is no grain-size discontinuity as required by the model favored by Swift et al. (1978). The limited geographic extent of the grain-size discontinuity, as evidenced in one core, strongly suggests that the process which brought about this discontinuity was limited in time and space. Interestingly, the mean grain-size distribution used by Swift et al. (1978; Fig. 2, this paper) to describe the estuarine sands in their core P1 is the same mean grain size distribution as for the surficial sands in their core P3.

In their discussion, Swift et al. comment on our lack of analysis of grain size and structure. Figure 9 (Stubblefield et al., this volume) includes grain size and structure, and includes grain-size gradients throughout the sand horizons. Our core data, as presented in Figure 9, compare favorably with the criteria suggested by Sanders and Kumar (1975) as indicative of shoreface deposition. The grain size of our mid-shelf ridge cores (1, 4, 5-8) is within the 2.25φ to 3φ range suggested by these workers (p. 66) as common in shoreface environments. The sand horizons in our cores are characterized by the absence of ripple cross-lamination, high-angle cross-stratification, and bioturbation. The only

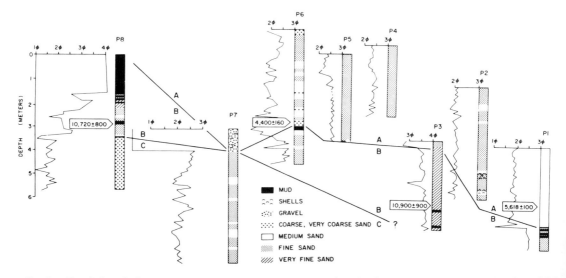

FIG. 2.—Correlation of vibracores in the Platt Massif, North Carolina (from Swift et al., 1978, Fig. 8b). Unit A is surficial sand. Unit B is estuarine sand. Unit C is older substrate.

bioturbation within the sand is restricted to our near-shore cores (A6, A7; Fig. 9). The structures observed in our cores are those common to shoreface environments. We feel that the grain size trends and sequence of sedimentary structures support our interpretation of the origin of these sediments.

Swift *et al.* (this volume) also state that Sanders and Kumar report interbedded layers of mud with thinly laminated sand of the shoreface. Swift *et al.* imply that our sandy units are too thick to satisfy the observations of Sanders and Kumar. The latter authors, however, suggest that sand horizons of 1 m thickness may result from a single storm (p. 67). By using their argument (p. 68) of sand from multiple storms events being stacked without an intervening coarse layer, an even thicker sand deposit may result. Thus, the thickness of our sand units is not in opposition to the values suggested for shoreface deposits by Sanders and Kumar.

Laminations within the sands are frequently difficult to recognize due to distortions of the unconsolidated sediment resulting from vibracoring operations. Unfortunately, our cores (Stubblefield *et al.*, this volume; Fig. 9) are no longer available for re-examination. However, in the companion study of an adjacent mid-shelf ridge (unpublished data), careful analysis indicates that thin, subhorizontal laminations occur throughout many of the sands interbedded within the muddy layer.

Finally, Swift *et al.* (this volume) use the grain-size distribution reported by Stubblefield and Swift (1981) as an argument against a barrier system. Swift *et al.* state that "if either of these were barriers, the high-energy coarse-grained size would be expected to be on the seaward side." However, our model requires ridge degradation with subsequent aggradation, thus the surficial grain size is not a reflection of barrier genesis. Stubblefield *et al.* (this volume) argue that the surficial size distribution is a function of recent modification by mid-shelf currents.

In summary, we believe that in addition to the conclusions drawn from the paleontology data that the grain size data, sedimentary structures, and sedimentary sequence agree sufficiently well with Sanders and Kumar's (1975) observations to also aid in supporting a model for shoreface deposition.

DEGRADED BARRIERS

The second major point raised by Swift *et al.* (this volume) is in regard to our concept of a degraded barrier. They termed our concept as "radical". We do, however, wish to make it clear that although Swift *et al.* equate our model with a "relict strandline", which is a subaerial coastline feature (Curray *et al.*, 1969; AGI, 1974, p. 696), we visualize on the New Jersey Shelf a barrier system with the dunes and upper shoreface removed. As such, the label of "relict strandline" is incorrect.

Swift *et al.* (this volume) focus their discussion of our degraded barrier model on two points: a stairstep criterion of Gilbert (1880) and certain "geometric problems". With Gilbert's criterion the shoreward barrier is higher in elevation than the adjacent seaward barrier. Swift *et al.* argue that our Figure 6 does not show the same stairstepping as does our schematic drawing (Stubblefield *et al.*, this volume, Fig. 11). As we stated in our previous paper, Figure 11 was not drawn to scale, but the geometry was representative of our cross-sectional profiles (Stubblefield *et al.*, this volume, Fig. 6). There is, in fact, a discordance between the two mid-shelf ridge sets over most of the area we studied. Lines, which better show this stairstep relationship between the two ridge sets, have been added to our Figure 6 (Fig. 3, this paper). Conservatively, a difference in height from 5 to 9 m is found. The exception is along profiles A and B, which are at the very northern limit of the well-developed mid-shelf ridge field (see Stubblefield *et al.*, this volume, Fig. 2). The stairstepping between ridge sets is much in evidence on the New Jersey mid-shelf, thus supporting Gilbert's criterion.

The second point raised by Swift *et al.* (this volume) in their discussion of degraded barriers is based on geometry. They question whether the mid-shelf ridges are truly coast-parallel. Obviously, there are several methods which can be used in interpreting the bathymetry. Our first approach in mapping the ridges was similar to that of McKinney *et al.* (1974), which was cited by Swift *et al.* (this volume, their Fig. 1b). We found this method to be too subjective because of the potential operator error involved in measuring the trend of the ridges. Instead we used the more objective approach of the closed-contour map (Stubblefield *et al.*, this volume, Fig. 3) as first suggested for the New Jersey shelf by McClennen and McMaster (1971). When this method is used, the mid-shelf ridges are approximately coast-parallel in alignment, rather than characterized by a large "Z" pattern as suggested by McKinney *et al.* (1974). We feel that the closed-contour map method is reproducible and objective, and we feel that this method correctly delineates the orientations of the ridges.

Swift *et al.*, (this volume) use the geometry of the sand ridges as a forum for discussing the origin of the larger mid-shelf ridges. We are somewhat confused by their arguments. Apparently, they suggest that the large, first-order ridges (McKinney *et al.*, 1974, Fig. 2a; Swift *et al.*, this volume, Ridges A-D, Fig. 1b) were cut into the massif (their term) by strong tidal currents. This is the model proposed by McKinney *et al.* (1974). Yet Swift *et al.* indicate, in their Figure 1b and their discussion, that one of these ridges, A, is a relict lower shoreface marking a period of near still-stand. It is unclear which of these two models they advocate for this particular ridge.

Equally unclear is their argument for the coast-oblique portions of the large ridges in their Figure 1b. Apparently, they argue that such orientations are a product of ridge migration and coastal retreat (Swift *et al.*, this volume, Fig. 12; Swift and Field, 1981). Their

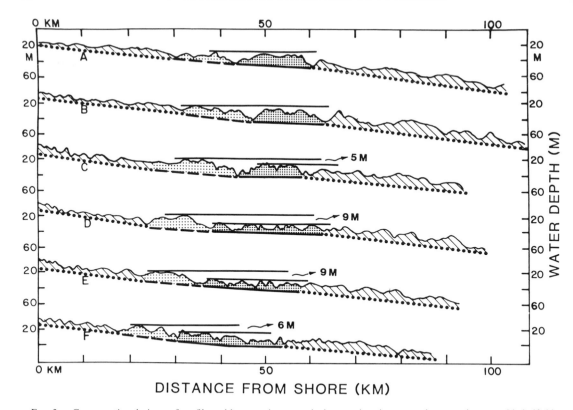

FIG. 3.—Cross-sectional views of profiles with approximate vertical separation, in meters, between the two mid-shelf ridge sets (modified after Stubblefield *et al.*, this volume, Fig. 6).

model implies that, with ridge migration and coastal retreat being equal, the nearshore ridges will intersect the coastline at a 45° angle. Similarly, with very little coastal retreat and periodic ridge migration, the nearshore ridges will assume a coast-parallel orientation. However, such a relationship remains to be documented. To the contrary, some evidence based on age dating questions this relationship. Radiocarbon dates from the "unpublished core data" indicate that a shoreface-connected ridge, in water depths of 10 m, was probably initiated within the past 1600 years. The sea-level curves for this general area indicate a maximum of 2 m rise during this time (Belknap and Kraft, 1977). This amount, which is considered extreme on other curves (e.g., Milliman and Emery, 1968), equates to approximately 350 m of horizontal transgression on the New Jersey coastline. In contrast, recent ridge migration has been documented on the coasts of the Middle Atlantic Bight of the eastern U.S. which greatly exceeds this amount of coastal retreat. Duane *et al.* (1972) reported 3600 m of net movement for ridges of Chincoteague, Virginia, in a 53-year period, whereas Moody (1964) documented 76 m of migration for some Delaware ridges during the single Ash Wednesday storm of 1962.

Swift and Field (1981) provide no measure for coastal retreat or ridge migration to account for various ridge orientations; therefore, it is difficult to evaluate their model in light of available data. However, one may argue that the apparent large amount of ridge migration relative to coastal retreat indicates that the relatively young nearshore ridge should be approaching a coast-parallel trend. Instead, when the ridge's trend is projected into the coastline, the resulting angle is 25°, an angle typical of many nearshore ridges along the U.S. Atlantic Coast (Duane *et al.*, 1972). If the relationship suggested by Swift and Field (1981) is not valid, then another mechanism must be sought to explain the coast-oblique and coast-normal portions of the large ridges in the interpretation advanced by McKinney *et al.* (1974) (Swift *et al.*, this volume, Fig. 1b). With our interpretation (Stubblefield *et al.*, this volume, Fig. 3), the origin of the ridge orientation is explained.

It is interesting that McKinney *et al.* (1974; Swift *et al.*, this volume, Fig. 1b) recognize a pattern of ridge orientations not unlike what we see in our closed-contour map (Stubblefield·*et al.*, this volume, Fig. 3). In the nearshore, the ridges intersect the coast at the well-recognized oblique angle. The ridges seaward of their ridge D have a trend which is at an oblique angle to the trend of the ridge D. Could these be the outer shelf coast-oblique ridges of Stubblefield *et. al.*, and possibly a partial counterpart of the present nearshore ridges barrier system?

CONCLUSIONS

Stubblefield *et al.* (this volume) proposed a model to explain the mid-shelf sand ridges on the New Jersey shelf. Swift *et al.* (this volume) contend that our model for the mid-shelf ridges is weak on two grounds: the muddy events in the mid-shelf ridges and our concept of degraded barrier. Macrofauna and microfauna data described in this paper from the muddy events and adjacent sands suggest that these muds were deposited on a shoreface and not in a lagoon as proposed by Swift *et al.* (1973). Likewise, the grain-size gradient for the New Jersey mid-shelf ridges does not show the consistent, sharp discontinuity which Swift *et al.* (1978) use as an argument for estuarine sands overlain by surficial sands on the North Carolina shelf. Similarly, the structure and sedimentary sequence found in cores from the mid-shelf ridges are compatible with deposition on a shoreface. As a result, we don't believe that Swift *et al.* (this volume) have presented conclusive arguments against the muddy horizons on the New Jersey shelf being part of a prograding barrier sequence.

The degraded barriers of our model are not relict strandlines as stated by Swift *et al.* (this volume). We propose a model of barrier degradation during the marine transgression followed by ridge aggradation on the mid-shelf (Stubblefield *et al.*, this volume, Fig. 11). The barrier degradation includes removal of the lagoonal muds by marine transgression prior to progradation of a more landward barrier. Our degraded barrier concept satisfies Gilbert's (1880) stairstep criterion. Figure 3 indicates that a stairstep occurs between two ridge sets over much of the New Jersey mid-shelf. Our method of determining the ridge orientation is an objective method and is at least equal to the approach of McKinney *et al.* (1974).

In summary, the data presented in Stubblefield *et al.* (this volume) and this paper support our interpretation of two ridge types on the New Jersey shelf. The mid-shelf ridges are degraded barriers which are being modified by shelf currents, whereas the nearshore ridges are post-transgressive features forming as shoreface-connected ridges.

REFERENCES

AGI, 1974, Glossary of Geology: American Geol. Inst., Washington, D.C., 805 p.

BELKNAP, D.F., AND J.C. KRAFT, 1977, Holocene relative sealevel changes and coastal stratigraphic units on the northwest flank of the Baltimore Canyon Trough geosyncline: Jour. Sed. Petrology, v. 47, p. 610-629.

BUZAS, M.A., 1965, The distribution and abundance of foraminifera in Long Island Sound: Smithsonian Misc. Coll., v. 149, 89 p.

CULVER, S.J., AND M.J. BUZAS, 1980, Distribution of recent benthic foraminifera off the North American Atlantic coast: Smithsonian Contrib. Marine Sci., No. 6, 512 p.

CURRAY, J.R., F.J. EMMEL, AND P.J.S. CRAMPTOM, 1969, Lagunas costeras, un simposio, *in* Memo. Simp. Int. Lagunas Costeras: UNAM-UNESCO, Nov. 28-30, 1967, Mexico, p. 63-100.

DUANE, D.B., M.E. FIELD, E.P. MEISBURGER, D.J.P. SWIFT, AND S.J. WILLIAMS, 1972, Linear shoals on the Atlantic inner shelf, Florida to Long Island, *in* D.J.P. Swift, D.B. Duane, and O.H. Pilkey (eds.), Shelf Sediment Transport: Process and Pattern: Dowden, Hutchinson, and Ross, Stroudsburg, Pennsylvania, p. 447-499.

GILBERT, G.K., 1880, Contributions to the history of Lake Bonneville: U.S. Geol. Survey, 2nd Ann. Rep., p. 169-200.

McCLENNEN, C.E., AND R.L. McMASTER, 1971, Probable Holocene transgressive effects on the geomorphic features of the continental shelf off New Jersey, United States: Maritime Sediments, v. 7, p. 69-72.

McKINNEY, T.F., W.L. STUBBLEFIELD, AND D.J.P. SWIFT, 1974, Large-scale current lineations, central New Jersey shelf: Investigations by sidescan sonar: Marine Geol., v. 17, p. 79-102.

MILLIMAN, J.D., AND K.O. EMERY, 1968, Sea levels during the past 35,000 years: Science, v. 162, p. 1121-1123.

MOODY, D.W., 1964, Coastal morphology and processes in relation to development of submarine sand ridges off Bethany Beach, Delaware: Unpublished PhD Dissertation, Johns Hopkins Univ., Baltimore, Maryland, 167 p.

MURRAY, J.W., 1968, Living foraminifers of lagoons and estuaries: Micropaleontology, v. 14, p. 435-455.

_____, 1969, Recent foraminifers from the Atlantic continental shelf of the United States: Micropaleontology, v. 15, p. 401-419.

SANDERS, J.E., AND N. KUMAR, 1975, Evidence of shoreface retreat and in place "drowning" during Holocene submergence of barriers, shelf off Fire Island, New York: Geol. Soc. America Bull., v. 86, p. 65-76.

STUBBLEFIELD, W.L., J.W. LAVELLE, D.J.P. SWIFT, AND T.F. McKINNEY, 1975, Sediment response to the present hydraulic regime on the central New Jersey shelf: Jour. Sed. Petrology, v. 45, p. 337-358.

_____, AND D.J.P. SWIFT, 1976, Ridge development as revealed by subbottom profiles on the central New Jersey shelf: Marine Geol., v. 20, p. 315-334.

_____, AND _____, 1981, Grain-size variation across sand ridges, New Jersey continental shelf, U.S.A.: Geo-Marine Letters, v. 1, p. 45-48.

SWIFT, D.J.P., D.B. DUANE, AND T.F. McKINNEY, 1973, Ridge and swale topography of the Middle Atlantic Bight, North America: Secular response to the Holocene hydraulic regime: Marine Geol. v. 15, p. 227-247.

_____, AND M.E. FIELD, 1981, Evolution of a classic sand ridge field: Maryland sector, North American inner shelf: Sedimentology, v. 28, p. 461-482.

_____, P.C. SEARS, B. BOHLKE, AND R. HUNT, 1978, Evolution of a shoal retreat massif, North Carolina shelf: Inference from areal geology: Marine Geol. v. 27, p. 19-42.

SAND BODIES ON MUDDY SHELVES: A MODEL FOR SEDIMENTATION IN THE WESTERN INTERIOR CRETACEOUS SEAWAY, NORTH AMERICA

DONALD J. P. SWIFT AND DUDLEY D. RICE

Exploration and Production Research, Arco Oil and Gas Company, Box 2819, Dallas, Texas 75221;
and U.S. Geological Survey, Box 25046, Denver, Colorado 80225

ABSTRACT

The continental shelf on the western margin of the Cretaceous western interior seaway was a muddy surface which bore abundant northwest-southeast trending sand bodies, as much as 20 m thick and many km long. Important examples are the Medicine Hat Sandstone, the Mosby Sandstone Member of the Belle Fourche Shale, the Shannon and Sussex Sandstone Members of the Cody Shale, and the Duffy Mountain sandstone and the Tocito Sandstone Lentil of the Mancos Shale. These deposits resemble the storm-built and tide-built sand ridges reported from the modern Atlantic Continental Shelf or from the Southern Bight of the North Sea. However, although modern sand ridges may protrude from the Holocene transgressive sand sheet through overlying Holocene mud deposits to be exposed on the present sea floor, no modern examples are known where sand ridges are completely encased in mud, as the Cretaceous examples seem to have been.

Hydrodynamical theory suggests that special circumstances may allow the formation of sand bodies from a storm flow regime whose transported load consists of sandy mud. Under normal circumstances, such a transport regime would deposit little clean sand. The sea floor is eroded as storm currents accelerate, but erosion ceases when the boundary layer becomes loaded with as much sediment as the fluid power expenditure will permit (flow reaches capacity). Deposition of a graded bed occurs as the storm wanes and a storm sequence is likely to consist of thin clay beds with basal sand laminae. However, slight topographic irregularities in the shelf floor may result in horizontal velocity gradients, so that the flow undergoes acceleration and deceleration in space as well as in time. Fluid dynamical theory predicts deceleration of flow across topographic highs as well as down their lee sides. The coarsest fraction of the transported load (sand) will be deposited in the zone of deceleration, and deposition will occur throughout the flow event. Relatively thick storm beds (2 to 10 decimeters) can accumulate in this manner. Enhancement of initial topographic relief results in positive feedback; as the bedform becomes larger, it extracts more sand from the transported load during each successive storm. Individual storm beds may tend to fine upward (waning current grading), but the sequence as a whole is likely to coarsen upwards, reflecting increasing perturbation of flow by the bedform as its amplitude increases.

Stability theory suggests that the end product of these processes should be a sequence of regularly spaced sand ridges on the shelf surface. However, sandstone bodies within the Cretaceous shelf deposits are quite localized in stratigraphic position and lateral distribution. Upward-coarsening sequences are a widespread phenomenon in the western interior Cretaceous system, and the sandstone bodies appear to constitute localized sand concentrations within more extensive sandy or silty horizons. Especially widespread upward-coarsening sequences appear to be due to the close coupling between activity in the overthrust belt to the west and sedimentation in the foreland basin. Each thrusting episode increased relief as well as the load on the crust. Initially, the increased relief as well as the load on the crust. Initially, the increased relief resulted in a flood of sediment transported to the shelf on the western margin of the basin so that the shelf became shallower. As it did so, wave scour on the shelf floor increased the amount of bypassing and resulted in the deposition of increasingly coarser sediment. As relief in the hinterland waned, subsidence overtook sedimentation and the shelf subsided. Renewed thrusting began the cycle anew. In a second mechanism for the formation of upward-coarsening sequences, tectonic uplift affect portions of the shelf as well. The initiation of Sevier or Laramide structural elements beneath the shelf and the remobilization of other, older structures created submarine topographic highs which caused slight sand enrichment over broad areas. The development of sand-enriched areas in the shelf floor by both mechanisms led to the flow-substrate feedback behavior that built large-scale, elongate bodies of clean sand.

INTRODUCTION

During Late Cretaceous time, a north-south trending subduction zone beneath the westward-moving North American Continent created a rising north-south trending land mass along the approximate extent of the present Rocky Mountains (Armstrong, 1968; see Fig. 1). An elongate epeiric sea (foreland basin) lay east of this landmass throughout the Late Cretaceous and extended from the Gulf of Mexico to the Arctic Ocean during portions of this time (Gill and Cobban, 1973). As the Cordilleran highlands underwent repeated deformation, uplift, and denudation, sediment was transported to the coast to form a clastic wedge extending into the western half of the Cretaceous seaway (see, for exam-

ple, Fig. 2). Tectonic movements interacted with sea level fluctuations so that several regressive-transgressive sequences were formed before the final withdrawal of the seaway in Late Cretaceous time (Jordan, 1982). In the eastern half of the sea, a carbonate lithotope moved east or west in response to shifts of the clastic lithotopes of the western margin (Rice and Shurr, 1983).

The coastal and shelf lithosomes of the clastic wedge were exposed by later Laramide movements in late Cretaceous and Tertiary time and have been well studied (summaries in Rice and Shurr, 1983; Shurr, this volume). More basinward deposits have been drilled during the coarse of petroleum exploration and production (Asquith, 1970, 1974). Recent studies have shown that a shelf-slope morphology with a well defined shelf break was generally present, although the slope was not extensive, beginning at perhaps a 100 m water depth and extending to a basin floor several hundred meters below. During the Campanian, the time interval considered in this report, the northern portion of the shelf (Montana) prograded eastward and eventually extended completely across the seaway.

The sediment load shed by the rising Cordilleran highlands was rich in fine sediment and probably high in volume; Gill and Cobban, (1973) report sedimentation rates of 0.15 to 8.0 x 10^{-3} cm yr on the basis of ammonite zonation. Gravel and much of the sand were trapped out on the coastal plain; more sand was removed from the sediment load by coastal and littoral environments. Sediment bypassing the coast to be deposited on the shelf and beyond consisted mainly of silt and clay with only subordinate amounts of sand. This sediment type built the massive shale accumulation of the Cretaceous seaway, consisting of such major units as the Mowry, Belle Fourche, Mancos, Pierre, Colorado, Cody, Gammon, and Carlile Shales (Fig. 3).

Subsurface exploration has shown, however, that the Upper Cretaceous shelf east of the Rockies also developed isolated sandstone bodies, particularly at times of maximum regression. These bodies occur within the several sandstone units shown in the correlation dia-

gram (Fig. 3) as the Medicine Hat Sandstone, the Mosby Sandstone member of the Belle Fourche Shale, the Tocito Sandstone lentil of the Mancos Shale, the Shannon and Sussex Sandstone members of the Cody Shale, and the Duffy Mountain Sandstone Member of the Mancos Shale (Rice, this volume; Spearing, 1976; Molenaar, 1973; Brenner, 1978; Boyles and Scott, 1982; Hobson et al., 1982). A detailed study of one of these units (Shannon Sandstone; Fig. 3) has suggested that it is organized into a hierarchy of sandstone bodies of five different sizes (Shurr, this volume; Fig. 4, this paper). Two important categories within this hierarchy are the lentil and elongate lens. Lentils are 15 to 23 m thick and up to 1500 km^2 in areal extent. They are composed of a series of elongate lenses, 12 to 20 m thick and about 50 km^2 in area. In Montana and northern Wyoming they are oriented northwest-southeast, with widths of several kilometers and lengths of 10 km or more (Rice and Shurr, 1983). In southern Wyoming and northwestern Colorado, the orientation of younger sandstone lenses swings to northeast-southwest (Boyles and Scott, 1982). As noted by Shurr (this volume), the geometry is strongly reminiscent of the storm-built sand ridges described from the Atlantic Continental Shelf of North America (Swift et al., 1973; Swift and Field, 1981), the shelves of southern Brazil and northern Argentina, and the west Frisian Coast of the North Sea (Swift et al., 1978a). It also resembles the tidal sand ridge topographies of Nantucket Shoals or the Southern Bight of the North Sea (Mann et al., 1981; Kenyon et al., 1981). A recent theoretical analysis by Huthnance (1982; see discussion by Figueiredo et al., 1981) suggests that both tide- and storm-built ridges may be formed by similar hydrodynamic mechanisms.

The Cretaceous shelf sandstone bodies possess a disconcerting characteristic; they are pods of sandstone enclosed entirely in shale and may occur several hundred kilometers from the time-equivalent paleoshoreline. Modern sand ridges, on the other hand, are molded into the upper surface of the transgressive sand sheet that constitutes the base of the Holocene section on all modern shelves (Review of Swift, 1976b). For instance, on the Atlantic Continental Shelf and in the Southern Bight of the North Sea, the ridged sand sheet is fully exposed. Mud patches may form in the swales between ridges during the quiescent summer period, and these may persist in very nearshore areas where the suspended sediment load is relatively high as on the New Jersey Coast (Swift, 1976a; Fig. 34). Elsewhere, on the southern New England Coast (Twichell et al., 1981), Spanish Mediterranean Coast (Maldonado et al., in press) or outer Amazon Shelf (Kuehl et al., 1982), modern mud deposits are burying or have buried sand ridge topographies. However in these cases, the sand ridges still protrude up into the mud from an underlying sand sheet. The sand, commonly accompanied by a basal gravel, rests disconformably on older Quaternary deposits. Because of innate

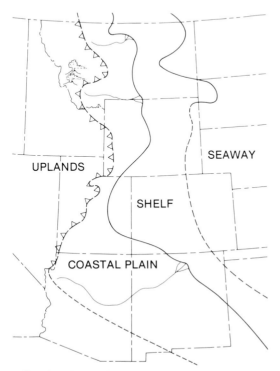

FIG. 1.—Campanian paleogeography of the Northern Rockies. Hachured line denotes the Absaroka and related thrusts (after Armstrong, 1968; Rice and Shurr, 1983).

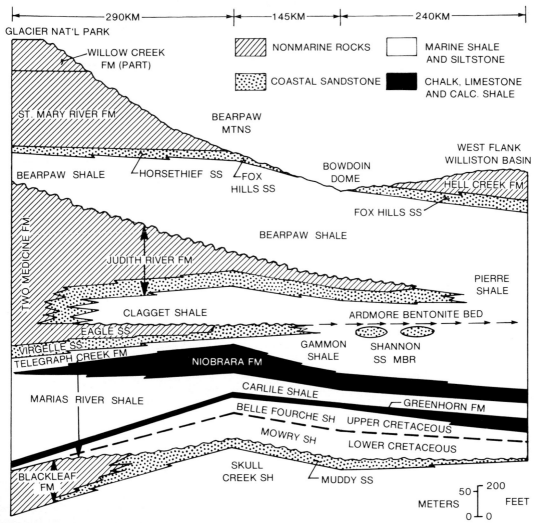

FIG. 2.—Schematic cross-section of the Upper Cretaceous section east of the Rocky Mountains in Montana demonstrating the geometry of the clastic wedge associated with the Sevier Orogeny (modified from Rice and Shurr, 1980).

differences between Mesozoic and Holocene sedimentation, or more probably because of our lack of knowledge of modern shelves, we have yet to locate a modern analog for the shale-enveloped sandstone bodies of the Cretaceous western interior sequence. In order to understand the facies relationships within this deposit, it is necessary to understand how sand can be bypassed across a muddy shelf surface and concentrated into sand bodies. In this paper, we review current fluid dynamical concepts that pertain to the problem and propose a model for sand body formation on a muddy shelf.

SEDIMENT TRANSPORT ON MUDDY SHELVES

Storm-Dominated Flow on Shelves

While we lack a uniformitarian model for sand bodies forming on a mud shelf, we probably do have a model for the shelf flow regime which created the sandy shale deposits of the Cretaceous western interior shelf. The area in question was a north-south-trending, eastward-facing shelf surface in the northern hemisphere and was thus in gross geometry similar to the modern Atlantic Shelf of North America, whose circulation is probably the best understood of any in the world (Beardsley et al., 1976; Mayer et al., 1979). Sediment transport on this shelf is storm-dominated (Vincent et al., 1981a). The tidal component of circulation on most parts of the Middle Atlantic Continental Shelf is mesotidal in nature and is inadequate to drive the observed sediment transport (Swift et al., 1981). The Cretaceous seaway similarly appears to have experienced microtidal to mesotidal regime (Slater, 1981), although tidal behavior on its shelves is poorly understood.

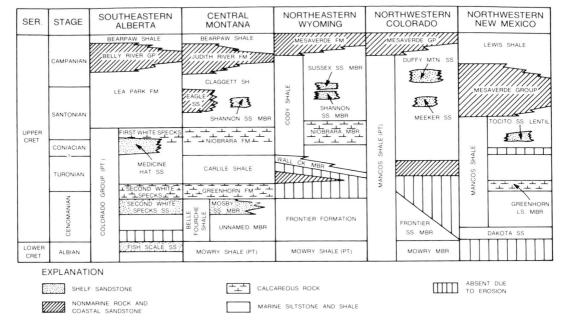

SER.	STAGE	SOUTHEASTERN ALBERTA	CENTRAL MONTANA	NORTHEASTERN WYOMING	NORTHWESTERN COLORADO	NORTHWESTERN NEW MEXICO

FIG. 3.—Correlation chart of Rocky Mountain Cretaceous sections (modified from Rice and Shurr, 1980).

The controlling mechanism for sediment transport on most of the central Atlantic Shelf is geostrophic storm flow (Boicourt and Hacker, 1976; see review of Ekman three-layer current system in Neuman and Pierson, 1966). The efficiency of coupling between wind and water flow depends on the trajectory of the storm with respect to the crescent-shaped shelf segment between Cape Cod and Cape Hatteras. The best "tuning" occurs when the storm crosses the shelf so that the western half of the storm nests into the curve of the shoreline. The contours of atmospheric pressure then parallel the contours of water depth, and northeast winds blow down the arc of the shelf (Beardsley and Butman, 1974; Beardsley et al., 1976). Wind stress causes surface water to move landward, i.e., to the right of the wind (Coriolis effect), until the sea surface slopes upwards toward the beach, creating a "storm tide" of 50 to 100 cm. Water movements associated with the growth of this "coastal set-up" are not particularly significant. The important effect is the response of the entire water column to the pressure gradient caused by the sea surface slope. Water parcels in the fluid interior attempt to flow seaward down the pressure gradient (Fig. 5). As they accelerate, they are deflected steadily to the right by a Coriolis force which intensifies as the parcel accelerates. Eventually, the Coriolis term and the pressure term balance each other in the equation of motion. At this point the entire shelf water mass is flowing parallel to the coast at sustained speeds on the order of 30 cm sec^{-1}. The flow is in the direction of the wind and is caused by the wind, but the nature of the geostrophic coupling between wind and

water is more complex than it appears.

Such storm flows are especially efficient in entraining bottom sediment, because associated storm waves produce a high-frequency component of motion (wave orbital current). Because of its rapidly fluctuating nature, this component increases bottom shear stress in nonlinear fashion (Vincent et al., 1981b; Grant and Madsen, 1979). Storm currents capable of transporting fine-to-medium sand for periods of hours or days occur repeatedly on the Atlantic Continental Shelf during the stormy winter season (Vincent et al., 1981a). While the frequency, intensity, and trajectories of Cretaceous mid-latitude storms may not have been the same as those of the modern Atlantic (Barron and Washington, in press), it is difficult to see how the response of the western shelf of the Cretaceous western interior seaway could have been qualitatively different.

A Temporal Acceleration Model of Shelf Sediment Transport

Storm flows that move over the shelf floor entrain and transport sediment. For steady, uniform flows (ignoring the high-frequency, reversing component due to wave orbital motion), sediment suspension can be described by the Rouse equation (Rouse, 1938). In its simplest form, the Rouse equation states that the mass of sediment particles carried downward by gravitational settling must equal the mass carried upward by turbulent motion:

$$\text{(down)} \; \omega N_z = -\frac{A_s}{\rho} \frac{\partial N}{\partial z} \; \text{(up)},$$

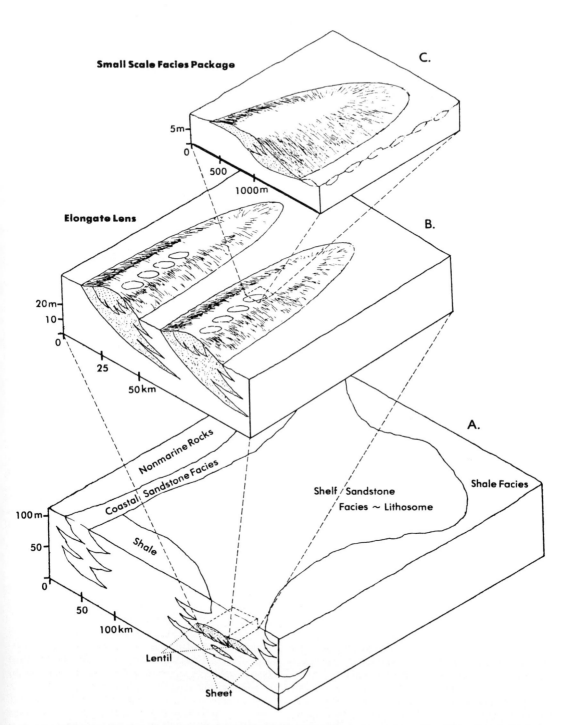

FIG. 4.—Hierarchical nature of sandstone bodies in Shannon Sandstone Member according to Shurr (this volume). Sandstone body classes in decreasing order of size are: lithosome, sandstone sheet, lentil, elongate lens, and small scale facies package.

where ω is the settling velocity, ϱ is the density of the fluid, N_z is the sediment concentration by volume at distance Z above the bottom, and A_S is the eddy viscosity applicable to sediment transfer.

Equation (1) can be integrated from a distance z_1 just above the bed to some distance to give:

$$\frac{N_z}{N_1} = \left[\frac{D-z}{z} \cdot \frac{z_1}{D-z_1} \right]^a , \qquad (2)$$

where D is total depth, a is $\omega/ku*$ and K is the Von Karman constant (0.4 for dilute suspensions; see for instance Ghosh et al., 1981). The concentration at height Z, N_z, can be predicted if the sediment settling velocity ω, the concentration at a reference level N_1, and the shear velocity $u*$, are known ($u*$, a function of the depth-averaged velocity, controls the eddy viscosity, A). Sediment concentration profiles described by the Rouse equation are exponential in nature, decreasing rapidly from the bottom to the first few centimeters above the bottom, and then more slowly as higher levels are attained. For transport loads of finer grain sizes, the profile becomes flatter and is stretched over a greater vertical distance.

If a sand fraction is also present, then some of the sediment will travel as bedload. On the continental shelf, where the mean flow component is strongly wave-modified, the volume bedload concentration, $C^*(t)$, is proportional to the value of the Shields number Ψ (dimensionless shear stress) in excess of that needed to entrain sediment (Vincent et al., 1981b):

$$C^*(t) = 0.09\Psi_{EX^{(t)}} , \qquad (3)$$

where the descriptor (t) refers to the time-varying nature of the parameter over the wave period. Bedload and suspended load concentrations tend to be correlated, since the power available to transport bedload is proportional to the power available to suspend sediment. The relation is even closer if the bottom is sandy; the amount of the interstitial fine sediment released to suspension then depends on the depth to which the bed has been stirred by bedload-transport (Vincent et al., 1981b).

The concept of capacity is implicit in equations (1) through (3); fluid power is proportional to the square of velocity (u^2 or $u*^2$), and in these equations, concentration (N of C^*) is a function of $u*$ (or u). As a storm current accelerates, the bottom is eroded. During the period of peak flow, however, the sediment load in the bottom boundary layer of the flow rapidly adjusts to the available fluid power. As long as velocity remains constant and the sediment load is in equilibrium with it, capacity has been attained and further erosion cannot occur; particles will fall out as fast as they hop up into the load.

The product of accelerating and decelerating storm currents may be observed in the Cretaceous shelf facies of the western interior seaway (Fig. 6). This lithofacies

association is made up of three characteristic lithofacies (Spearing, 1976; Rice and Shurr, 1980; Boyles and Scott, 1982; Rice, this volume). An extensive sandy shale lithofacies contains elongate sandstone lenses (sandstone lithofacies). The sandstone lenses grade into the surrounding shale through thin-bedded shales and sandstones (transitional or thin-bedded lithofacies). The thin-bedded lithofacies was first recognized by Spearing (1976) and has been subdivided by Tillman and Martinsen (1979), whose genetic classification corresponds in general to the spectrum of stratification types described by Reineck and Wunderlich (1968) as flaser bedding, wavy bedding and lenticular bedding (Fig. 7, 8). The thin-bedded lithofacies and especially the sandstone lithofacies is characterized by glauconite (up to 30%) and siderite clasts (Spearing, 1976).

Studies of strata formation on modern shelves, where accumulation rates can be established by radioisotope chronology (Nittrouer and Sternberg, 1981), suggest a "temporal acceleration" model for the formation of the sandy clay and thin-bedded lithofacies. The model is based on the premise that muddy shelves tend to be very flat; fine sediment does not support high marine slopes (Swift, 1976b). Storm currents in this model do not undergo spatial accelerations in response to the rise and fall of the bottom beneath them; they accelerate and decelerate only in time, as the storm current waxes and wanes. The model could in fact be called the square wave model; it is a permissible simplification to describe a current that shifts abruptly from low fair-weather levels of velocity and sediment load to high storm levels, then shifts as abruptly back again. In such a regime it is possible to mix a subordinate quantity of fine sand into a silt and clay load without giving rise to extensive sand deposits, even when storm flows are very frequent and very intense. Each storm can only erode a few centimeters down into the sea bed before capacity is reached; the bed is then effectively armored. Even a 10-year or 100-year storm need not cause more than several decimeters of erosion. As the flow wanes, the transported load may fall out as a size-graded sand and mud couplet (Reineck and Singh, 1972) which may be partly remixed by bioturbation during the ensuing fair-weather period.

In a pure temporal acceleration model, the velocity history of the flow is the same at every place. The same bed must be resuspended again and again so that net long-term aggradation is impossible. In reality, however, the shelf surface over which the water flows has broad, gentle topographic gradients, and the forcing winds also are characterized by gentle regional gradients. The resulting currents must therefore experience slight regional variations in velocity, and therefore more deposition can occur in some areas than in others. Sedimentation under such circumstances is described by the sediment continuity equation (Exner, 1925):

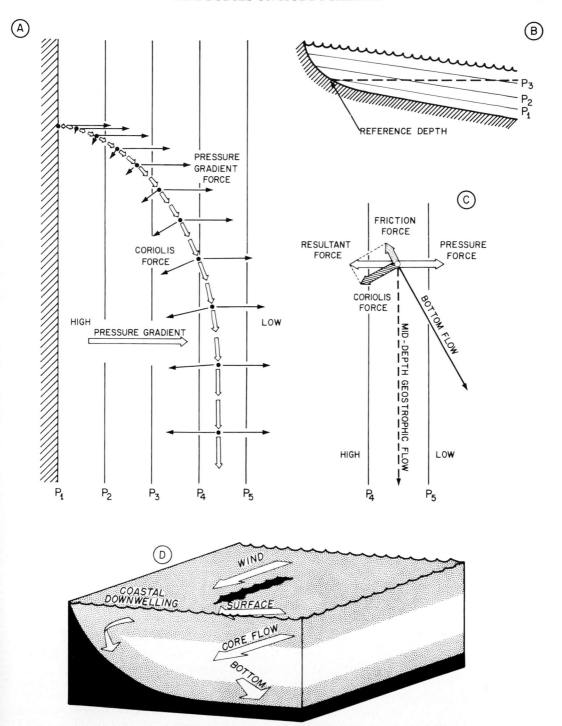

FIG. 5.—Geostrophic flow on the continental shelf. *A*, In this plan view, a parcel of water at a reference depth moves seaward (to the right) in response to pressure-gradient force. As it accelerates, it experiences a Coriolis force impelling it to the right of its trajectory. *B*, Cross-section of hypothetical shelf experiencing geostrophic flow illustrating relation of sea surface slope, isobaric surfaces, and reference depth. *C*, Relation between geostrophic flow and flow in bottom boundary layer. *D*, Block diagram of geostrophic flow, showing landward moving upper boundary layer (stippled), fluid interior (clear), and seaward veering boundary layer (stippled). Modified from Strahler (1963) and Swift (1976a).

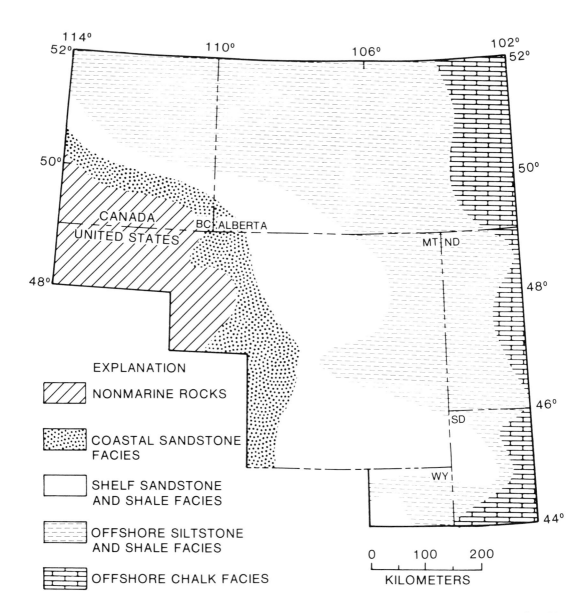

F$_{IG}$. 6.—Lithofacies map of Upper Cretaceous Eagle Sandstone and equivalent rocks in the northern Great Plains (from Rice and Shurr, 1980).

$$\frac{\partial \eta}{\partial t} = - \epsilon \frac{\partial q}{\partial X}, \qquad (4)$$

where q is sediment discharge in gm cm^{-1} normal to the transport direction X, η is bed height relative to a reference level, t is time and ϵ is a dimensional constant related to sediment porosity.

The equation states that the rate of erosion or deposition of the bed is controlled by the intensity of the horizontal transport gradient. Because very broad topographic and storm velocity gradients exist on even very flat and muddy shelves, there may be long-term trends towards deposition. Hence, we may imagine a shelf deposit in which a series of eroded remnants of storm beds and laminae are capped by the full sequence of the last storm. In this sequence, there is only partial fractionation with respect to the bulk grain-size composition of the shelf sediment input. Some of the fine sand is segregated into bedload laminae, but much of it has traveled in suspension together with silt and clay and is disseminated throughout the mud.

Nittrouer and Sternberg (1981), have reviewed calculations of Smith (1977) and have incorporated them

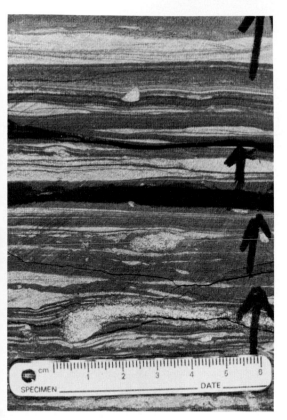

FIG. 7.—Wavy-bedded sandstone (from a core of the basal part of the Shannon Sandstone Member; Getty Oil 20-11 Ione, Sec. 20, T44N, R75W, Hartzog Draw Field, Campbell County, Wyoming).

into a more general model of strata formation by the interaction of biological and physical processes. In this model the character of the sediment formed is defined by the ratio of the sediment mixing rate to the sediment accumulation rate. Mixing may be induced by either burrowing organisms or storm currents. For strata for which physical processes are dominant, the ratio is described by a non-dimensional parameter H:

$$H = \frac{L_p \; F_p}{A}, \qquad (5)$$

where L_p is the depth of erosion by the storm current, F_p is the frequency of storm events, and A is the sediment accumulation rate. L_p and F_p together define storm mixing. Nittrouer and Sternberg (1981) note that as the value of H increases (storm mixing increases) relative to the accumulation rate, the deposit becomes more massive, since only the sandy base of the graded storm bed is preserved for burial and each such base is very similar in grain size to the preceding one (Fig. 9). However, the equation does not parameterize the grain-size characteristics of the sediment. Since the sediment load of the Cretaceous Western Interior seaway was

relatively sand-poor, the stratification sequences are skewed toward the fine-grained, well-stratified end, rather than the massive, coarse end, regardless of the value of H.

Our model for the formation of storm-stratified clay and clay-sand sequences is presented in Figure 10. For simplicity, mud with a symmetrical size-frequency distribution is shown as undergoing fractionation into symmetrically distributed mud and sand deposits. More probably, fractionation would result in asymmetric daughter populations (Ghosh *et al.*, 1981). Studies of shale samples from the Cretaceous Western Interior seaway (Tourtelot, 1962) show that at least some massive shales (parent populations) are bimodal to begin with (silt-clay mode with a secondary sand mode).

The traction and suspension deposits in Figure 10 are of equal modal height, but their frequency curves are scaled by weight percent, not absolute weight. The mud lithofacies is sand poor, and the suspension deposits comprise 95 percent or more of the deposit; sand laminae deposited from the traction load are volumetrically unimportant.

Spatial Flow Gradients and Sediment Transport

Recent theoretical studies by Smith (1970) and Huthnance (1982) suggest a variant model for shelf sedimentation, in which a storm deposit can grow to a much greater thickness. These authors have investigated bedform formation by undertaking a form of numerical flow-modeling known as stability analysis. Water flowing over a boundary such as the sea floor is sheared flow; the lowermost layer of water molecules is motionless, while each successively higher layer in the flow moves at a higher speed. Sheared flows may experience gradients not only in velocity, but also in pressure, viscosity, temperature and density or sediment concentration in the direction normal to the shear force and may become unstable in response to any of these gradients. Secondary flow components may appear as regularly repeated patterns of velocity variation superimposed on the mean flow component. The flow satisfies the three continuity laws of mass, energy and momentum, but in such a way that any small disturbance is initially self-aggravating. These instabilities may arise over plane beds or may be initiated by slight initial irregularities in the bed. Eventually, however, as the perturbed flow and the bed deform in response to each other, a new stable state is attained.

In stability analysis, an algorithm is constructed from modifications of the equations of motion that describe the fluid motion of interest. These equations are solved to discover whether a small sinusoidal disturbance of one variable will be damped or amplified under the chosen range for other variables in the system. If it is amplified, the analysis seeks the limiting condition of the amplification set by the other variables, at which a new state of quasi-equilibrium is attained.

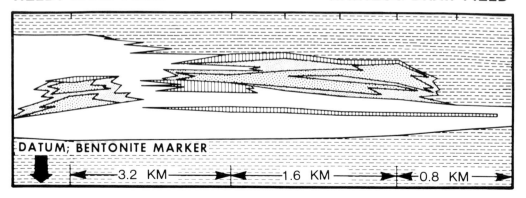

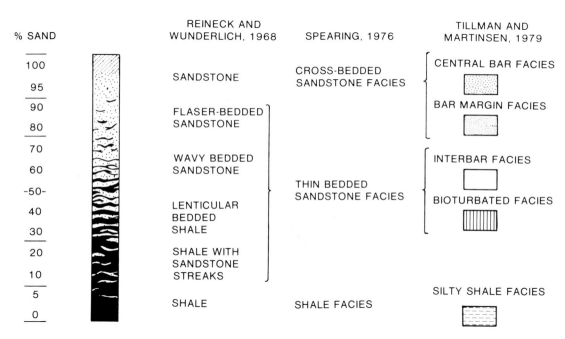

FIG. 8.—Cross-section of sandstone bodies in the Shannon Sandstone member of the Cody Shale at Heldt and Hartzog Draw fields, northern Wyoming (after Tillman and Martinsen, 1979) and comparison of facies classifications.

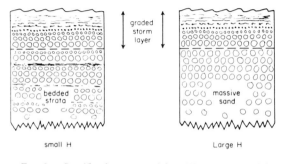

FIG. 9.—Stratification expected for different values of the physical parameter H (Nittrouer and Sternberg, 1981).

Smith's (1970) stability analysis was concerned with the role of inertial terms in the equations of motion. He has shown that as water flows over a sand bed with high and low areas, bottom shear stress will be out of phase with the topography. Flow must accelerate over the bottom high in order to maintain continuity. However the greatest bottom shear stress will be experienced not at the crest of the bottom high but on its up-current flank, where the more rapidly moving water, normally a short distance above the bottom, converges with the bottom as the bottom rises (Fig. 11). Bottom shear stress and sediment discharge increase to a maximum on the up-current flank, then decrease from that point over the crest; hence according to the continuity equa-

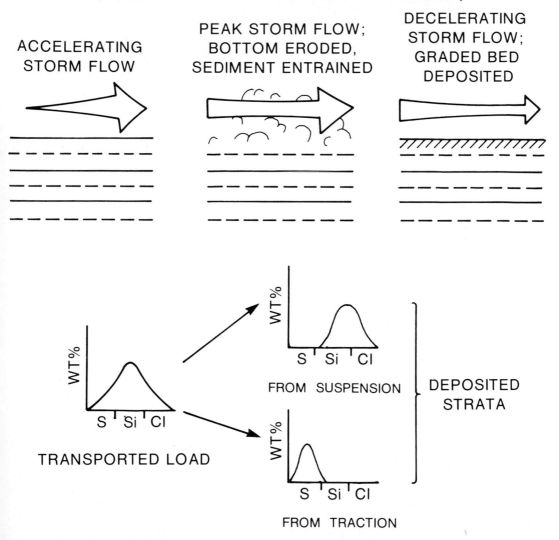

NORMAL SHELF SEDIMENTATION
(TEMPORAL ACCELERATION MODEL)

FIG. 10.—Model for the production of the "normal" Cretaceous shelf facies (as in Fig. 7) by flows whose velocities vary primarily in time rather than in space. S, Si, Cl refer to sand, silt, and clay.

tion (4), the up-current flank must erode while the crest and down-current flank upgrade. Figure 11 is drawn to show how a minor topographic high is maintained, but Smith's (1970) work leads to a more fundamental conclusion: a sand bed is inherently unstable in response to flow in excess of the threshold velocity; any initial unevenness of the bottom, no matter how small, must be amplified. The resulting bottom features will extend themselves in a cross-flow direction as ripples or waves and will adjust their along-flow spacings to conform to the separation and attachment of turbulent wakes. The

result is an ordered array of bedforms.

This theory has been modified by Richards (1980) to describe the formation of current ripples (spacings of centimeters) and megaripples (spacing of meters or tens of meters), and could be extended to describe the larger sand waves (spacing of 100 m or more; see Swift et al., 1979, for review of nomenclature). Sand ridges, a yet larger scale of shelf bedform, consist of sand bodies with spacings of several kilometers (Swift et al., 1973). Unlike ripples, megaripples, and sand waves, which are flow-transverse bedforms, sand ridges tend to be aligned at small angles to flow. This is true for both the storm-built variety (Swift *et al.*, 1978a) and the tide-built variety (Kenyon *et al.*, 1981). A stability analysis by Huthnance (1982) suggests that at this large spatial scale the phase lag between bottom shear stress and topography is no longer sufficient to cause crestal aggradation. Instead, if the flow crosses the bedform at right angles to the crest, it accelerates in order to maintain continuity, and deposition cannot occur. However, if the flow is oblique to the ridge crest, the flow may be viewed as consisting of a cross-ridge and an along-ridge component (Fig. 12). The cross-ridge component of flow accelerates to satisfy continuity as before, but the along-ridge component of flow loses energy to the bottom by frictional drag and deposition occurs.

Huthnance (1982) plotted the angle between flow direction and sand ridge orientation against the wave number (a function of ridge spacing) for a given set of flow parameters. His stability analysis allows him to contour the resulting field for the rate of ridge growth. Huthnance's (1982) analysis implies that a sandy substrate is innately unstable in response to flow at very large spatial scales, as well as at the intermediate scales investigated by Smith (1970) and Richards (1980). Huthnance's analysis shows that initial topography with a broad spectrum of orientations and spacings must, given a few millennia in which to grow, evolve into an orderly array of sand ridges.

If spatial scale is a valid criterion, the elongate sandstone bodies on the Cretaceous shelf of New Mexico, Colorado, Wyoming, Montana and Alberta have more in common with modern shelf sand ridges (spacings of several kilometers) then they do with sand waves (spacings of hundreds of meters). Studies by Shurr (this volume) indicate that, in fact, elongate sandstone bodies and small-scale facies packages resembling sand waves occur together in Cretaceous shelf deposits (Fig. 4) as sand ridges and sand waves do on modern shelves (Swift *et al.*, 1978b).

In the Huthnance flow model for sand ridge formation, very slight topographic variation can result in spatial velocity gradients that induce deposition. If the velocity history at any point on a shelf with initial topographic irregularities is viewed for convenience as a square wave, as suggested earlier, the deposition across the region due to temporal deceleration can only occur when the storm current ceases. Deposition at each initial topographic high, however, can occur throughout the duration of the storm in response to spatial deceleration. Thus, the storm deposit will be thicker over such initial highs, and will more efficiently induce deposition during the next storm. Furthermore, spatial decelerations result in lateral fractionation of the deposited loads; the deposit becomes finer-grained downstream. Thus the topographic high will become capped with well sorted sand, while the associated mud is deposited in a downstream shadow zone. As the sand body grows upwards, storm

flows over the crest are intensified. Megaripples and sand waves will migrate across the upper surface of the sand body and enhance the winnowing process. This self-exciting sequence for sand body formation is presented diagrammatically in Figure 13, where it occurs because near-bottom water slows (undergoes spatial deceleration) as it moves over the tops and down the down-current side of topographic highs. As indicated by equation 4, particles are deposited not because velocity drops below the critical value (this is not necessary), but simply because there is less power available to suspend sediment per unit area of bottom.

The authigenic component of the sandstone facies (glauconite and siderite) is compatible with this scenario of prolonged concentration of the coarse fraction of a sediment load with a broad grain-size range. The glauconite would have been formed from the alteration of fecal pellets and as foraminiferal casts within the shale lithofacies and would have been concentrated along with the quartz clasts. The siderite is most common where the glauconite concentration is especially high. It is present both as "concretions" and as clasts, testifying to the duration and intensity of the winnowing process.

DISCUSSION: TIME AND SPACE CONTROLS OF SAND BODY FORMATION

Sourceland Tectonics and Sedimentation

Our review of fluid dynamical concepts leads us to conclude that two types of sedimentary regime may have existed on the western continental shelf of the Cretaceous western interior seaway. During "normal" sedimentation, storm-suspended fine sand and mud moved intermittently across an essentially flat shelf surface, slowly building a storm-stratified sandy mud sequence. However, under appropriate conditions, the shelf surface became locally unstable; slight topographic irregularities became self-perpetuating zones of sand accumulation during successive storm flows. Feedback of this sort has clearly transpired between the sea floor and flows on modern sandy shelves; examples are the tide-dominated North Sea (Kenyon *et al.*, 1981) and the storm-dominated Maryland Shelf (Swift and Field, 1981). The stability analysis of Huthnance (1982) provides a plausible model for such flow-substrate interactions.

However, these explanations by themselves cannot constitute a full and sufficient model for shelf sandstone body distribution in the Cretaceous western interior seaway, because they would require sandstone bodies at every horizon from the shoreline to the shelf edge spaced at intervals determined by Huthnance theory . Instead, sandstone bodies are confined to relatively narrow time-rock intervals, 30 m or less thick, which are in many cases periods of maximum regression (Fig. 3). Sandstone bodies within these intervals are also somewhat laterally restricted into "lentils" up to 1500 km^2 in extent (Fig. 4). We infer that the sand

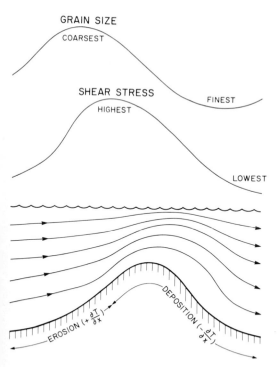

GRAIN SIZE

COARSEST

SHEAR STRESS

FINEST

HIGHEST

LOWEST

$EROSION (+ \frac{\partial T}{\partial x})$

$DEPOSITION (- \frac{\partial T}{\partial x})$

FIG. 11.—Model for the growth of a bedform from a slight initial topographic high, due to phase lag between topography and bottom shear stress (from Swift and Field, 1981, based on Smith, 1970).

content of the storm-transported load must reach a certain threshold before the "normal" model for shelf sedimentaion (Fig. 10) is no longer valid, and the sand-accumulating model (Fig. 13) becomes applicable. In other words, the pump must be primed by increasing the sandiness of the horizon on a regional basis before sand bodies can be triggered and grow.

There appears to be a pervasive phenomenon within the Cretaceous western interior section of North America known as the upward-coarsening sequence. In the Upper Cretaceous Wapiabi and Cardium Formations of Alberta, for example, as many as four such cycles can sometimes be seen in a single riverbank outcrop sequence (Walker *et al.*, 1982). Here 100 m sequence of silty shale may culminate in a few decimeters of fine sandstone or pebbly fine sandstone. Two such cycles characterize the Turonian Turner Sandy Member of the Carlile Shale in the Black Hills, South Dakota section, each culminating in a meter or less of fine, hummocky or cross-bedded sandstone, capped in turn by a decimeter of coarse pebbly sandstone (Merewether *et al.*, 1979). As noted above, sand bodies coarsen upwards (because as they grow they increasingly perturb near-bottom flow) but the upward-coarsening cycles described above occur over much broader areas than a single sandstone body, although sandstone bodies may occur within such intervals.

A number of clues exist which allow us to recon-

struct the events which lead to these upward-coarsening cycles. Foremost among these is the superb well log correlation through the Pierre Shale of eastern Wyoming by Asquith (1970), which in its density of data and resolution of detail rivals a seismic line. The well log correlations (Fig. 14) clearly show that the Campanian shelf edge grew upward and eastward into the seaway in a series of pulses, and that the shelf subsided beneath it as it did so. Each pulse is recognizable on well logs as an upward-coarsening cycle which consists primary of shale, but which is terminated by a few meters of sandstone with reduced self-potential value. The tops of two of these cycles are correlated with sandstone body horizons (Sussex and Shannon Sandstone Members).

These upward-coarsening cycles resemble the "asymmetrical cyclothems" described by Kauffman (1977) as "reflecting mainly progradation ("regression") of nearshore marine and marginal marine clastics along the tectonically active margin of the basin" versus the "symmetrical cyclothems" which he reported from the basin center. Kauffman (1977) recognizes two eustatic regressions during the interval represented by Figure 14 (R7, during the Niobrara cyclothem, and R8 during the Claggett cyclothem), but the cycles of Figure 14 appear to be a higher frequency phenomenon.

Recent studies (Beaumont, 1981; Jordan, 1982) suggest that tectonic, rather than eustatic, sea-level fluctuations controlled sedimentation in the Cretaceous seaway. The seaway is seen as a foreland basin which subsided in response to the loading of successive thrust plates from the Sevier orogenic belt to the west onto the adjacent cratonic crust. The relationship suggests a model (Fig. 15) in which the movement of a thrust plate increases relief, resulting in a flood of sediment onto the east-facing coastal plain and presumably a seaward advance of the shoreline. However, the thrust plate also loads the underlying crust and induces subsidence, as does the redistributed sediment. As the relief caused by the tectonic event undergoes erosional reduction, the rate of subsidence decreases more slowly than the rate of sedimentation; a balance point is passed and the shoreline begins to retreat. Eventually tectonic activity resumes, and the cycle begins again. In its simplest form, this model implies that the stratigraphy of the Cretaceous seaway is a record of the thrusting history of the Sevier Overthrust Belt, much as the stratigraphy of the Atlantic Continental Shelf is a record of the thermal history of the underlying crust. During such a period of thrust-controlled subsidence, eustatic sea-level flucions would only cause a second order modulation of the rapidly accumulating sediment pile ($\sim$ 5,000 m in 60 million years).

We suggest that the successive sediment cycles revealed in Figure 14 constitute a record of Campanian tectonics in the overthrust belt, probably due to repeated movement on the Absaroka Thrust (Jordan, 1982). Each cycle coarsens upward because water

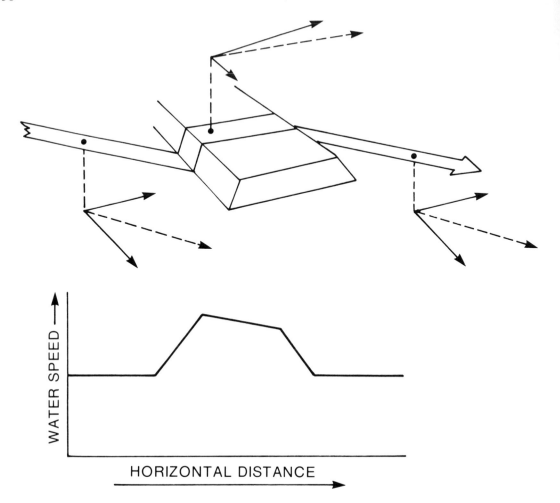

FIG. 12.—Model for growth of a large scale (2 km spacing) sand ridge, based on analysis of Huthnance, 1982. See text for explanation.

depths decrease throughout its history. The sediment fraction selected for permanent disposition during intermittent storm transport across the storm surface must get coarser because of increasing wave agitation of the sea floor (Fig. 16). At the beginning of such a cycle, fines would be trapped primarily on the shelf surface. Towards the end, they would be largely bypassed over the shelf edge.

Recent studies by Curry (1973, 1976) of Upper Cretaceous sequences in northeastern Wyoming have provided a glimpse of the shoreline process analogous to the shelf progradational sequences (Fig. 17). The sequences are attributed to the seaward growth of a delta, although the control is not adequate to distinguish a deltaic bulge from a straight shoreline. In Figure 17, the successive progradational units are thinner, five m or less in thickness, as compared to the 10 to 20 m of the shelf edge units. The difference must in part be attributed to the greater efficiency of bypassing in the shallow wave-agitated inner shelf water column. It is also probable that these nearshore progradational units

represent a smaller time-scale phenomenon such as river avulsion, rather than variations in the rate of tectonic activity, and Curry's entire sequence may in fact be equivalent to a single cycle within the Asquith section.

Progradational sequences at both spatial scales, whatever the causative factor, have resulted in upward-coarsening sequences that pass from shale to siltstone, sandy siltstone or fine sandstone near the top. We suggest that these sand-enriched substrates became capable of a second stage of flow-substrate feedback, leading to the growth of narrow (5 to 10 km) bodies of well-sorted sand (sandstone facies, Fig. 8) whose spacings and orientations would have been controlled by the mechanisms described by Huthnance (1982; see Fig. 12).

Shelf Tectonics and Sedimentation

Orogenic activity appears to have controlled the distribution of sand on the Cretaceous western interior

SAND BODY FORMATION (SPATIAL ACCELERATION MODEL)

BOTTOM CURRENT ACCELERATES
UP FORWARD SLOPE OF HIGH,
DECELERATES OVER CREST

SAND LENS DEPOSITED OVER HIGH;
FINES DEPOSITED DOWN CURRENT

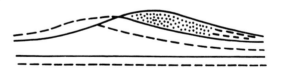

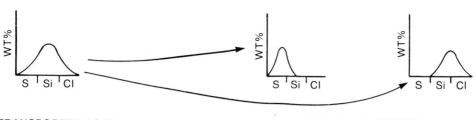

TRANSPORTED LOAD

DEPOSITED STRATA

FIG. 13.—Process model for sand body formation on the muddy Cretaceous shelf of the western interior seaway, due to the occurrence of spatial as well as temporal velocity variations. Compare with Figure 4 as a product model, with Figure 10 as a model for sedimentation with temporal variation only. (S = sand, Si = silt, Cl = clay).

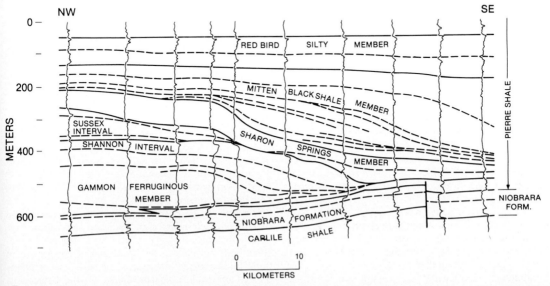

FIG. 14.—Resistivity-log section through the Pierre Shale, eastern Wyoming, revealing the progradational structure of the Campanian Shelf edge. Sandstone bodies are prominent in the Shannon and Sussex intervals (from Brenner, 1978, after Asquith, 1970).

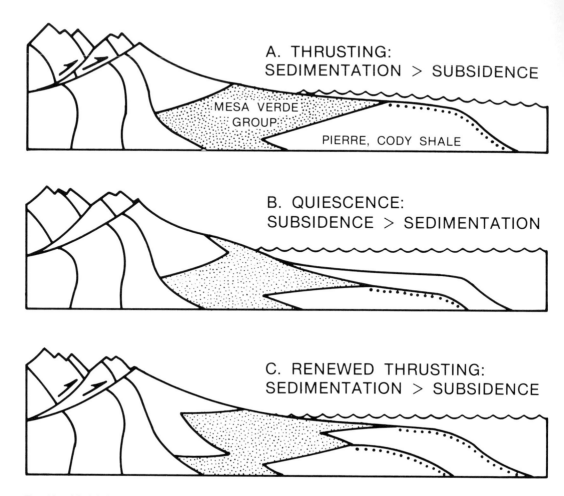

A. THRUSTING:
SEDIMENTATION > SUBSIDENCE

MESA VERDE GROUP

PIERRE, CODY SHALE

B. QUIESCENCE:
SUBSIDENCE > SEDIMENTATION

C. RENEWED THRUSTING:
SEDIMENTATION > SUBSIDENCE

FIG. 15.—Model showing the generation of upward-coarsening, progradation shelf sequences. *A*, Renewed thrusting in the source area increases rate of sedimentation; tectonic loading initiates subsidence. *B*, Sedimentation at first exceeds subsidence. *C*, As relief decreases, subsidence exceeds sedimentation.

shelf in a second and more direct way. During the late Cretaceous, the mode of crustal deformation shifted from overthrusting in western Alberta, Montana, Wyoming and Utah (Sevier orogeny) to basin and uplift deformation that involved the foreland area as well as the overthrust belt (Laramide orogeny). During this phase, basement highs east of the Laramide Front, such as the Bowdoin Dome and the Central Montana Uplift, were reactivated. Their stratigraphy reveals a history of tectonic activity that extends back into the Paleozoic. They were intermittently active during Late Cretaceous time as indicated by the thinning of major units across their crest, and Shurr (1983) and Rice (1983) have shown that they are the locus of large-scale groupings of Cenomanian and Campanian sandstone bodies (Fig. 18). In Wyoming, a related phenomenon appears; subsiding Laramide basins (Powder River Basin, Hanna Basin) developed second-order flank folds (Salt Creek Anticline, Lost Soldier Anticline). A

study by Reynolds (1976) has shown that the Lost Soldier Anticline began to grow in Campanian time, and that as it did so, shelf sandstone bodies (Haystack Mountain Formation) formed across it. The sandstone bodies of the Shannon Sandstone Member exposed on the west flank of the Powder River Basin appear to bear a similar relation to the Salt Creek Anticline (R. Martinsen, personal communication, 1982; Tillman and Martinsen, this volume).

The correlation between the growth or rejuvenation of structural elements and sandstone deposition suggests a second two-stage model for Cretaceous sandstone body formation. In this scenario, the growth of flank folds and the reactivation of basement elements created regional submarine highs on the Cretaceous western interior shelf. The relief was sufficient to cause spatial acceleration and deceleration of flow during intermittent movement of the storm-transported load across these zones, so that a slight enrichment of the

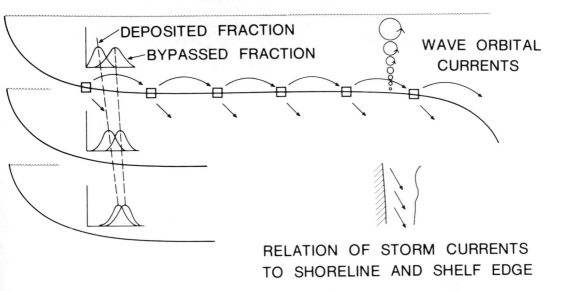

FIG. 16.—Model depicting the generation of an upward-coarsening sequence. Upbuilding of shelf surface leads to greater fluid power expenditure, and a shift of the partition between deposited fraction and bypassed fraction toward the coarse end.

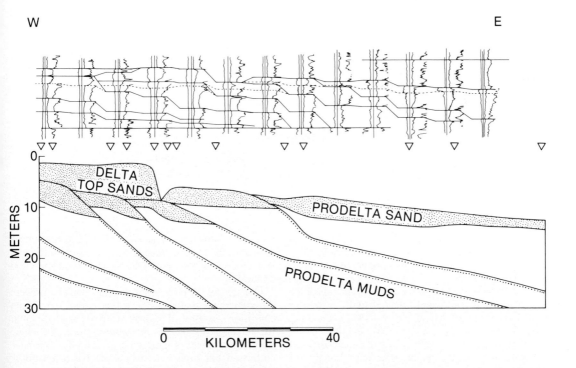

FIG. 17.—Upward-coarsening progradational sequences in the middle Campanian Teapot Sandstone Member of the Mesa Verde Formation, western margin of the Powder River Basin, Wyoming (after Curry, 1976). Triangles indicate location of wells in the lower panel.

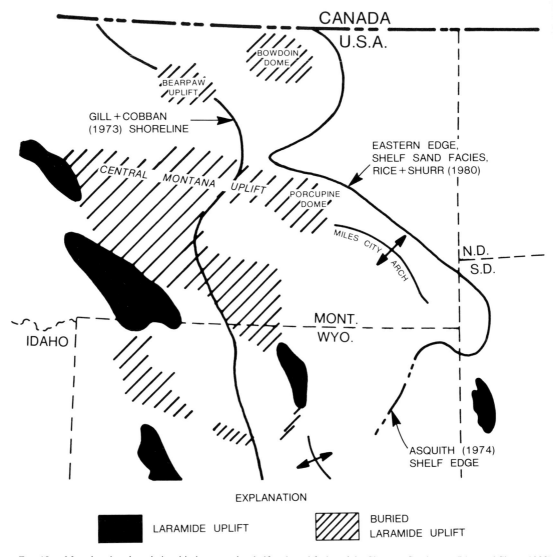

Fig. 18.—Map showing the relationship between the shelf and sand facies of the Shannon Sandstone (Rice and Shurr, 1983) and the northern edge of the Central Montana Uplift; compare with Figure 1.

substrate occurred over an area many tens of kilometers wide. The enrichment of the substrate was a consequence of the mechanisms summarized in Figure 13. Modern analogs of bottom response to flow at this regional scale exist as large-scale relief elements of the modern North Atlantic Shelf. Such features as the Virginia Beach Massif (Swift *et al.*, 1977), the Platt Shoals Massif, (Swift *et al.*, 1978b), and the Hudson Divide (Freeland *et al.*, 1981) cause regional variations in storm flow intensity that are reflected in the pattern of sediment deposition around them. On the Cretaceous shelf this kind of substrate-flow interaction was apparently sufficient in some areas to result in the deposition of a thin-bedded sandstone facies (Fig. 8). Attainment of this level of sand enrichment seems to have permitted the second stage of flow-substrate feedback, that elsewhere was caused by shelf progradation and upbuilding.

Our review of fluid- and sediment-dynamic process on continental shelves suggest that under appropriate conditions, a flux of sediment across a continental shelf floor in response to intermittent storm flows may lead to the growth of large-scale sand bodies. We have buttressed our argument with some basic functional relationships between sediment concentration and fluid power, but we have not been able to pursue quantitative aspects of our model very far because the entrainment process for fine cohesive sediment that constituted the bulk of the sediment load on the Cretaceous shelf is poorly understood. However, we can turn to the Cretaceous stratigraphy of the western interior seaway as circumstantial evidence. The localization of sandstone body arrays in space and time indicates that the feedback process which led to sand body formation was not easily triggered, but once initiated could lead to sandstone bodies 18 to 20 m thick and up to 50 km^2 in areal extent.

REFERENCES

ARMSTRONG, R. I., 1968, Sevier Orogenic Belt in Nevada and Utah: Geol. Soc. America Bull., v. 79, p. 425–458.

ASQUITH, D. O., 1970, Depositional topography and major marine environments, Late Cretaceous, Wyoming: Am. Assoc. Petroleum Geologists Bull. v. 54, p. 1184–1224.

_____, 1974, Sedimentary models, cycles, and deltas, Upper Cretaceous, Wyoming: Am. Assoc. Petroleum Geologists Bull., v. 58, p. 2274–2283.

BEARDSLEY, R. C., W. C. BOICOURT, AND D. V. HANSEN, 1976, Physical oceanography of the Middle Atlantic Bight, in M. Gross (ed.), Middle Atlantic Continental Shelf and the New York Bight: Am. Soc. Limnology and Oceanography, Spec. Symp. 2, p. 20–34.

_____, AND B. BUTMAN, 1974, Circulation on the New England continental shelf: Response to strong winter storms: Geophys. Res. Lett., v. 1, p. 181–184.

BEAUMONT, C., 1981, Foreland basins: Geophys. Jour. Royal Astron. Soc., v. 65, p. 241–329.

BOICOURT, W. C., AND P. W. HACKER, 1976, Circulation in the Atlantic continental shelf of the United States, Cape May to Cape Hatteras: Mem. Soc. Roy. des Sci. de Liege, 6 ser., v. 10, p. 187–200.

BOYLES, M. J., AND A. J. SCOTT, 1982, A model for migrating shelf bar sandstones in upper Mancos Shale (Campanian), northwestern Colorado: Am. Assoc. Petroleum Geologists Bull., v. 66, p. 441–508.

BRENNER, R. L., 1978, Sussex Sandstone of Wyoming—Example of Cretaceous offshore sedimentation: Am. Assoc. Petroleum Geologists Bull., v. 62, p. 181–200.

CURRY, W. H., III, 1973, Parkman Delta in central Wyoming: Earth Sci. Bull., v. 6, no. 4, p. 5–18.

_____, 1976, Late Cretaceous Teapot Delta of the southern Powder River Basin, Wyoming: Wyoming Geol. Assoc., Guidebook 28th Ann. Field Conf., p. 21–28.

EXNER, F. M., 1925, Über die Wechselwirkung zwischen Wasser und Geschiebe in Flussen: Setzungber. Acad. Wiss. Vienna, H. 3–4, p. 165–180.

FIGUEIREDO, A. G., JR., D. J. P. SWIFT, W. L. STUBBLEFIELD, AND T. CLARKE, 1981, Sand ridges on the inner Atlantic shelf of North America; Morphometric comparisons with Huthnance stability model: Geo–Marine Letters, v. 1, p. 187–191.

FREELAND, G. C., J. D. STANLEY, D. J. P. SWIFT, AND D. M. LAMBERT, 1981, The Hudson Shelf Valley: Its role in shelf sediment transport: Marine Geol., v. 42, p. 399–427.

GILL, J. P. AND W. A. COBBAN, 1973, Stratigraphy and geologic history of the Montana Group and equivalent rocks, Montana, Wyoming, and North and South Dakota: U.S. Geol. Survey Prof. Paper 776, 37 p.

GHOSH, J.R., R.S. MOZENDU, AND S. SENGUPTA, 1981, Methods of computation of suspended load from bed materials and flow parameters: Sedimentology, v. 28, p. 781–791.

GRANT, W. D., AND O. S. MADSEN, 1979, Combined wave and current interaction with a rough bottom: Jour. Geophys. Res., v. 84, p. 1797–1808.

HOBSON, J. P., JR., M. L. FOWLER, AND E. A. BEAUMONT, 1982, Depositional and statistical exploration models, Upper Cretaceous offshore sand complex, Sussex Member, Horse Creek Field, Wyoming: Am. Assoc. Petroleum Geologists Bull., v. 66, p. 689–707.

HUTHNANCE, J. M., 1982, On one mechanism forming linear sand banks: Estuarine, Coastal and Shelf Sci., v. 14, p. 79–89.

JORDAN, T. E., 1982, Thrust loads and foreland basin evolution, Cretaceous, western United States: Am. Assoc. Petroleum Geologists Bull., v. 65, p. 2506–2520.

KAUFFMAN, E. G., 1977, Geological and biological overview: Western Interior Cretaceous basin: Mountain Geologist, v. 14, p. 75–99.

KENYON, N. H., B. H. BELDERSON, A. H. STRIDE, AND M. A. JOHNSON, 1981, Offshore sand banks as indicators of net sand transport and as potential deposits, in S. C. Nio, R. T. E. Schuttenhelm, and T. C. E. van Weering (eds.), Holocene Sedimentation in the North Sea Basin: Int. Assoc. Sedimentologists Spec. Pub. 5, p. 257–268.

KUEHL, S. A., C. A. NITTROUER, AND D. J. DEMASTER, 1982, Modern sediment accumulation and strata formation on the Amazon continental shelf: Marine Geol., v. 49, p. 274–300.

MANN, R. G., D. J. P. SWIFT, AND R. PERRY, 1981, Size classes of flow–transverse bedforms in a subtidal environment, Nantucket Shoals, North American Atlantic shelf: Geo–Marine Letters, v. 1, p. 39–43.

MAYER, D. W., D. V. HANSEN, AND D. A. ORTMAN, 1979, Long term current and temperature observations on the Middle Atlantic Shelf: Jour. Geophys, Res., v. 84, p. 1776–1792.

MEREWETHER, E. A., W. A. COBBAN, AND E. T. CAVANAUGH, 1979 Frontier Formation and equivalent rocks in eastern Wyoming: Mountain Geologist, v. 16, p. 67–101.

MOLENAAR, C. M., 1973, Sedimentary facies and correlation of the Gallup Sandstone and associated formations, North Western New Mexico, in Cretaceous and Tertiary rocks of the Southern Colorado Plateau: Four Corners Geol. Soc. Mem., p. 85–110.

NEUMANN, G., AND W. J. PIERSON, JR., 1966, Principles of Physical Oceanography: Prentice Hall Inc., Inglewood Cliffs, New Jersey, 545 p.

NITTROUER, C. E., AND R. W. STERNBERG, 1981, The formation of sedimentary strata in an allochthonous shelf environment: The Washington Continental Shelf: Marine Geol., v. 42, p. 201–232.

REINECK, H. E., AND F. WUNDERLICH, 1968, Classification and origin of flaser and lenticular bedding: Sedimentology, v. 11, p. 99–104.

_____, AND I. B. SINGH, 1972, Genesis of laminated sand and graded rhythmites in storm–sand layers of shelf mud: Sedimentology, v. 18, p. 123–128.

REYNOLDS, M. W., 1976, Influence of recurrent Laramide structural growth on sedimentation and petroleum accumulation, Lost Soldier area, Wyoming: Am. Assoc. Petroleum Geologists Bull., v. 60, p. 11–33.

RICE, D. D., 1980, Coastal and deltaic sedimentation of Upper Cretaceous Eagle Sandstone: Relation to shallow gas accumulations, north–central Montana: Am. Assoc. Petroleum Geologists Bull, v. 64, p. 316–338.

RICE, D. D., AND G. W. SHURR, 1983, Patterns of sedimentation and paleogeography across the Western Interior Seaway during time of deposition of Upper Cretaceous Eagle Sandstone and equivalent rocks, northern Great Plains: in M. W. Reynolds, and E. D. Dolly, (eds.) Mesozoic paleogeography of the West-Central United States: Rocky Mountain Section SEPM Symposium 2, p. 337-358.

_____, AND G. SHURR, 1980, Shallow, low permeability reservoirs of northern Great Plains—Assessment of their natural gas resources: Am. Assoc. Petroleum Geologists Bull., v. 64, p. 969–987.

RICHARDS, K. J., 1980, The formation of ripples and dunes on an erodable bed: Jour. Fluid Mechanics, v. 94, p. 592–618.

ROUSE, H., 1938, Mechanics for Hydraulic Engineers: Dover Pub. [Dover edition issued in 1961], New York, 322 p.

SLATER, R. D., 1981, A numerical model of tides in the Cretaceous seaway of North America: Unpub. M.S. Thesis, Univ. Chicago, Chicago, Illinois, 72 p.

SMITH, J. D. 1977, Modeling of sediment transport on continental shelves, in E. D. Goldberg, I. N. McCave, J. J. O'Brien, and J. H. Steele (eds.), The Sea; Ideas and observations on progress in the study of the seas, v. 6, Marine modeling: John Wiley and Sons, New York, p. 539–577.

SMITH, J. P., 1970, Stability of a sand bed subjected to a shear flow of low Froude number: Jour. Geophys. Res., v. 75, p. 5428–5940.

SPEARING, D. R., 1976, Upper Cretaceous Shannon Sandstone: An offshore, shallow–marine sand body: Wyoming Geol. Assoc., Guidebook 28th Ann. Field Conf., p. 65–72.

STRAHLER, A., 1963, The Earth Sciences: Harper and Row, New York, 681 p.

SWIFT, D. J. P., 1976a, Coastal sedimentation, in D. J. Stanley and D. J. P. Swift (eds.), Marine Sediment Transport and Environmental Management: John Wiley and Sons, New York, p. 255–310.

_____, 1976b, Continental shelf sedimentation, in D. J. Stanley and D. J. P. Swift (eds.), Marine Sediment Transport and Environmental Management: John Wiley and Sons, New York, p. 311–350.

_____, D. B. DUANE, AND T. F. McKINNEY, 1973, Ridge and swale topography of the Middle Atlantic Bight, North American: Secular response to the Holocene hydraulic regime: Marine Geol., v. 15, p. 227–247.

_____, AND M. FIELD, 1981, Evolution of a classic sand ridge field: Maryland Sector, North American Inner Shelf: Sedimentology, v. 28, p. 461–482.

_____, G. L. FREELAND, AND R. A. YOUNG, 1979, Time and space distributions of megaripples and associated bedforms, Middle Atlantic Bight, North American Atlantic Shelf: Sedimentology, v. 26, p. 389–406.

_____, T. NELSEN, M. McHONE, B. HOLLIDAY, H. PALMER, AND G. SHIDELER, 1977, Holocene evolution of the inner shelf of southern Virginia: Jour. Sed. Petrology, v. 47, p. 1454–1474.

_____, G. PARKER, N. W. LANDREDI, G. PERILLO, AND K. FIGGE, 1978a, Shoreface–connected sand ridges on American and European shelves: A comparison: Estuarine and Coastal Marine Sci., v. 7, p. 257–273.

_____, P. C. SEARS, B. BOHLKE, AND R. HUNT, 1978b, Evolution of a shoal retreat massif, North Carolina Shelf: Inferences from areal geology: Marine Geol, v. 27, p. 19–42.

_____, R. A. YOUNG, T. CLARKE, AND C. E. VINCENT, 1981, Sediment transport in the Middle Atlantic Bight of North American: Synopsis of recent observations, in D. S. Nio, R. T. E. Schuttenhelm, and T. C. E. van Weering (eds.), Holocene Sedimentation in the North Sea Basin: Int. Assoc. Sedimentologists Spec. Pub. 5, p. 361–383.

TILLMAN, R. W., AND R. M. MARTINSEN, 1979, Hartzog Draw Field, Powder River Basin, Core Workshop, in R. F. Flory (ed.), Core Seminar, Rocky Mountain Section: Am Assoc. Petroleum Geologists—Soc. Econ. Paleontologists and Mineralogists, Convention, June 3–7, Casper Wyoming, p. 1-38.

TOURTELOT, H. A., 1962, Preliminary investigation of the geologic setting and chemical composition of the Pierre Shale, Great Plains region: U.S. Geol. Survey Prof. Paper 390, 42 p.

TWITCHELL, D. C., C. E. McCLENNEN, AND B. BUTMAN, 1981, Morphology and processes associated with the accumulation of the fine–grained deposit on the southern New England Shelf: Jour. Sed. Petrology, v. 51, p. 269–280.

VINCENT, C. E., D. J. P. SWIFT, AND B. HILLARD, 1981a, Sediment transport in the New York Bight, North American Atlantic Shelf: Marine Geol. v. 42, p. 369–398.

_____, R. A. YOUNG AND D. J. P. SWIFT, 1981b, Bedload transport under waves and currents: Marine Geol., v. 39, p. M71-M80.

WALKER, R. G., 1982, Clastic units of the front ranges, foothills and plains in the area between Field, B. C., and Drumheller, Alberta: Int. Assoc. Sedimentologists, 11th Ann. Congress Guidebook 21A, 160 p.

GEOMETRY OF SHELF SANDSTONE BODIES IN THE SHANNON SANDSTONE OF SOUTHEASTERN MONTANA

GEORGE W. SHURR

U.S. Geological Survey and St. Cloud State University, St. Cloud, Minnesota 56301

ABSTRACT

The Shannon Sandstone Member of the Gammon Shale in southeastern Montana is one of several Upper Cretaceous sandstones in the western interior of the United States which are interpreted as shelf sediments. Paleogeographic reconstruction suggests that the site of Shannon deposition in southeastern Montana was near the edge of a broad shelf more than 322 km (200 mi) from the shoreline. The Shannon is early Campanian in age and is equivalent to the Eagle Sandstone of central Montana. Exposures on the north flank of the Black Hills and subsurface data from oil and gas test holes reveal that the Shannon is made up of a hierarchy of sandstone bodies. From largest to smallest, the hierarchy is made up of five elements of different sizes: lithosome (level I), sheet (level II), lentil (level III), elongate lens (level IV), and small-scale facies packages (level V).

The largest elements of the geometric hierarchy are documented in regional subsurface studies. The total sandstone lithosome is 121-182 m (400-600 ft.) thick, covers 117,000 km^2 (45,000 mi^2), and is enclosed in sandy shale. In the center of the lithosome, there is a sandstone sheet which is 30 m (100 ft) thick and has an area of 18,200 km^2 (7,000 mi^2). The sheet is, in turn, composed of four coalesced lentils 15-23 m (50-75 ft) thick, each covering 1500 km^2 (600 mi^2).

The sandstone in one of these lentils can be traced in closely spaced bore-holes and in nearby exposures. Detailed subsurface studies indicate that the lentil consists of at least two smaller elongate lenses; each lens is about 12-18 m (40-60 ft) thick and is about 52 km^2 (20 mi^2) in areal extent. In outcrop, a series of four lithologic units form a coarsening-upward cycle within a single elongate lens. Within the lenses, the uppermost sandstone units of the coarsening-upward cycle contain three different lithologies in a facies relationship. The sandstone facies are arranged in individual packages which are small-scale, covering areas of about .12 km^2 (.05 mi^2) and having thicknesses of 3-6 m (10-20 ft). The individual small-scale facies packages have a distinct geometry and are imbricate.

Each of the five levels of the geometric hierarchy recognized in the Shannon shelf sandstone represents different formative processes. The larger elements of the hierarchy, specifically the lithosome (level I) and sheet (level II), were probably produced by long-term processes, such as episodic paleotectonism. The smaller elements may have been a response to short-term depositional processes on the ancient shelf. The lentils (level III), elongate lenses (level IV), and small-scale facies packages (level V) constitute an ancient response model that is very similar to a hierarchy of morphologic elements observed on modern continental shelves. Comparison with the modern hierarchical response model suggest that the lentils may be interpreted as complex fields of sand ridges, that the elongate lenses may be individual sand ridges, and the the small-scale facies packages may represent shelf sand waves located on the crest and flanks of the sand ridges.

INTRODUCTION

Upper Cretaceous marine sediments in the western interior of the United States contain excellent, well-described examples of shelf sandstones. These sandstones have been employed as models for the recognition of shelf sandstones in other areas. (Johnson, 1978; Walker, 1979). During the past decade, the work of several investigators has converged toward a picture of facies variation and geometry that is essentially the same for different stratigraphic units in widely-separated areas. Some of these units are: the Hygiene Sandstone Member (Porter, 1976; Kiteley, 1977) of the Pierre Shale in the Denver Basin of Colorado, the Shannon Sandstone Member of the Steele Shale in the Powder River Basin of Wyoming (Davis, 1976; Spearing, 1976; Seeling, 1978; Tillman and Martinsen, 1979), and the Sussex Sandstone Member of the Steele Shale in the Powder River Basin of Wyoming (Berg, 1975; Brenner, 1978). Most of these detailed studies have been carried out because the shelf sandstones are important reservoirs for hydrocarbons.

In Montana, Upper Cretaceous shelf sandstones

have been interpreted from electric-log characteristics and mapped regionally in five separate stratigraphic intervals (Rice and Shurr, 1980). Shallow biogenic natural gas is produced from conventional, high-porosity and high-permeability reservoir rocks and from low-porosity and low-permeability, "tight" reservoirs within the shelf sandstones. Although the tight reservoir rocks are distributed throughout wide areas of eastern Montana, the conventional reservoir rocks are largely confined to sandstone bodies with distinctive geometries. These sandstone bodies have genetic significance and are believed to represent morphologic elements of the ancient shelf.

This report describes the geometry of shelf sandstone bodies found within the Shannon Sandstone Member of the Gammon Shale in southeastern Montana. Data are summarized from regional subsurface correlations and from local surface and subsurface studies (Fig. 1). The sandstone bodies of this study are equivalent to the Shannon Sandstone Member of the Steele Shale in eastern Wyoming, to the Eagle Sandstone in central Montana, and to the Milk River Formation in southeastern Alberta (Rice, 1980).

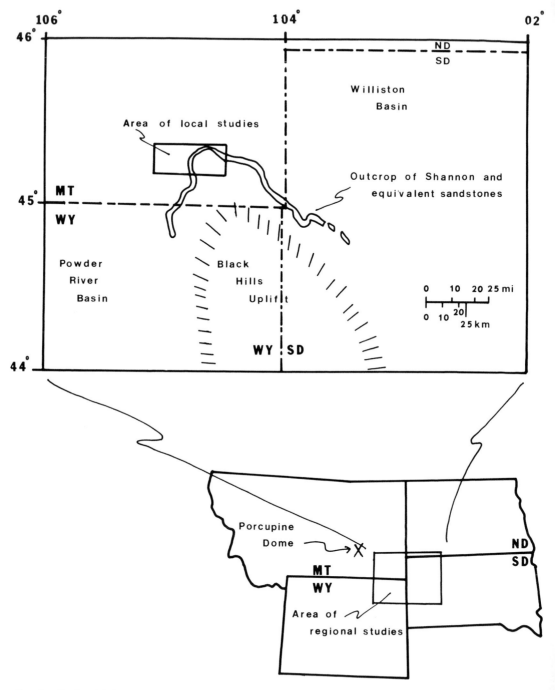

FIG. 1.—Regional index map indicating area of local studies, major tectonic elements and outcrop belt of Shannon and equivalent sandstones.

GEOLOGIC SETTING

The Shannon Sandstone Member of southeastern Montana lies about 45 m (150 ft) below the top of the Gammon Shale. The Gammon is composed of non-calcareous gray shale as much as 300 m (900 ft) thick and overlies the calcareous shale and marls of the Niobrara Formation. The regional stratigraphic setting of Shannon sandstone bodies is summarized in Figure

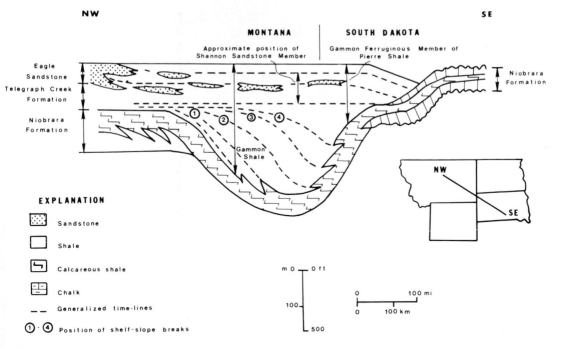

Fig. 2.—Generalized regional stratigraphic setting of shelf sandstones in the Shannon Sandstone Member of the Gammon Shale in southeastern Montana and South Dakota.

2, which illustrates the stratigraphic equivalence of the Shannon Sandstone Member of the Gammon Shale, the Eagle Sandstone, and Gammon Member of the Pierre Shale, and the upper part of the Niobrara Formation. The form of the generalized time lines is based upon correlation of subsurface markers that are probably thin siltstone beds and bentonite beds. The geometric interpretation of the time lines is analogous to correlations in equivalent units in the southern Powder River Basin (Asquith, 1970) and the facies relationships between the Gammon and Niobrara incorporate regional biostratigraphic syntheses (Cobban, 1964).

Exposures of rocks equivalent to the Shannon Sandstone Member occur in an arcuate band around the north flank of the Black Hills Uplift in southeastern Montana, northeastern Wyoming, and northwestern South Dakota. Although in outcrop the sandstone is known locally as the Groat Sandstone Bed (Rubey, 1930), the term "Groat" has not commonly been employed in subsurface studies. The term "Groat" is restricted to outcrop usage, where it is a bed in the Gammon Ferruginous Member of the Pierre Shale. In the subsurface north and west of this outcrop belt, the sandstone is termed the Shannon Sandstone Member of the Gammon Shale. This usage conforms to the recognition of the Gammon Shale at Porcupine Dome in central Montana (Gill, Cobban, and Schultz, 1972) and to the use of the Shannon Sandstone Member as it is applied in the subsurface of Wyoming's Powder

River Basin. The Groat and Shannon lie in the zones of *Scaphites hippocrepis* III and *Baculites* sp. (smooth), which are assigned to the lower Campanian (Gill and Cobban, 1973; Rice, 1980).

The structural setting of southeastern Montana was described by Knechtel and Patterson (1962) and by Robinson, Mapel, and Bergendahl (1964). The Black Hills Uplift separates the Williston Basin of western North Dakota and northwestern South Dakota from the Powder River Basin of northeastern Wyoming and southeastern Montana (Fig. 1). Arcuate outcrop belts define the overall structure in the study area as a broad anticlinal nose plunging to the north and northwest. Many small, elongate anticlines and synclines are mapped with individual axes parallel to the regional plunge of the uplift. Local domes are generally thought to be the result of igneous intrusion. A few faults occur, but are limited in extent and show small displacements.

Structural trends have been integrated with linear features mapped on satellite images to outline a grid of structural blocks bounded by lineaments in southeastern Montana (Shurr, 1975; 1979). The concept of lineament-block tectonics has been applied in the region by Thomas (1974), and the lineaments he mapped on aerial photographs are very similar, though not identical with those mapped on satellite images. Lineament-bound blocks were probably the sites of paleotectonism because sedimentation patterns seem to reflect block geometries. Within the study area, lineaments apparently influenced deposition of Paleszoic sediments (Maughan, 1966; Brown 1978) and Upper Cretaceous sediments (Shurr, 1975; 1979), including the Shannon.

The paleogeographic setting of the northern Great Plains during the Late Cretaceous was dominated by a seaway that extended north and south through the western interior of North American (Gill and Cobban, 1973). Deposition in the seaway was controlled by cycles of transgression and progradation (Weimer, 1960; Rice, 1980) and the Shannon Sandstone is a part of one of these cycles. The cycle began with deposition of the Niobrara Formation during the transgression; the remainder of the cycle was deposited during progradation (Shurr, 1975). The Shannon was apparently deposited in the later phases of the progradation.

An interpretation of the paleogeography during the progradation that produced the Shannon is presented in Figure 3. Clastic sediments from source areas in western Montana and Wyoming were deposited on the western shelf and chalks accumulated on the eastern shelf (Shurr and Rice, 1979). As the strandline migrated seaward (eastward), the western shelf prograded into an area of the basin previously occupied by deeper water along the eastern border of Montana and Wyoming (Asquith, 1970; Shurr, 1975). This progradation to the southeast is shown by the interpreted migrating position of the shelf-slope break (indicated schematically by inflection points in the time lines labeled 1 through 4, Fig. 2). Figure 3 illustrates how the geometry of the basin changed from the beginning (A) to the end (B) of the cycle. The marine basin was progressively filled from north to south by this southeastward progradation and at the end of the cycle, the only area of deeper water that remained was in Wyoming and South Dakota (Fig. 3). Asquith (1970) has demon-

strated that progradation during subsequent cycles filled this area of deeper water. As the basin was filled from the northwest, clastics engulfed the eastern shelf and carbonate deposition was terminated. The position of the eastern shelf margin remained fairly constant, as indicated by the western termination of two chalk tongues which are shown in Figure 2.

This speculation on the paleogeography assumes that the tectonic basin approximated an area of deeper water; that is, episodic rates of tectonism exceeded rates of sedimentation. Alternatively, it could be assumed that sedimentation kept pace with tectonism so that a relatively deep bathymetric basin was never produced. Under this assumption, the "shelf-slope break" would probably represent a minor topographic feature, such as an edge of a marine terrace, and the central area filled by progradation would be an area of slightly deeper water in the central part of a relatively continuous shelf. However, under both assumptions, the geometry shown in Figure 3 would be similar.

Initial recognition of the Shannon in southeastern Montana as a shelf sandstone facies (Rice and Shurr, 1980) was based upon electric log characteristics and upon the paleogeographic interpretations discussed in the preceding paragraphs. Interpretation of the Shannon as a shelf sandstone is supported by the recognition of specific sandstone geometries that are very similar to morphologic features on modern continental shelves.

HIERARCHY AS A DESCRIPTIVE DEVICE

The geometry of sandstone bodies in the Shannon can be described in terms of a hierarchy. Hierarchies are collections of specific elements arranged into levels; each level consists of several smaller elements from the next lower level (Simon, 1962). For example, if the largest elements are assigned to level I, each element of level I is a composite of several smaller level II elements. Similarly, each level II element is made up of still smaller lever III elements.

Applying the concept of hierarchy to describe the Shannon, sandstone bodies of different sizes are taken as elements; these elements are arranged into five levels. The levels are, from largest to smallest: the total sandstone lithosome (level I), sandstone sheets (level II), sandstone lentils (level III), elongate lenses (level IV), and small-scale facies packages (level V). The lithosome, sheet, and lentils are regional sandstone bodies and are described in the next section of this report. The lentils and small-scale facies changes are local sandstone bodies and are each described in separate portions of the report.

GEOMETRY OF REGIONAL SANDSTONE BODIES (LEVELS I, II AND III)

The Shannon Sandstone Member in southeastern Montana is located in a facies belt of shelf sandstones about 370 km (230 mi) east of equivalent nonmarine rocks in western Montana. The shelf sandstone facies

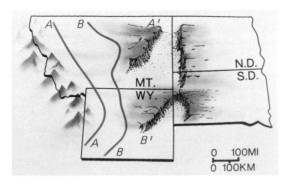

FIG. 3.—Paleogeography of the Late Cretaceous seaway during the progradation that culminated in deposition of the Shannon Sandstone Member. Strand position A is early in the progradation (*Desmoscaphites bassleri*) and position B is late in the cycle (*Baculites* species, smooth) (Gill and Cobban, 1973). The exaggerated relief shown at A' and B' are the western shelf margins which are time-equivalents of the respective strand lines. Modified from Shurr (1975).

is found throughout eastern Montana; it passes west-ward into a facies of coastal sandstone and eastward into a shale facies (Fig. 4). The coastal sandstone facies has recently been described in north-central Montana (Rice, 1980). The shelf sandstone facies described by Rice and Shurr (1980) includes volumes of sandstone, that is sandstone lithosomes, surrounded by larger lithosomes composed of interbedded shale, siltstone, and sandstone. The western boundary of the shale facies is mapped where the interbedded shale, siltstone, and sandstone pass laterally into shale. The belt of shelf sandstone facies includes three regional sandstone bodies: lithosome (level I), sheet (level II), and lentil (level III).

Within the regional study area shown in Figure 1, the Shannon encompasses a large and complex sandstone lithosome (level I) that covers about 117,500 km^2 (45,000 mi^2)(Table 1). The sandstone lithosome is enclosed by a unit 121 to 182 m (400 to 600 ft) thick, composed of interbedded shale, siltstone, and sandstone, representing the upper part of the Gammon Shale. The total Gammon is as much as 300 m (900 ft) thick in the regional study area. A diagrammatic cross section through the sandstone lithosome is shown in Figure 5.

The sandstone sheet (level II, Table 1) occupying the center of the lithosome (Fig. 5) has been mapped using subsurface data (Fig. 6), and covers approximately 18,200 km^2 (7,000 mi^2) in southeastern Montana and adjacent parts of South Dakota and Wyoming. The Groat Sandstone Bed of outcrop usage is equivalent to the sandstone sheet mapped in subsurface. The sheet is composed of a series of coalesced sandstone lentils; areas of the sheet intervening between individual lentils are mostly about 7.6 m (25 ft) thick.

Individual lentils (level III, Table 1) of sandstone, generally less than 23m (75 ft) thick and covering approximately 1,500 km^2 (580 mi^2), can be indentified within the lithosome and sheet. Four regional lentils coalesce in the central part of the lithosome to produce the sandstone sheet (Fig. 6) and four more lentils are found at other stratigraphic positions within the lithosome (Fig. 5). The thickness and distribution of the four isolated lentils have been mapped regionally (Shurr, 1975). The four lentils that are a part of the sheet are shown on Figure 6 as discrete pods (A through D, Figs. 5 and 6).

The external aspects of lentils (level III) are shown in Figures 5 and 6. In general, the length and width of a lentil are approximately equal. The base of a lentil is usually gradational and the upper contact is sharp. Lentils A and B (Figs. 5 and 6) have recently been mapped in the subsurface (John T. Jenkins, Jr., personal communication, 1981). This study shows that the Montana portion of lentil A consists of at least 30 northwest oriented elongate sandstone lenses and lentil B is composed of about 10 such elongate lenses. The length and width of lentil B are not as equal as shown in Figure 6

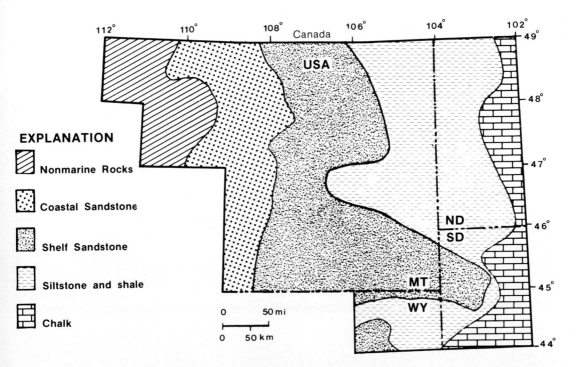

FIG. 4.—Regional distribution of Shannon shelf sandstone and laterally equivalent facies. Map summarizes distributions within the stratigraphic interval shown in Figure 2.

TABLE 1.—HIERARCHY OF ANCIENT SHELF SANDSTONE BODIES

Level	Elements, Informal Description	Area in Km² (mi²)		Thickness in m (ft)		Genetic Interpretation
I	Sandstone lithosome and enclosing sediments	117,000	(45,000)	121-182	(400-600)	Progradational deposits over entire shelf
II	Sheets or "benches"	18,200	(7,000)	30	(100)	Areas of sandstone deposition confined to lineament-bound blocks
III	Regional lentils	1,500	(580)	15-23	(50-75)	Sand-ridge fields
IV	Elongate lenses	52	(20)	12-18	(40-60)	Sand ridges
V	Facies packages	0.12	(0.05)	3-6	(10-20)	Sand waves

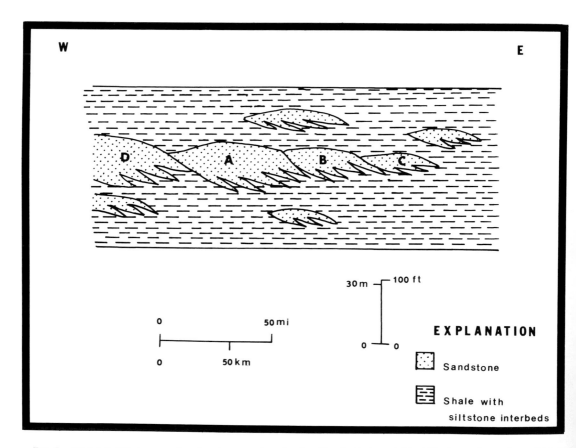

FIG. 5.—Diagrammatic cross section showing geometry of regional lentils (A through D) within the sandstone sheet which is encompassed by the lithosome that constitutes the Shannon Sandstone Member. The location of individual lentils A through D are shown in Figure 6 and the line of section lies approximately along latitude 45°30′ (Fig. 6). The base and top of the lithosome are the subsurface markers A and D, respectively, in Figure 9. Generalized from Shurr (1975).

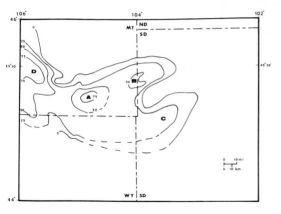

FIG. 6.—Net sandstone thickness of the sandstone sheet composed of coalesced lentils. Individual lentils are labeled A through D. Dashed contours are reconstructed in areas where sandstone was removed by erosion. Contour interval 25 ft (7.6 m). From Shurr (1975).

and it has an overall trend to the northwest. A few isolated elongate lenses are found within the sandstone sheet near the margins of lentils A and B. This is analogous to the isolated lentils which are found within the lithosome near the margins of the sandstone sheet (Fig. 5).

Lithology and thickness variations within an individual lentil are related to the component elongate lenses. Thus, the description of elongate lenses presented in the next section of this report constitutes a description of the internal aspects of a lentil. For example, the gradational base is shown in the correlation of outcrop measured sections (Fig. 7).

GEOMETRY OF LOCAL SANDSTONE BODIES:
ELONGATE LENSES (LEVEL IV)

Elongate lenses of sandstone (level IV) are found within regional lentils (Table 1). The elongate lenses have areas of about 52 km² (20 mi²) and thicknesses of 12 to 18 m (40 to 60 ft.). This generalization is based

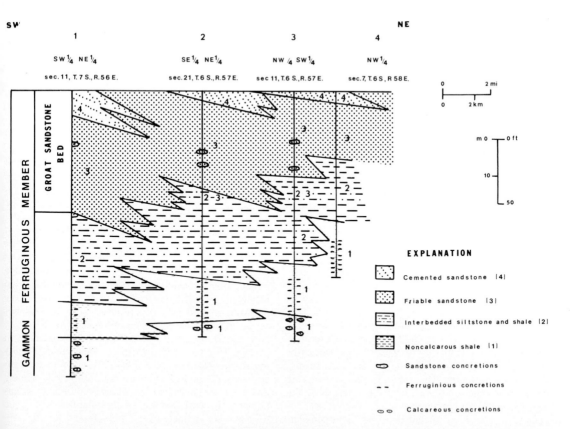

FIG. 7.—Correlation of measured sections through the Groat Sandstone Bed (equivalent to the Shannon Sandstone Member of the subsurface) of the Gammon Ferruginous Member of the Pierre Shale. See Figure 8 for locations of sections. Small-scale facies changes within units 3 and 4 are shown in Figures 12, 13, and 15.

upon investigation of the internal geometry of lentil A (Fig. 6) in southeastern Montana. The study employs data collected in areas near the Montana-Wyoming border where the Shannon-equivalent, the Groat Sandstone Bed, is exposed and where subsurface control points are closely spaced.

Outcrop Observations

In four townships in the eastern half of the area of local studies (Fig. 1), the Groat Sandstone Bed is exposed where beds of cemented sandstone form resistant caps on the top of buttes. Eight stratigraphic sections were examined in this area, lithologies in the Groat and Gammon were described, and stratigraphic thicknesses were measured (Fig. 7). Four lithologic units are recognized on the basis of grain size, bedding, and weathering characteristics. These units form a coarsening-upward sequence and from bottom to top are:

Unit 1.—Noncalcareous dark, gray shale lies at the base of each section. Generally about 15 m (50 ft) is exposed; the maximum thickness observed in outcrop is 60 m (200 ft). Large calcareous, septarian claystone concretions are found below an interval that has small, moderate-red to dark-reddish-brown, ferruginous claystone concretions. Unit 1 grades upward into unit 2 by the addition of siltstone beds and by an increased silt content in the shale.

Unit 2.—Interbedded siltstone and shale 15 to 21 m (50 to 70 ft) thick overlies unit 1. Grayish-orange siltstone occurs in dark-gray, noncalcareous, silty shale as interbeds less than 0.3 m (1.0 ft) thick. Weathered surfaces are medium light gray to very light gray.

Unit 3.—Nearly structureless, bioturbated, friable, fine-grained sandstone interbedded with siltstone and claystone is found above unit 2 and forms unit 3. Unit 3 is 14 to 24 m (45 to 75 ft) thick and weathers to a moderate-yellow to grayish-orange soil. Fresh surfaces are grayish orange to medium light gray. Sandstone interbeds are 30 to 50 cm (12 to 20 in.) thick. The proportion and thickness of claystone and siltstone interbeds decrease upward from the gradational base. Calcareous siltstone concretions are found in the lower part of the unit. The upper part is dominantly sandstone and includes concretionary masses of cemented sandstone. The contact with the overlying butte caps is gradational.

Unit 4.—Crossbedded and burrowed, well-cemented sandstone forms butte caps generally less than 3.3 to 4.5 m (10.0 to 15.0 ft) thick. Three distinctive lithologies within the sandstone are arranged in discrete facies packages. The lithology and geometry of these small-scale sandstone facies are described in the next section of this paper.

Units 1 and 2 are included in the Gammon Ferruginous Member of the Pierre Shale and units 3 and 4 are the Groat Sandstone bed as mapped in the area by Robinson, Mapel, and Bergendahl (1964). Strata above unit 4 are rarely preserved on the buttes, but in the outcrop belt west of the buttes, westward-dipping units include shales similar to unit 1 that overlie the cemented sandstone with a sharp, nongradational contact. In total, the stratigraphic succession of the four lithologic units, when combined in a variety of proportions, represents a single upward coarsening cycle as much as 45 m (150 ft) thick.

The stratigraphic relationships among these lithologic units are shown in Figure 7. In general, the shale and interbedded siltstone and shale (units 1 and 2) appear to rise stratigraphically, becoming younger to the northeast; in addition, the friable and well-cemented sandstones (units 3 and 4) thin to the northeast. Although the facies shown in Figure 7 are not easily traced laterally in outcrop because of vegetation cover and separation of the buttes, the discontinuous nature of the well-cemented sandstone (unit 4) between sections 1 and 2 can be observed in exposures that are a part of the outcrop belt of Groat located in the eastern part of T. 6 S., R. 56 E. (Fig. 8).

Facies changes are very obvious in the stratigraphic section oriented northeast-southwest (Fig. 7), which is perpendicular to the trend of clustered elongate buttes (Fig. 8). In contrast, essentially no variation in lithology or thickness was observed in correlations parallel to the butte trend (for example, northwest and southeast of measured-section 1, Fig. 8). The butte clusters are believed to indicate sandstone bodies having elongate axes trending N. 30° W. In the correlations perpendicular to this trend (Fig. 7), sections 1, 2, and 4 may be located in separate elongate sandstone bodies. The clusters of buttes may represent the preservational limits of thick parts of the sandstone bodies. Based upon outcrop observations only, these comments on sandstone geometry may appear to be speculative. However, subsurface studies establish the existence of elongate sandstone lenses trending parallel to the buttes.

Subsurface Studies

In four townships adjacent to and west of the area of Groat outcrops (Fig. 8), bentonite beds and sandstone units provide useful subsurface markers (Fig. 9). The Ardmore Bentonite Bed is a distinctive subsurface marker often considered as the top of the Eagle in Montana (for example, Rice, 1976). A marker below the Ardmore in the upper part of the Gammon Member of the Pierre Shale has been correlated (Shurr, 1975) with the H-bentonite bed of Knechtel and Patterson (1962). This bentonite bed is the uppermost electric-log marker (D, Fig. 9) employed throughout the subsurface study area. The lowest marker (A, Fig. 9) may also be a bentonite bed, but its exact lithology is unknown. The base and top of the Shannon Sandstone Member are defined by markers B and C (Fig. 9), respectively.

Lithologic interpretations of electric-log properties were established by correlation of a typical log with lithologic units described in a nearby stratigraphic sec-

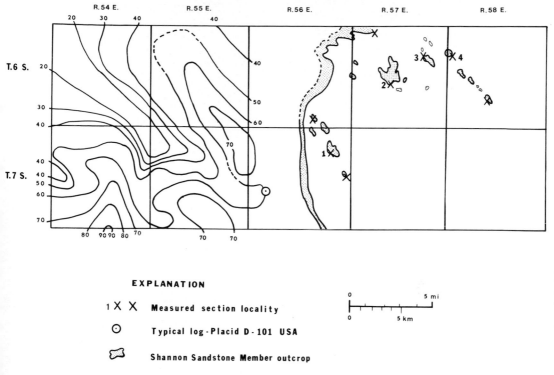

EXPLANATION

1 X X   **Measured section locality**

⊙   **Typical log - Placid D - 101 USA**

   Shannon Sandstone Member outcrop

FIG. 8.—Map showing location of buttes capped by Groat Sandstone and sandstone isopach map for the Shannon Sandstone Member of subsurface usage. Contour interval is 10 ft (3 m). Dashed contours and contacts are approximate. Approximately 50 subsurface control points were used. A typical subsurface log is shown in Figure 9. See Figure 1 for location of map area.

tion (Fig. 9). Relatively low resistivity and spontaneous potential above marker A represent the shale of unit 1. The upward increase in resistivity reflects the upward increase in grain size observed in outcrop, as shale grades upward into interbedded siltstone and shale. The relatively sharp base of the well-developed sandstone used in subsurface mapping (marker B, Fig. 9) has little expression in outcrop. It is possible that marker B reflects the top of the interval in which claystone and siltstone interbeds, rather than friable sandstone, dominate unit 3. Increased log values above marker B are associated with the part of unit 3 dominated by friable sandstone. The highest resistivity and spontaneous potential are observed just below marker C and these are characteristic of the crossbedded, well-cemented sandstone of unit 4. Diminished log values above marker C represent shale.

The distribution and thickness of the Shannon Sandstone Member in the subsurface is shown in Figure 8. Elongate lenses (level IV) cover approximately 52 km^2 (20 mi^2) and parallel the trend of buttes in the area of outcropping Groat. The elongate lenses are the elements of the geometric hierarchy that compose the thickest part of regional lentil A in Figure 6. Thicknesses of 12 to 18 m (40 to 60 ft) are common within the lenses. Where the sandstone is thickest, as in the

southern part of T. 7 S., R. 54 E., breaks in the spontaneous-potential curve and multiple peaks in the resistivity curve suggest a stacking of separate units of cemented sandstone similar to the facies shown between measured sections 2 and 3 (Fig. 7).

The geometry of parallel and perhaps prograding elongate sandstone lenses has been observed in the Eagle Sandstone at several localities in Montana. The rimrock above Billings has been shown to display this geometry and subsurface study reveals the orientation of the lenses (Shelton, 1965). Exposures of the Eagle near Winnett and Mosby in central Montana show gently inclined stratigraphic datums similar to those at Billings, although the existence of lenses in the subsurfaces has not been established.

Comparison with Wyoming Shelf Sandstones

The lithologies and sandstone geometries found in the Shannon Sandstone Member of the Gammon Shale of southeastern Montana are similar to the shelf sandstones described in the Upper Cretaceous of Wyoming.

The Shannon Sandstone Member of the Steele Shale in the Powder River Basin is a well-documented example of a shelf sandstone. Spearing (1976) described three facies in the Shannon at Salt Creek in the western part of the basin: his "silty marine mudstone facies" is

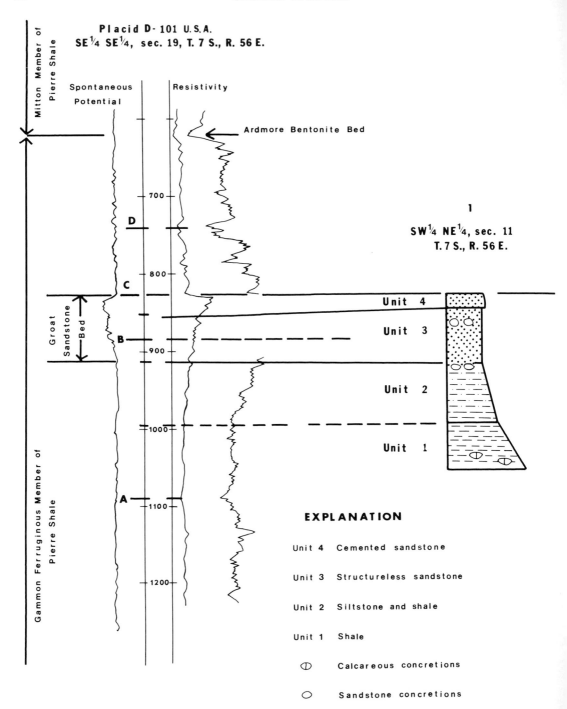

FIG. 9.—Subsurface log and lithologic section showing correlation of subsurface parastratigraphic units with outcropping lithologic units. Vertical log scale is in feet. See Figure 8 for location of log and measured section 1.

analogous to the interbedded siltstone and shale in the Groat (unit 2); his "thinly-bedded sandstone facies" is analogous to the friable sandstone interbedded with siltstone and claystone in the Groat (unit 3); and his "crossbedded sandstone facies" is analogous to the crossbedded cemented sandstone in the Groat (unit 4). Based on Spearing's (1976) descriptions and my own reconnaissance at Salt Creek, the Shannon at Salt

Creek seems less burrowed than the Groat, which has fairly extensive bioturbation in unit 3 and parts of unit 4. Spearing (1976) presented isopach maps of sandstone bodies in the Shannon that have the approximate dimensions of the regional lentils recognized in the subsurface unit equivalent to the Groat (Fig. 6).

Heldt Draw and Hartzog Draw are oil fields producing from the Shannon in the central Powder River Basin about 64 km (40 mi) northwest of Salt Creek. Sandstone facies similar to those described by Spearing (1976) were recognized at Heldt Draw and Hartzog Draw by Davis (1976), Seeling (1978), and Tillman and Martinsen (1979). In addition, elongate lenses with dimensions similar to the elongate lenses shown in Figure 8 are documented in all three of these papers. Regional studies of the Shannon (Crews, Barlow, and Haun, 1976) showed that elongate lenses at Heldt Draw and Hartzog Draw are components of larger regional sandstone lentils. Thus, the Shannon Sandstone Member in the Powder River Basin appears to have regional lentils and constituent elongate lenses similar to those found in the Shannon of southeastern Montana.

The Sussex Sandstone Member of the Steele Shale, which is a shelf sandstone about 122 m (400 ft) above the Shannon, has been described by Berg (1975) at the House Creek oil field. House Creek is located in the central Powder River Basin, approximately 16 km (10 mi) east of Hartzog Draw and is elongate northwest to southeast parallel with Heldt Draw and Hartzog Draw. The facies described by Berg (1975) seems only broadly analogous to those recognized in the Groat outcrop: his "shale with sandstone lenses" may approximate the interbedded siltstone and shale (unit 2) and the interbedded sandstone and claystone (lower part, unit 3); the "pebbly sandstone", "crossbedded sandstone", and rippled sandstone" facies in the Sussex represent the crossbedded, well-cemented sandstone (unit 4) in the Groat outcrops that are extensively bioturbated (unit 3 and parts of unit 4).

Regional studies of the Sussex Sandstone Member in the central Powder River Basin document sandstone lentils and sheets of coalesced lentils (Crews, Barlow, and Haun, 1976; Brenner, 1978) that have the same dimensions as regional lentils in the Shannon in southeastern Montana. The lentils in the Sussex coalesce in a manner very similar to the configuration observed in the central part of the shelf sandstone lithosome in the Shannon (Figs. 5 and 6). Within the Sussex regional lentils, studies of individual oil fields reveal small elongate lenses similar to those shown in Figure 8. For example, net sandstone maps of the Sussex at House Creek oil field (Berg, 1975) and Triangle U oil field (Anderman, 1976) document such elongate lenses.

The Shannon and Sussex in the central Powder River Basin are well-documented examples of shelf sandstones and they have sandstone geometries and lithologies very similar to the Shannon in the northern Powder River Basin in southeastern Montana.

GEOMETRY OF LOCAL SANDSTONE BODIES:
SMALL-SCALE FACIES PACKAGES (LEVEL V)

Detailed outcrop studies in southeastern Montana reveal facies variations and sandstone bodies that to date either have not been recognized or do not exist in other shelf sandstones. These variations and geometries are important because they generally occur over distances of less than 300 m (1000 ft) and would be difficult to recognize even in studies of core from close-spaced wells. Outcrop studies suggest that the elongate sandstone lenses (level IV) are, in turn, made up of sets of smaller, discrete facies packages (level V, Table 1).

Lithologic Units

Small-scale sandstone bodies are visible in exposures of units 3 and 4 (Fig. 7) in the area of local studies. The bodies are defined by three laterally equivalent sandstone types that are termed structureless sandstone, burrowed sandstone, and crossbedded sandstone. These three facies are recognized on the basis of gross lithology, bedding, sedimentary structures, and degree of burrowing (Table 2). From bottom to top they are:

Structureless sandstone.—Friable, structureless, glauconitic sandstone mottled with clay, characterizes the upper part of unit 3. The sandstone is poorly sorted and fine grained, possibly grading into silt particle sizes. Ripples and parallel laminations 1 cm (0.4 in) thick are rarely preserved. Bioturbation is commonly so complete that only a few identifiable trace fossils remain and a structureless appearance in outcrop is the result (A, Fig. 10). Only two trace fossils are recognized in the structureless sandstone. One is *Chondrites* and the second resembles *Cylindrichnus*. The latter trace fossil is a series of conical and concentric sheaths of claystone and sandstone ranging from 15 to 20 cm (5.9 to 7.9 in.) in diameter (Fig. 11). Concentrations of ammonites and wood fragments are often found associated with concretionary sandstone masses. This facies is poorly exposed and grades laterally and downward

Fig. 10.—Typical outcrop of bioturbated, structureless sandstone (A) and burrowed, cemented sandstone (B) in the butte cap at measured-section 4 (Fig. 8).

Fig. 11.—Trace fossil resembling *Cylindrichnus* which is found in structureless and in burrowed sandstone.

into interbedded siltstone and claystone.

Burrowed sandstone.—Butte caps (B, Fig. 10) consist of medium-to fine-grained glauconitic sandstone that commonly has clay drapes 1.0 to 3.0 cm (0.4 to 1.2 in.) thick. Beds are usually less that 25 cm (10 in.) thick and have flat, locally erosioinal bottoms. Burrowing commonly extends down into the bed from the upper surface. Where there is little burrowing, symmetric interference ripple sets and smaller tabular cross sets are observed. Body fossils are rare, but trace fossils are diverse and abundant. At least five trace fossils

are fairly common. In addition to the *Chondrites* and the *Cylindrichnus* forms also observed in the structureless sandstone, the burrowed sandstone contains *Ophiomorpha* and small vertical tubes resembling *Trichichnus*, which are about 1 mm (0.04 in.) in diameter and generally less than 5 cm (1.9 in.) long. The fifth trace fossil is a series of horizontal, smooth-walled burrows that are both straight and branching. This facies has a calcareous cement and weathers to moderate brown and grayish orange. Resistant butte capps (unit 4) consist of this well-cemented sandstone and the better-cemented parts of the overlying crossbedded sandstone.

Crossbedded sandstone.—The crossbedded sandstone is medium to coarse grained and in places it has small chert clasts and an associated salt and pepper aspect. It is better-sorted than the burrowed sandstone below it. Clay clasts are found locally, but continuous clay drapes are very rare. Sandstone beds are less than 50 cm (20 in.) thick; these contain megaripples and rhomboid ripples near the top and are locally channeled at the base. Within the beds, well-preserved laminations are 1.0 cm (0.4 in.) thick and form open trough to tabular crossbeds. Fossils are rare and burrowing is minimal. The only trace fossils found in the crossbedded sandstone are the small vertical *Trichichnus*-like tubes also found in the burrowed sandstone. On weathered surfaces, the sandstone is generally well cemented and is light-gray to white.

TABLE 2.—SANDSTONE FACIES WITHIN SMALL-SCALE FACIES PACKAGES

Characteristic	Crossbedded Sandstone	Burrowed Sandstone	Structureless Sandstone
Gross Lithology	Medium to coarse grained, clean sandstone having local salt and pepper and clay clasts.	Medium to fine grained, moderate to poorly sorted, glauconitic sandstone. Clay drapes.	Fine-grained sandstone to siltstone, poorly sorted and glauconitic. Mottled with clay.
Bedding	Beds less than 50 cm thick having internal laminations 1 cm thick. Lower bedding surfaces slightly erosional and show channels locally.	Beds less than 25 cm thick having flat, sharp bases and burrowing near the tops.	1-cm. thick lamination, where preserved.
Sedimentary structures	Large, open, low-relief trough to tabular crossbedding. Some rhomboid ripples, megaripples, and channels.	Local preservation of symmetric interference ripple sets and small tabular crossbedding.	Rare ripples and crossbedding.
Burrowing	Minimal. Few trace fossils. Primary bedding well preserved.	Moderate. Trace fossils common. Some primary bedding preserved.	Extreme. Few recognizable trace fossils and little primary bedding.
Fossils	Rare large *Baculites*.	Rare *Baculites* and bivalves.	Common *Baculites*, including large coiled *Placenticeras*(?), and wood fragments.
Energy level	High	Moderate	Low
Position on sand wave	Crest	Margin	Trough between waves.

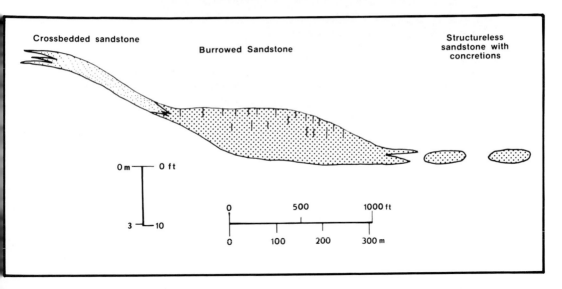

Fig. 12.—Idealized geometry of lateral relationships within small-scale facies packages found in sandstone exposures. Sketch is perpendicular to the elongate axis of the package. Elongation is commonly north-south. Vertical lines in burrowed sandstone are representations of burrows.

Individual Facies Packages

The vertical succession of three sandstones is also seen in lateral variations within resistant butte caps. These lateral facies relationships define discrete sandstone bodies here termed small-scale facies packages. Within a single facies package, crossbedded sandstone grades laterally to burrowed sandstone, which in turn grades into structureless sandstone containing local concretionary sandstone masses (Fig. 12). Thus, there is an overall lateral and downward decrease in grain size and sorting associated with changes in bedding and sedimentary structures (Table 2). Along the same gradient there is an increase in burrowing and an increase in the number of body fossils. Directional data from sedimentary structures suggest that sediment transport was both parallel and perpendicular to the gradient shown in Figure 12 and Table 2.

The complete lateral sequence of facies within a package is rarely preserved or exposed in its entirety. The geometry shown in Figure 12 is a composite based on correlation of several sets of closely spaced sections in different packages. However, the lateral facies relationship between two of the three sandstone types can be seen in many exposures. Instances of burrowed cemented sandstone passing into structureless sandstone are more frequently observed than crossbedded sandstone passing into the burrowed cemented sandstone. This is probably because the stratigraphic position of the structureless sandstone, where it is observed in outcrop, is below the resistant butte cap composed of burrowed sandstone. The crossbedded sandstone is less frequently preserved as the highest unit on the butte cap because it is not well cemented. The generalized correlations do give some estimates of the dimensions of the facies package. Additional inferences on size and shape can be drawn from the configuration of butte tops.

Differences in texture and cementation among the facies within a single package appear to influence the configuration of butte tops. Most commonly parts of butte tops are elongated at right angles to the gradient shown in Figure 12 and Table 2. Parallel to the elongation, units are easily traced for 300 to 450 m (1000 to 1500 ft) (Fig. 13). However, perpendicular to the elongation, facies changes of the type shown in Figure 12 make individual units difficult to trace; sandstone facies interfinger over distances of 150 to 300 m (500

Fig. 13.—Process model for sand body formation on the muddy Cretaceous shelf of the western interior seaway, due to the occurrence of spatial as well as temporal velocity variations. Compare with Figure 4 as a product model, with Figure 10 as a model for sedimentation with temporal variation only. (S = sand, Si = silt, Cl = clay).

to 1000 ft). The axes of oblate parts of the butte tops seem to mark the thickest part of the burrowed cemented sandstone that is the central part of the facies packages. Crossbedded sandstone locally lies above the cemented sandstone rim and structureless sandstone usually lies below the rim. This geometry suggests a progradation arrangement of sets of packages, because the vertical succession of lithology reflects the lateral variation.

Sets of Facies Packages

Individual facies packages appear to be stacked laterally in sets of progradational prisms similar to those recognized in coastal sandstones of the Upper Cretaceous (Asquith, 1970; Shurr, 1975). The vertical succession between lenses generally repeats the lateral variation of facies within a single package. Thus, structureless sandstone usually grades upward into burrowed sandstone and this, in turn, is overlain by crossbedded sandstone. Figure 14 shows the idealized geometry that would result from the imbricate arrangement of a set of facies packages. Such imbricate arrangements are not easily photographed, because the lithologic contrasts are not conscpicuous from a distance. However, field sketches (Fig. 15) emphasizing the lithologies do illustrate the lateral and vertical stacking of sets of facies packages.

Sets of facies packages show changes in specific characteristics from southwest to northeast in the correlation section shown in Figure 7. These changes may indicate differences among the separate elongate sandstone lenses suggested to be present in the section. For example at locality 1 (Fig. 7), progradational facies packages are closely packed and all three internal sandstone lithologies are well developed. In contrast at locality 2, the packages are more widely separated vertically and horizontally, and much bioturbated sandstone is present. Greater changes were observed at localities 3 and 4, where the packages cannot be easily identified and the crossbedded sandstone facies is dominant. Although the geometry of sets of facies packages may be somewhat speculative, it is nonetheless clear that the characteristics of these sand bodies do change over distances of 9.6 to 19.0 km (6 to 12 mi).

GEOMETRIC HIERARCHY OF SANDSTONE BODIES

Summary

The preceeding descriptions of regional and local shelf sandstone bodies can be summarized in a classification that is a geometric hierarchy. The geometric hierarchy shown in Table 1 has sandstone bodies of different sizes as the elements and these elements are arranged into five levels. From largest to smallest, the levels are: lithosome, sheet, lentils, elongate lenses, and small-scale facies packages. The hierarchy is summarized diagrammatically in Figure 16.

In southeastern Montana the total sandstone lithosome (level I) in the Shannon (Fig. 5) is the highest, that is largest, level of the hierarchy. The lithosome is made up of a regional sandstone sheet (level II) that is itself composed of four lentils (A, Fig. 16; Fig. 6). Each sandstone lentil (level III) is an element of the next level in the hierarchy and an individual lentil contains many elongate lenses (B, Fig. 16; Fig. 8). The elongate lenses, in turn are a fourth level in the hierarchy. The fifth-level elements are the small-scale facies packages that are found in a single elongate lens (C, Fig. 16). This hierarchical classification is based upon observations in the Shannon of southeastern Montana, but it also appears to have application in well-documented shelf sandstone bodies of the Shannon and Sussex in the Wyoming Powder River Basin.

Genetic Interpretations

Interpretation of the processes that produced the geometric hierarchy shown in Table 1 is facilitated by employing an intrinsic property of hierarchical organizations. Specifically, the largest elements in any hierarchy require a longer time to be produced; that is, they are the result of long-term processes, when compared with smaller elements (Simon, 1962). Thus, the geometric hierarchy of shelf sandstone bodies (Table 1) can be related to a process model that has a hierarchical structure with respect to time. Although the fundamental nature of the processes may be different from one level to the next, (for example, tectonism versus sedimentation) the larger sandstone bodies require longer times to be built than the smaller sandstone bodies.

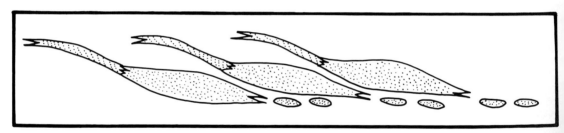

Fig. 14.—Imbricate lateral stacking of several small-scale facies packages (level V). No horizontal or vertical scale. See Figure 12 for description of facies.

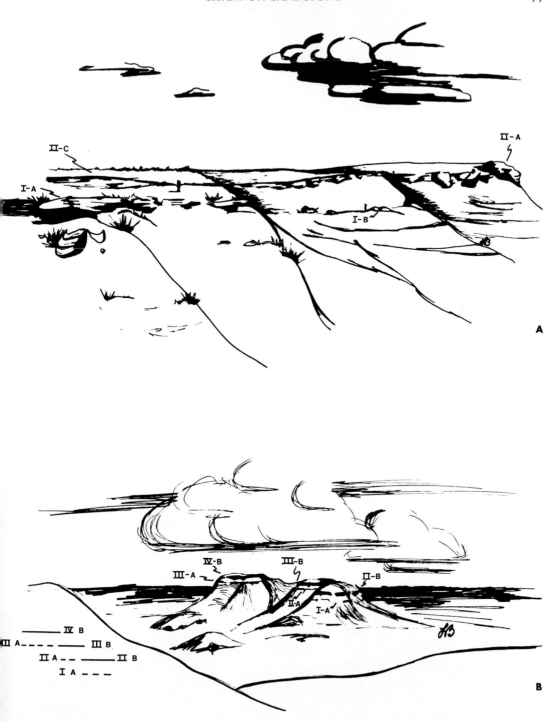

FIG. 15.—Field sketches of small-scale facies packages; prepared by Laurie Butler. A, Butte top at measured section 2 (Fig. 8). Lower facies package has burrowed sandstone (I-A) and structureless sandstone (I-B) preserved. In the overlying package crossbedded sandstone (II-C) burrowed sandstone (II-A) are preserved. B, Butte top at section measured southeast of section 1 (Fig. 8) showing four stacked packages (I through IV). Only burrowed sandstone (I-A) is preserved in the lowest facies package. The two middle packages include both burrowed sandstone (II-A and III-A) and structureless sandstone (II-B and III-B). The upper package is marked by structureless sandstone (IV-B). Generalized succession of facies packages is shown in lower left.

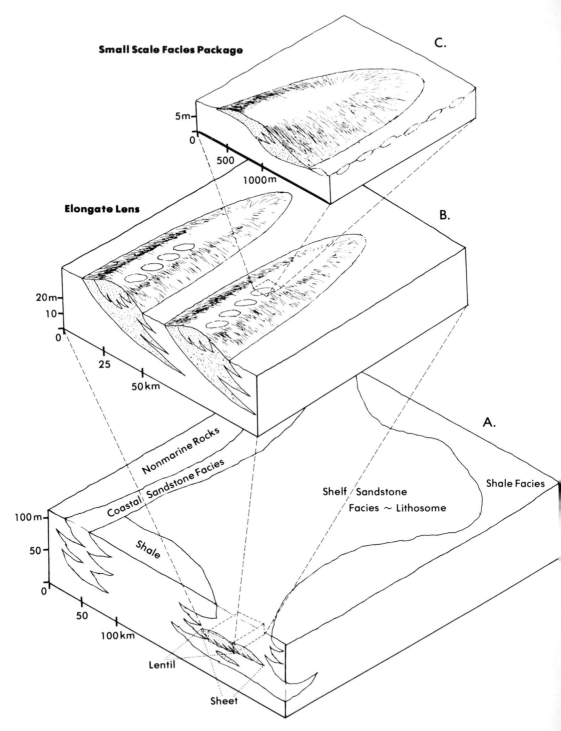

Fig. 16.—Summary of hierarchical nature of shelf sandstone bodies. A, Lithosome (level I), sandstone sheet (level II), and lentils (level III). B, Elongate lens (level IV). C, Small-scale facies package (level V). See text for discussion.

The largest element of the geometric hierarchy is the total lithosome (level I, Table 1) and the processes producing this large element acted over long time intervals. In particular, the lithosome is a part of a cycle of transgression and progradation that lasted about four million years. The lower levels of the hierarchical response model are smaller and are produced by shorter term processes. The sandstone sheet (level II) and the constituent sandstone lentils (level III) have been shown to be controlled, at least in part, by paleotectonic events on structural blocks bounded by lineaments (Shurr, 1975; 1979). The paleotectonism influencing these two smaller elements of the geometric hierarchy was a series of distinct events separated by time intervals of one-half to one million years.

The smaller elements in the shelf-sandstone hierarchy (levels III, IV, and V, Table 1) are the result of short-term sedimentation processes on the shelf. The origin of these smaller elements will be explored in some detail because they provide support for the interpretation of the Shannon as a shelf sandstone.

MORPHOLOGIC ELEMENTS OF THE SHELF

The smaller elements (levels III, IV, and V) of the hierarchy in Table 1 constitute a hierarchical response model of ancient shelf sandstone bodies. A very similar hierarchical response model is found in modern shelf sands (Table 3). The hierarchical aspect of modern shelf sand bodies has been emphasized recently by Walker (1979) in a review of published work dealing with shelf sedimentation patterns. Large sand bodies, sometimes termed "retreat massifs", are complex fields of linear sand ridges (Swift, 1975). In turn, the sand ridges include smaller sand bodies known as sand waves. The large ridge fields, which are the first-order (largest) elements of Walker's (1979) hierarchy, probably correspond with the regional lentils of the Shannon (level III, Table 1). Second-order linear sand ridges correlate well with elongate lenses (level IV). And third-order sand waves are equivalent to small-scale facies packages (level V). Comparisons between the ancient and modern hierarchical response models at each level of the hierarchy are discussed below.

Ridge Fields

Regional sandstone lentils (level III, Table 1) in the Shannon of southeastern Montana may be related to the large, complex ridge fields (Table 3) of Walker's (1979) hierarchy. The general size of the modern sand bodies, specifically 72 km (45 mi) long, 21 km (13 mi) wide, and 10 to 30 m (33 to 100 ft) thick (Walker, 1979), is similar to the size of the lentils in the Shannon which are 15 to 23 m (50 to 75 ft) thick and cover about 1500 km² (580 mi²). Two other characteristics of the modern ridge fields can be compared with the lentils in the Shannon: first, the large modern sand bodies tend to be elongate in a direction almost perpendicular to the coast, and second they lie offshore from capes and estuaries. Lentils in the Shannon do lie at a high angle to the strand line (compare Figs. 3 and 6). However, capes and estuaries have not been specifically documented in the Upper Cretaceous strand lines, and this test of the ridge-field model cannot be applied at the present time. As mentioned in the previous section of this report, the distribution of the lentils, that is of the ridge fields, was controlled at least in part by paleotectonism.

The term "retreat massif" has been applied to the large modern sand bodies here correlated with ancient sandstone lentils. The Atlantic shelf off North America (Swift, Stanley, and Curray, 1971) and the North Sea (Swift, 1975) are examples of areas in which the term "retreat massifs" has been used. However, there is no agreement on the relict nature of these modern sand bodies and hence, the more descriptive term "ridge field" is used in this paper. Use of "ridge field" emphasizes the hierarchical nature of the modern response model; ridge fields are complexes of linear sand ridges.

Sand Ridges

Elongate lenses recognized in the Shannon (level IV, Table 1) are equivalent to linear sand ridges found on modern continental shelves. Recent studies show that sand ridges on the Maryland sector of the North American inner shelf have lengths of 8.3 to 13.9 km (5.1 to 8.6 mi), widths of 1.4 to 2.3 km (0.9 to 1.4 mi), and relief values of 6.1 to 9.1 m (20.0 to 29.8 ft) (Swift and Field, 1981). On the central Atlantic shelf of the United States, sand ridges have spacings of 2.0 to 5.0 km (1.2 to 3.0 mi) and lengths of 12 to 54 km (7 to 32 mi) (Swift, Kofoed, Saulsbury, and Sears, 1972). Detailed studies of the sublittoral sand sheet on the outer part of the shelf (Knebel and Spiker, 1977) revealed that the average thickness of 5 to 7 m (19 to 32 ft)

TABLE 3.—HIERARCHY OF MODERN SHELF SAND BODIES

Elements	Length in km (mi)	Width in km (mi)	Relief in m (ft)	Reference
Ridge fields	72 (45)	21 (13)	10-30 (33-100) Thickness	Walker, 1979
Sand ridges	8-14 (5-9)	1-2 (0.6-1)	6-9 (20-30)	Swift and Field, 1981
Sand waves		350 m (1150 ft) Spacing	4 (13)	Knebel and Folger, 1976

approximates the relief commonly observed on sand ridges. In addition, these studies show that medium to coarse sand in the upper part of the sheet is underlain by muddy sands.

The dimensions (level IV, table 1) and the spacing (Fig. 8) of the elongate lenses are approximately the same as modern sand ridges (Table 3). Furthermore, the lithologic units observed in outcrop can be easily interpreted as having been deposited at different topographic positions on a single ridge. Interbedded siltstone and shale (unit 2, Fig. 7) may represent a trough position. Structureless bioturbated sandstone containing interbedded siltstone and shale (unit 3, Fig. 7) may have been deposited on the ridge flank. The crossbedded and burrowed, cemented sandstone (unit 4, Fig. 7) was probably formed on the ridge crest. The interpretation of these lithologies as topographic positions on sand ridges agrees with observations of vibracores taken from trough to crest on modern sand ridges (Strubblefield and Swift, 1976).

The term "sand ridge" has been previously applied by Seeling (1978) to elongate lenses in the Shannon at Heldt Draw in the central Powder River Basin. Sand ridge is probably more appropriately applied to the ancient shelf sand bodies than the term "bar", which has been frequently employed in the past. Sand ridges are morphologic elements of modern shelves, whereas bars are usually attached to the strand.

Sand Waves

Small-scale facies packages (level V, Table 1) are very similar to modern shelf sand waves (Table 3). Shelf sand waves are sometimes called dunes or megaripples, but should not be confused with bedforms described in flume studies that are also called sand waves (Walker, 1979). Shelf sand waves are much larger than the features observed in flumes. A recent study of sand waves on a part of the outer continental shelf off the eastern United States (Knebel and Folger, 1976), reveals that these asymmetric bedforms have an average spacing of 350 m (1150 ft) and an average relief of 4 m (13 ft). The upper part of the sand unit having these bedforms is medium sand and the lower part is medium to fine sand containing clay and silt. They also reported that no sedimentary structures are present in a vibracore from a trough between sand waves and extensive bioturbation was observed during submersible dives.

The geometry and sediments in small-scale facies packages are comparable with those observed in sand waves. Sandstone facies variation at right angles to the elongate axis of a facies package (Fig. 12 and Table 2), may reflect sedimentologic variations from crest to trough on a sand wave. Coarser-grained, crossbedded sandstone was deposited near the crest and finer-grained sandstone was deposited in the trough. This type of deposition suggests a decrease in energy levels from crest to trough. A gradient in bioturbation is also present from crest to trough: the little-burrowed, crossbedded sandstone grades to burrowed sandstone which may characterize the flank of the sand wave and the burrowed sandstone grades to thoroughly churned, structureless sandstone that probably was formed in the trough. The gradient of bioturbation off the crest and into the trough is probably a consequence of increased organism activity associated with the decreased energy levels from crest to trough.

The stratigraphic thickness between the base and top of a single facies package (3 to 6 m [10 to 20 ft], Fig. 12) should approximate the relief on the prograding bedforms that produced the unit. This generalization is a consequence of the imbricate geometry shown in Figure 14; it also explains the similarity between sand sheet thickness and bedform relief observed on sand ridges (Knebel and Spiker, 1977). The relief deduced from the thickness of a facies package in the Shannon is approximately that observed on modern sand waves.

Shelf Processes

Sediment transport on continental shelves could be the result of waves, tidal currents, wind-driven currents, internal waves, or currents related to submarine canyons (Knebel and Folger, 1976). There is no prevailing consensus on the relative importance of these energy sources and indeed different sources may affect different parts of a shelf. For example, it has been suggested that a spectrum of conditions exist between the extremes of fair-weather, tide-dominated sedimentation and storm-influenced sedimentation (Anderton, 1976). However, it has also been suggested that the pure end members of such a spectrum do not exist and that most sand bodies are influences by both tides and storms (Walker, 1979). Although a detailed interpretation of the shelf sandstone hierarchical response model must await the development of widely accepted process models for sedimentation on modern continental shelves, some preliminary generalizations can be made regarding conditions on the ancient shelf at the time of Shannon deposition.

The area of the shelf located in southeastern Montana was probably characterized by overall energy levels that were lower than in the shelf area located farther south in the west-central part of Wyoming's Powder River Basin. This generalization is based upon the fact that in southeastern Montana the Shannon has a smaller proportion of crossbedded sandstone and has more extensive bioturbation than in the west-central Powder River Basin. The specific energy source or combination of sources responsible for the energy levels cannot be identified at the present time. However, some speculations are possible on the relative strength of one energy source in different parts of the Upper Cretaceous Shelf. That energy source is tidal currents.

The concept that tidal currents are stronger on wide shelves has been applied to several ancient examples, including the Upper Cretaceous of the western interior (Klein and Ryer, 1978). The paleogeography illustrated in Figure 3 shows a very wide shelf in Montana. By the end of the regression that produced the Shannon, clastics from a western source area had filled the northern part of the basin and engulfed the eastern shelf to form a single wide shelf. The remnant of the western shelf margin nearest to the Black Hills study area is located farther from the strand than in Wyoming. A narrow shelf is indicated in Wyoming by the smaller distances from strand to shelf margin. This difference in shelf width suggests that tidal currents might have been stronger on the wider Montana part of the shelf. Thus, the higher energy levels postulated for the Wyoming shelf in the previous paragraph probably were not due to tidal currents alone. It must be emphasized that the overall energy levels were most likely produced by tidal current augmented by an unknown combination of energy sources.

The recognition and discussion of morphologic elements provides a preliminary synthesis of conditions on the Montana part of the shelf in the study area. Sand waves (level V) were most active on the crests and flanks of sand ridges (level IV). The sand ridges themselves were concentrated in ridge fields (level III) that were distributed over tracts of the shelf in which depositional patterns were controlled by paleotectonism on lineament-bound blocks (Shurr, 1975). Studies of ancient (Goldring and Bridges, 1973) and modern (Swift and Freeland, 1978) shelf sands suggest that processes of sedimentation on the outer shelf may be distinct from those on the inner shelf, near the strand. Thus, differences in shelf-sandstone geometry and in the hierarchical response model, may exist between the study area and localities farther west.

CONCLUSIONS

The Shannon Sandstone Member in southeastern Montana is a shelf deposit which displays a series of sandstone bodies of different sizes. A hierarchical classification is useful in describing the geometries, because large sandstone bodies are composed of progressively smaller sandstone bodies (Table 1). From largest to smallest, the observed elements are:

1) Sandstone lithosome (level I)—volume of sandstone enclosed in a large lithosome of shale interbedded with siltstone and sandstone.

2) Sandstone sheet (level II)—well-developed sandstones located in the center of the lithosome.

3) Sandstone lentils (level III)—thick "pods" within the sheet defined by contouring subsurface data on net sandstone thickness.

4) Elongate sandstone lenses (level IV)—components of lentils documented in closely spaced boreholes and in correlation of measured exposed sections.

5) Small-scale facies packages (level V)—laterally equivalent sandstone facies observed in outcrops on butte tops.

The largest sandstone bodies are interpreted to be the result of long-term processes; the lithosome is part of a slow progradation and the depositional patterns in the sheet sandstone are controlled by episodic paleotectonism. Lentils may also be influenced by paleotectonism. The smaller sand bodies are produced by short-term processes. The lentils, elongate lenses, and small-scale facies packages constitute a response model in ancient shelf sediments that appears to correspond with specific morphologic features on modern shelves. Lentils may be the remains of complex sand-ridge fields. Elongate lenses probably represent individual sand ridges. And the small-scale facies packages may correspond with modern shelf sand waves.

Although there is a strong correspondence between the ancient Shannon and modern hierarchical response models, there is a problem with the process model. At the present time, the processes producing morphologic elements on modern shelves are not well understood. As a result the process model responsible for the observed hierarchy of Upper Cretaceous sandstone bodies cannot be simply borrowed from studies of modern shelf sands. Instead, it may be more productive to deduce formative processes by the detailed investigation of ancient shelf sandstone bodies.

ACKNOWLEDGEMENTS

The cooperation and assistance of ranchers in the area, particularly the Ed Gardner family, is gratefully acknowledged. Technical assistance was provided by Terri Ericson and Jayne Sieverding. Artwork by Laurie Butler and Jim Bertram is included. Jan Maslonkowski typed the manuscript. The study was carried out with funding from the Department of Energy.

REFERENCES

ANDERMAN, G. G., 1976, Sussex Sandstone production, Triangle U Field, Campbell County, Wyoming: Wyoming Geol. Assoc., Guidebook 28th Ann. Field Conf., p. 107-113.

ANDERTON, R.,1976, Tidal-shelf sedimentation: An example from the Scottish Dalradian: Sedimentology, v. 24, p. 429-458.

ASQUITH, D. D., 1970, Depositional topography and major marine environments, Late Cretaceous, Wyoming: Am. Assoc. Petroleum Geologists Bull., v. 54, p. 1184-1224.

BERG, R. L., 1975, Depositional environment of Upper Cretaceous Sussex Sandstone, House Creek Field, Wyoming: Am. Assoc. Petroleum Geologists Bull., v. 59, p. 2099-2110.

BRENNER, R. L., 1978, Sussex Sandstone of Wyoming—Example of Cretaceous offshore sedimentation: Am. Assoc. Petroleum Geologists Bull., v. 62, p. 181-200.

BROWN, D. L., 1978, Wrench-style deformational patterns associated with a meridional stress axis recognized in Paleozoic rocks in parts of Montana, South Dakota, and Wyoming: Montana Geol. Soc. 24th Ann. Conf. Symposium, p. 17-31.

COBBAN, W. A., 1964, The Late Cretaceous cephalopod *Haresiceras* Reeside and its origin: U.S. Geol. Survey Prof. Paper 619, 29p.

CREWS, G. C., J. D. BARLOW, JR., AND J. D. HAUN, 1976, Upper Cretaceous Gammon, Shannon, and Sussex Sandstones, central Powder River Basin, Wyoming: Wyoming Geol. Assoc., Guidebook 28th Ann. Field Conf., p. 9-20.

DAVIS, M. J. T., 1976, An environmental interpretation of the Upper Cretaceous Shannon Sandstone, Heldt Draw Field, Wyoming: Wyoming Geol. Assoc., Guidebook 28th Ann. Field Conf., p. 125-138.

GILL, J.R., AND W. A. COBBAN, 1973, Stratigraphy and geologic history of the Montana Group and equivalent rocks, Montana, Wyoming, and South Dakota: U.S. Geol. Survey Prof. Paper 776, 37 p.

_____, _____, AND L. G. SCHULTZ, 1972, Correlation, ammonite zonation, and a reference section for the Montana Group, central Montana: Montana Geol. Soc., Guidebook 21st Ann. Field Conf., p. 91-97.

GOLDRING, R., AND B. BRIDGES, 1973, Sublittoral sheet sandstones: Jour. Sed. Petrology, v. 43, p. 736-747.

JOHNSON, H. D.,, 1978, Shallow siliciclastic seas, in H. G. Reading (ed.), Sedimentary Environments and Facies: Blackwell, Oxford, p. 207-258.

KITELEY, L. W., 1977, Shallow marine deposits in the Upper Cretaceous Pierre Shale of the northern Denver Basin and their relation to hydrocarbon accumulation: Rocky Mountain Assoc. Geologists Field Conf. Guidebook, p. 197-211.

KLEIN, G. D., AND T. A. RYER. 1978, Tidal circulation patterns in Precambrian, Paleozoic and Cretaceous epeiric and mioclinal shelf seas: Geol. Soc. America Bull., v. 89, p. 1050-1058.

KNEBEL, H. J., AND D. W. FOLGER, 1976, Large sand waves on the Atlantic outer continental shelf around Wilmington Canyon, off eastern United States: Marine Geol., v. 22, p. M7-M15.

_____, AND E. SPIKER, 1977, Thickness and age of surficial sand sheet, Baltimore Canyon Trough area: Am. Assoc. Petroleum Geologist Bull., v. 61, p. 861-871.

KNECHTEL, N. M., AND S. H. PATTERSON, 1962, Bentonite deposits of the northern Black Hills district, Wyoming, Montana, and South Dakota: U.S. Geol. Survey Bull. 1082-M, p. 893-1030.

MAUGHAN, E. K., 1966, Environment of deposition of Permian salt in the Williston and Alliance Basins: Northern Ohio Geol. Soc. 2d Symp. on Salt, p. 35-47.

PORTER, K. W., 1976, Marine shelf model, Hygiene Member of Pierre Shale, Upper Cretaceous, Denver Basin, Colorado, in R. C. Epis and R. J. Weimer (eds.), Studies in Colorado Field Geology: Prof. Contrib. Colorado School of Mines No. 8, p. 251-263.

RICE, D. D., 1976, Stratigraphic sections from well logs and outcrops of Cretaceous and Paleocene rocks, northern Great Plains, Montana: U.S. Geol. Survey Oil and Gas Invest. Chart OC-71.

_____, 1980, Coastal and deltaic sedimentation of Upper Cretaceous Eagle Sandstone—Relation to shallow gas accumulations, north-central Montana: Am. Assoc. Petroleum Geologists Bull., v. 64, p. 316-338.

_____, AND G. W. SHURR, 1980, Shallow, low-permeability reservoirs of northern Great Plains—Assessment of their natural gas resources: Am. Assoc. Petroleum Geologists Bull., v. 64, p. 969-987.

ROBINSON, C. S., W. J. MAPEL, AND M. H. BERGENDAHL, 1964, Stratigraphy and structure of the northern and western flanks of the Black Hills Uplift, Wyoming, Montana, and South Dakota: U.S. Geol. Survey Prof. Paper 404, 134 p.

RUBEY, W. W., 1930, Lithologic studies of fine-grained Upper Cretaceous sedimentary rocks of the Black Hills region: U.S. Geol. Survey Prof. Paper 165-A, p. 1-54.

SEELING, A., 1978, The Shannon Sandstone, a further look at the environment of deposition of Heldt Draw Field, Wyoming: Mountain Geologist, v. 15, p. 133-144.

SHELTON, J. W., 1965, Trend and genesis of lowermost sandstone of Eagle Sandstone at Billings, Montana: Am. Assoc. Petroleum Geologists Bull., v. 49. p. 1385-1397.

SHURR, G. W., 1975, Marine cycles in the lower Montana Group, Montana and South Dakota: Unpub. PhD Dissertation, Univ. Montana, Missoula, 310 p.

_____, 1979, Lineament control of sedimentary facies in the northern Great Plains, United States, in M. H. Podwysocki and J. L. Earle (eds.), Proceedings of Second International Conference on Basement Tectonics: Basement Tectonics Committee, Denver Colorado, P. 413-422.

_____, AND D. D. RICE, 1979, Stratigraphic studies of the Upper Cretaceous in eastern North and South Dakota [Abs.]: Geol. Soc. American Abstracts with Programs, v. 11, p. 256.

SIMON, H. A., 1962, The architecture of complexity: Am. Philosophical Soc. Proc., v. 106, p. 467-482.

SPEARING, D. R., 1976, Upper Cretaceous Shannon Sandstone—An offshore, shallow-marine sand body: Wyoming Geol. Assoc., Guidebook 28th Ann. Field Conf., p. 65-72.

STUBBLEFIELD, W. L., AND D. J. P. SWIFT, 1976, Ridge development as revealed by subbottom profiles on the central New Jersey shelf: Marine Geol., v. 20, p. 315-354.

SWIFT, D. J. P., 1975, Tidal sand ridges and shoal-retreat massifs: Marine Geol., v. 18, p. 105-134.

_____, AND M. E. FIELD, 1981, Evolution of a classic sand ridge field: Maryland sector, North American inner shelf: Sedimentology, v. 28, p. 461-482.

_____, AND G. L. FREELAND, 1978, Current lineations and sand waves of the inner shelf, Middle Atlantic Bight of North America: Jour. Sed. Petrology, v. 48, p. 1257-1266.

_____, J. W. KOFOED, F. P. SAULSBURY, AND P. SEARS, 1972, Holocene evolution of the shelf surface, central and southern Atlantic shelf of North American in D. J. P. Swift, D. B. Duane, and O. H. Pilkey (eds.), Shelf Sediment Transport—Process and Pattern: Dowden, Hutchinson, and Ross, Stroudsburg, Pennsylvania, p. 621-644.

_____, AND D. J. STANLEY, AND J. R. CURRAY, 1971, Relict sediments on continental shelves: A Reconsideration: Jour. Geol., v. 79, p. 322-346.

THOMAS, G. E., 1974, Lineament-block tectonics—Williston-Blood Creek basin: Am. Assoc. Petroleum Geologists Bull., v. 58, p. 1305-1322.

TILLMAN, R. W., AND R. S. MARTINSEN, 1979, Hartzog Draw Field, Powder River Basin, Wyoming, *in* R. F. Flory (ed.), Core Seminar: 28th Ann. Meeting, Rocky Mountain Sec., Am. Assoc. Petroleum Geologists, p. 1-38.

WALKER, R. G., 1979, Facies models 7. Shallow marine sands, *in* R. G. Walker (ed.), Facies Models: Geosci. Canada Reprint Series no. 1, p. 75-88.

WEIMER, R. J., 1960, Upper Cretaceous stratigraphy, Rocky Mountain area: Am. Assoc. Petroleum Geologists Bull., v. 44, p. 1-20.

THE SHANNON SHELF-RIDGE SANDSTONE COMPLEX, SALT CREEK ANTICLINE AREA, POWDER RIVER BASIN, WYOMING

R. W. TILLMAN AND R. S. MARTINSEN

Exploration and Production Research Laboratory, Cities Service Company, Tulsa, Oklahoma 74102,
and Consultant 3901 Grays Gables Road, Laramie, Wyoming 82070

ABSTRACT

Two vertically stacked shelf-ridge (bar) complexes in the Shannon Sandstone member of the Cody Shale (designated upper and lower sandstones) crop out in the Salt Creek anticline of the Powder River Basin, Wyoming. The shelf-ridge complexes are composed primarily of moderately to highly glauconitic, fine- to medium-grained lithic sandstone and attain thicknesses of over 70 feet. The shelf-ridge complexes were deposited at least 70 miles from shore at middle to inner shelf depths by south to southwest-flowing shore-parallel currents intensified periodically and frequently by storms. Ridges in each sequence trend north-south, slightly oblique to current flow. A possible source of sediments for the shelf ridges was the Eagle Sandstone shoreline and deltaic deposits of southern Montana 200 miles to the northwest.

Eleven facies were defined in outcrop on the basis of physical and biologic sedimentary structures and lithology. Vertical and lateral changes in facies are relatively abrupt where observed in closely spaced outcrop sections, and, in general, facies are stacked in coarsening-upward sequences with Central Bar Facies commonly immediately overlying Interbar Sandstone Facies. Porous and permeable potential reservoir facies include: Central Bar Facies, a clean cross-bedded sandstone; Bar Margin Facies (Type 1), a highly glauconitic, cross-bedded sandstone containing abundant shale and limonite (after siderite) rip-up clasts and lenses; and Bar Margin Facies (Type 2), a cross-bedded to rippled sandstone. These facies were formed by sediment transported and deposited in the form of medium- to large-scale planar-tangential troughs and sand waves on and across the tops of ridges by moderate to high energy shelf currents. Storm flow deposited Central Bar (planar laminated) Facies are rare.

Finer-grained, non- to marginal-reservoir quality facies include Interbar Sandstone Facies (rippled to ripple-form bedded sandstone), Bioturbated Shelf Sandstone Facies, Bioturbated Shelf Siltstone Facies, Interbar Facies (interlaminated rippled sandstone and shale), Shelf Sandstone and Shelf Siltstone Facies (sub-horizontally laminated sandstone and siltstone). Interbar Sandstone Facies were most commonly deposited lateral to the higher energy portions of the ridges as well as near the base of the shelf ridges during their initial development. The two bioturbated facies most commonly occur near the base of the ridge complex and between the two vertically stacked ridge complex sequences, and probably represent periods of slow deposition. The Shelf Silty Shale Facies is actually a facies of the Cody Shale of which the Shannon Sandstone is a member.

The most common vertical sequence of sandstone facies is one in which a coarsening-upward sequence is formed where Central Bar Facies overlie Interbar Sandstone Facies; this contrasts with the sequence observed at Hartzog Draw Field, 25 miles to the northeast, where the most common coarsening-upward sequence from bottom to top is Interbar Facies, Bar Margin Facies (Type 1) and Central Bar Facies. Relatively abrupt lateral changes in facies are observed in surface cross sections spaced from one-fourth to one mile apart. Thickness changes, but not facies changes, are readily observable on subsurface cross sections constructed using SP-resistivity logs.

The association of two vertically stacked shelf-ridge complexes at Salt Creek is atypical compared to other Shannon sequences in the Powder River Basin in several respects. The lower sandstone is correlative with productive Shannon sandstones in many of the Powder River Basin fields (e.g., Hartzog Draw); the upper sandstone sequence is only locally developed in the area of the Salt Creek anticline. Also, the spacing between ridges and the length to width ratios of the ridges at Salt Creek are much smaller than those in other areas. These differences are particularly apparent in the upper sequence wherein the sandstone bodies appear to have oblate geometries very unlike the strongly linear geometries typical of most shelf sandstone ridges. These differences are attributed to the presence during Shannon time of an actively growing paleo-high in the area of the present day Salt Creek anticline which localized sand deposition and ridge formation. Similar early structural growth and its influence on shelf sedimentation has been well-documented for the Lost Soldier anticline area by Reynolds (1976).

Baculites zones are commonly used for surface correlations in the study area and in Upper Cretaceous units throughout Wyoming. Subsurface cross sections paralleling surface sections at distances from one half to three miles away corroborate the surface sandstone correlations in the Salt Creek area. Bentonites above and below the Shannon form excellent subsurface correlation datums.

Foraminiferal data indicate that the shelf-ridge complexes were deposited in water depths ranging from the middle shelf to the outer part of the inner shelf. A wide diversity in size, orientation, and type of burrow-fill material suggests a relatively hospitable environment for burrowers in portions of the shelf-ridge complexes. Rare *Teichichnus*, *Thallasinoides*, *Chondrites*, and plural curving tubes were identified. Common Cretaceous shoreline traces such as *Ophimorpha*, *Asterosoma* and *Rhizocorallium* were not observed. The Bioturbated Shelf Sandstone Facies and the Bioturbated Shelf Siltstone Facies range from 75 to 95% burrowed. Burrowing in the other facies averages from 5 to 27%. Glauconite is present throughout and is most abundant in association with shale rip-up clasts and limonite (after siderite) lenses and rip-up clasts in Bar Bargin Facies (Type 1).

Transport directions, determined by abundant high angle cross beds (mostly planar-tangential), indicate a south-southwest transport direction (188°) for current deposition of the high energy facies. The range of variation in transport direction at individual outcrops and overall is relatively small (60°). Most current ripples in the Interbar Sandstone Facies also indicate a southerly transport direction. Only very locally, in the top foot or two of some Central Bar and Bar Margin Facies, trough orientations indicate transport directions strongly oblique (northeast) to the general south-southwest flow direction.

In both outcrop and in Hartzog Draw Field, ridge complexes trend nearly north-south, slightly oblique to current flow. Detailed subsurface correlations of the Shannon sand ridges throughout the Powder River Basin, using well-developed bentonite markers, show the reservoir facies to "rise and fall" parallel to their elongation, indicating that the ridges were not deposited in layer-cake fashion. In the Salt Creek area, the reservoir facies generally gradually rise in section to the south parallel to the direction of current flow. In the Hartzog Draw area, the ridges rise to the north, opposite to current flow; they also rise in paired-fashion laterally east and west. These stratigraphic patterns of development of the higher energy shelf-ridge facies are interpreted to reflect sea-floor topography during their deposition.

INTRODUCTION

This study of the Shannon Sandstone is based primarily on detailed measured outcrop sections in the area of Salt Creek Field in the Central Powder River Basin, Wyoming (Figure 1). Closely spaced, detailed measured sections described during 1977-1981 were utilized to construct a series of surface cross sections that were compared with subsurface sections which include Salt Creek, Shannon, Teapot, and Meadow Creek fields, all of which produce from the Shannon.

Stratigraphy

The Shannon Sandstone is one of three sandstones (Fishtooth sandstone, Shannon Sandstone, and Sussex Sandstone) deposited in the Powder River Basin during Lower Campanian time (Figure 2).

The Shannon Sandstone is included in the Montana Group, first formally proposed as a rock-stratigraphic unit by Eldridge (1889, p. 313). As originally defined, it included all the formations from the Fort Benton to the Fox Hills (Figure 2). In 1973 Gill and Cobban divided the Montana Group, as originally defined, into two formal rock-stratigraphic Groups. The older formations in the Group were assigned to the Colorado Group, and the Montana Group was restricted to the Telegraph Creek and younger formations (Figure 2). In its type area of southwestern Montana, the basal Montana Group formation is the Telegraph Creek Formation which is transitional from the Colorado Shale into the overlying, Shannon equivalent, Eagle Sandstone. The boundaries of the Montana rock-stratigraphic Group are time transgressive through at least nine ammonite zones (4-5 million years) and for this reason Gill and Cobban (1973) recommend that the Montana Group be used only as a rock-stratigraphic unit.

The name Shannon Sandstone was first used for a 170-foot thick sandstone series consisting of two resistant beds separated by 100 feet of softer sandstone that crops out in the Salt Creek Field area (Wilmarth, 1938). According to Baker (1957), who discussed the origin of the name Shannon, the first oil well drilled in the area of Salt Creek Field was drilled by M. P. Shannon in 1889 to a depth of 700 feet. The oil field and the sandstone which forms the escarpment around the Salt Creek Anticline were both named Shannon. Interestingly, Mr. Shannon then built his own refinery in Casper to refine oil from his oil pool; the capacity of this small refinery, built in 1894, was 100 barrels per day.

During the early part of this century, additional small fields were discovered in the Salt Creek area. These included the Meadow Creek (North, East and West) and Teapot Fields (Figure 3), all of which had part of their production from the Shannon.

Early Facies Analyses

Parker (1958) was the first to provide evidence that the Shannon Sandstone was a shelf deposit, presumably deposited in the middle to outer shelf in water depths exceeding 100 feet. Spearing (1975, 1976) corroborated Parker's conclusions and presented more supporting evidence of shelf sand deposition for the Shannon; he recognized two sandstone facies types, a "thin-bedded sandstone facies" and a "cross-bedded sandstone facies". These facies are significantly subdivided in this study. He also drew generalized isopach maps and measured transport directions for the upper and lower Shannon sandstones. The general wide-shelf interior-seaway model of Asquith (1970, 1974) was found to be applicable for the Shannon based on subsurface correlations (Tillman and Martinsen, 1979). The environmental facies model utilized in this study is modified from that of Porter (1976) which was developed for the Upper Cretaceous Hygiene Sandstone in the Denver Basin of Colorado.

Shannon Isopach Maps

A top of Sussex to top of Frontier interval (Figure 4) isopach map (Figure 5) in the Powder River Basin was constructed by Crews et al. (1976). Significant east-to-west thickening of the interval is shown. In contrast to the regional isopach, the Shannon, which is a small portion of the Sussex to Frontier interval, does no show an east-to-west thickening (Figure 6A). Instead, thickening of the Shannon interval occurs near the south end of the Powder River Basin and in southern Montana. As will be discussed later, the thickening in southern Montana may be related to the presence of an Eagle "delta" in that area.

A detailed isopach of approximately 2600 square miles of the Powder River Basin north of the Salt Creek study area shows the Shannon interval to range from 20 to just over 70 feet in thickness (Figure 6B). Included in this area are what Crews et al. (1976) term "relatively clean resistive sandstone facies"; these "clean" areas correspond with the areas of Shannon production in numerous fields including Hartzog Draw and Heldt Draw Fields.

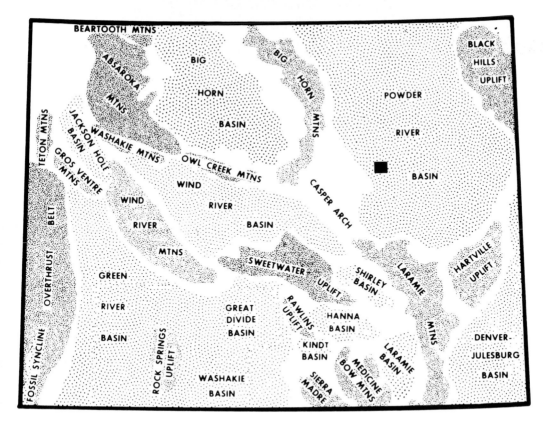

FIG. 1.—Basins, uplifts and mountain ranges of Wyoming. The location of the Salt Creek outcrop area is shown (black square) in the Powder River Basin.

The Shannon Sandstone crops out in the Salt Creek area of this study and at Parkman northwest of the study area. To the east, equivalent age rocks are either absent (Figure 4) or are shale. Shannon equivalents in the Hanna Basin (Figure 1) are included in the lower part of the Haystack Mountain Formation (Figure 2). In southern Montana, in the northwest part of the Powder River Basin, shoreline and nearshore lower Campanian sediments of the Telegraph Creek-Eagle Sandstone are observed. In central Wyoming, shoreline and continental deposits of the Mesa Verde form the western boundary of the seaway (Figure 7). Recent work by Barratt and Scott (1982) on the Phayles Sandstone documents the shoreline aspects of Shannon age equivalents in the Wind River Basin. In the northeastern part of the Powder River Basin in Montana, offshore sandstone deposits are present in the lower Campanian; these are variously termed Groat or Shannon (Shurr, this volume).

MEASURED SECTIONS

Fourteen outcrop sections in the Salt Creek area were measured, described, and interpreted for this study. The locations of the sections are shown in Fig-ure 3. Outcrop sections were measured using a Jacob's staff and Brunton compass. Field parties were organized so that each section was measured and scaled off, while a second group described and interpreted the facies. Using the scaled-off outcrop, detailed sketches and descriptions were made for each genetic unit. A major emphasis was placed on determining characteristics which would define individual facies. Once facies were defined, specific facies boundaries were determined and the stratigraphic sketch section was completed. In order to define more precisely the components of each facies, detailed checklist-percentile forms (Table 1) were completed for many of the facies. The detailed check lists were used conscientiously during the first phase of the study. Information from the checklist was used as a format to compile the brief summary of each facies included on the stratigraphic sections. Check list forms such as shown in Table 1 were utilized to quantitatively describe individual units in many of the measured sections. On the check list form are tabulated percentages and other numerical data under the headings: (1) *lithology*, (2) *physical structures*, (3) *biogenic sedimentary structures*, and (4) *bed thickness* and *contact types*.

Stratigraphic correlation chart (Upper Cretaceous stages and substages).

Columns: Southeastern Alberta 3; Central Montana 2; Hardin-Billings 1; Judith River 1; Hanna Basin 1; Powder River Basin (Red Bird and vicinity 1, Parkman 1, Salt Creek 1).

Stages: Maestrichtian (Upper, Lower); Campanian (Upper, Lower); Santonian; Coniacian.

Southeastern Alberta 3: Bearpaw; Judith River; Lea Park; Pakowki; Milk River; Alderman Member; Deadhorse Coulee; Virgelle Sandstone; Telegraph Creek; Colorado; Lloyd-Minster.

Central Montana 2: Bearpaw; Judith River; Claggett Shale; Eagle (Upper Member, Middle Member, Virgelle Sandstone); Telegraph Creek; Niobrara Shale.

Hardin-Billings 1: Hell Creek Formation; Fox Hills Sandstone; Bearpaw; Shale; Parkman Sandstone; Claggett Shale; Eagle Sandstone; Virgelle Sandstone Member; Telegraph Creek Formation; Niobrara Shale Formation (part); Cody Shale (part).

Judith River 1: Hell Creek Formation; Fox Hills Sandstone; Bearpaw Shale; Judith River Formation; Parkman Sandstone; Claggett Shale; Unnamed nonmarine member; Eagle Sandstone; Virgelle Sandstone Member; Telegraph Creek Formation; Colorado Shale (part).

MONTANA GROUP; COLORADO.

Hanna Basin 1: Ferris Formation (part); Medicine Bow Formation; Fox Hills Sandstone; Lewis Shale; Dad Ss Mbr; Almond Formation; Pine Ridge Sandstone; Marine Member; Allen Ridge Formation; Hatfield Sandstone Member; O'Brien Spring Ss Mbr; Haystack Mountains Formation; Tapers Ranch Ss Mbr; Steele Shale; Niobrara Formation; Frontier Formation (part); Mesaverde Group.

Red Bird and vicinity 1: Lance Formation; Fox Hills Sandstone; Upper unnamed shale member; Kara Bentonitic Member; Lower unnamed shale member; Red Bird Silty Member; Mitten Black Shale Member; Sharon Springs Member; Gammon Ferruginous Member; Niobrara Formation; Pierre Shale.

Parkman 1: Lance Formation; Fox Hills Sandstone; Bearpaw Shale; Parkman Sandstone; Sandstone; Niobrara Shale Member (part); Judith River Formation; Cody Shale (part).

Salt Creek 1: Lance Formation; Fox Hills Sandstone; Lewis Shale; Sandstone; Teapot Ss Mbr; Marine shale member; Nonmarine member; Parkman Sandstone Member; Stray sandstone; Shale; Sussex Sandstone Member; Shannon Ss Mbr; Fishtooth sandstone; Shale; Niobrara Shale Member (part); Mesaverde Formation; Cody Shale (part).

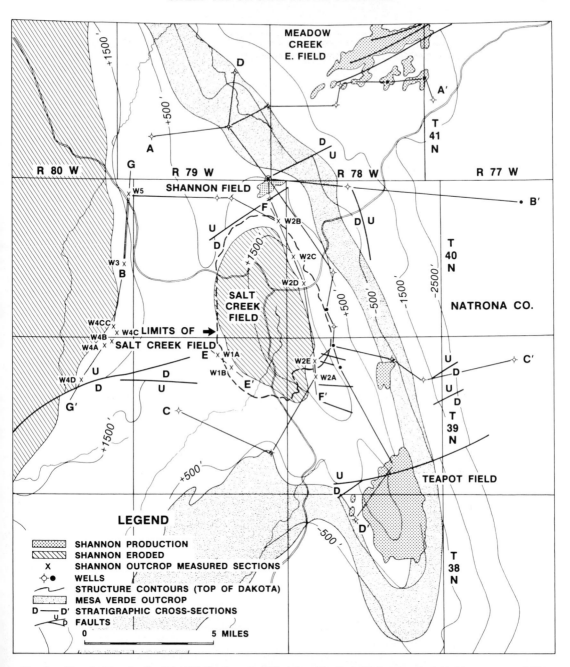

FIG. 3.—Map showing details of the Salt Creek area of Wyoming. The Cody Shale outcrop area (white area west of Mesa Verde) includes the Shannon and the Sussex Sandstones in the Salt Creek area of Natrona County, Wyoming. Areas indicated by diagonal lined pattern are areas where the Shannon has been eroded. Production from the Shannon is shown at three locations: Meadow Creek, Shannon, and Teapot Fields. The outcrop pattern on the post-Shannon Mesa Verde Formation, which overlies the Cody and outcrops to the east, is shown. Locations of surface and subsurface cross-sections are shown. Subsurface sections are A-A', B-B', C-C', and D-D'; surface sections are E-E', F-F', and G-G'. Only wells utilized in the cross-sections are plotted. Structure contours for the top of the Dakota Sandstone are indicated.

FIG. 2.—Correlation of the Upper Cretaceous of Wyoming, Montana, and southern Alberta. Ardmore Bentonite (above Sussex) indicated by lazy Y pattern. Sources of sections indicated by number: from (1) Gill and Cobban, 1973; (2) Rice and Cobban, 1977; and (3) Meijer Drees and Mhyr, 1981.

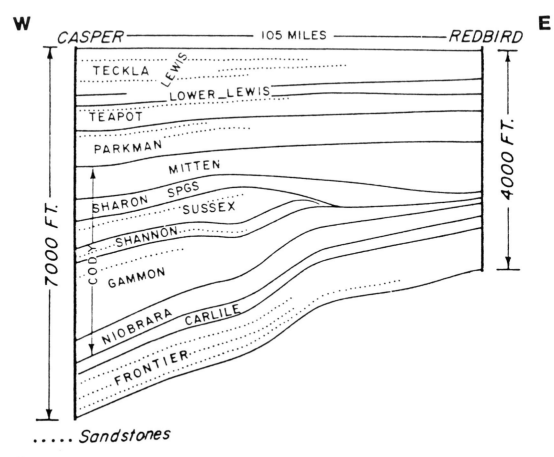

FIG. 4.—Upper Cretaceous formations in the Powder River Basin. Cody Formation (arrow) includes all units from the Niobrara through the Mitten Shale. The Gammon-Shannon-Sussex package includes apparently similar units, all of which thin significantly to the east. Sandstones are indicated by dotted lines. (After Gill and Cobban, 1966).

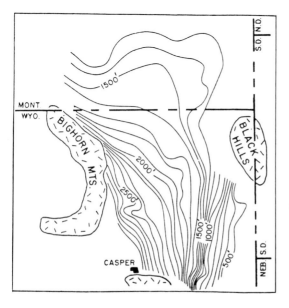

The measured sections labeled W1A (Figure 8), W1B (Figure 9), W2A (Figure 10A), W2B (Figure 10B), and W3 (Figure 11) through W2E rim the outcrop forming the topographic reflection of the Salt Creek Anticline (Figures 3 and 12). The sections numbered from W3 through W5 lie on a series of low hills west of the anticlinal outcrops on west facing cliffs. The lengths of the measured sections vary from 60 ft. (W4B) to 193 ft. (W2E). Measured section W4A (Figure 13) is 230 ft. long, but it includes 130 ft. of silty shale at the base.

FIG. 5.—Isopach map of the interval from the top of the Frontier Sandstone to the regional (Ardmore) bentonite marker at the top of the Sussex. Included are all the sandstone-bearing units of the Cody Formation in the Powder River Basin. The section thickens progressively from 500 feet on the east to nearly 3,000 feet on the west. (After Crews *et al.*, 1976).

Facies of the Shannon offshore sand-ridge complexes were determined from the study of the detailed outcrop measured sections; ten facies were recognized. These include several facies which are included in Spearing's (1976) "cross-bedded sandstone facies", the Central Bar and the Bar Margin Facies. Spearing's "thin-bedded sandstone facies" are subdivided in this study to include Interbar and Interbar Sandstone Facies, Bioturbated Shelf Sandstone Facies (Table 2) and locally a rather nondescript Shelf Sandstone Facies. The facies are described in this paper in an order which is inferred to correspond to successive decreases in energy of the depositing currents.

Central Bar Facies

This facies is one of the highest energy facies observed at Salt Creek and is dominated by high angle cross-bedded sets which commonly have a truncated horizontal to subhorizontal upper surface (Figure 14A). Where outcrop continuity allowed, single beds in this facies were traced laterally for up to 100 yards. The cross beds are formed by migrating medium- to large-scale troughs, mega-ripples, and sand waves. Internally, these features may best be described in cross section as troughs (curved lamina, tangential base), planar (lamina)-tangential (base), and planar-tabular (planar laminated with an acute angle at the base). Where bedding plane outcrop surfaces are visible, distinction between the planar-tabular bedding and the trough cross bedding is relatively straightforward, since the planar-tabular beds have a nearly straight crest, and the troughs have a sinuous crest with indentations along the crest of up to several feet. Planar-tangential bedding ranges from moderately straight crested to nearly trough-like in aspect when viewed on a bedding plane surface. This type of bed-form is most easily recognized in vertical section (Figure 14B). This type of bedding is gradational into the trough-type bedding, and presumably conditions for its formation are more similar to those which form the troughs than those active during periods of formation of the planar-tabular beds. Based on flume work carried on by Costello and Southard (1981) and others, the planar-tabular cross bedding is formed during lower energy conditions than those that form the troughs; it is assumed that the planar-tangential form is transitional between the two forms.

Individual sets in this facies commonly range from 6 to 18 inches in thickness, and only very rarely exceed two feet in thickness. Commonly the top of a set is truncated; however, based on the geometry of the cross beds, it appears that less than 30% of individual sets are eroded and more commonly considerably less appears to have been removed by erosion which preceded the deposition of the next bed (Figure 15A). Locally at the top of some units the uppermost cross beds show bidirectional flow directions which probably are the result of variation in flow or possibly reworking during the last stages of bar formation (Figure 15B). Recumbent folding, interpreted to be the result of soft sediment deformation caused by subsequent currents flowing over the bar (Figure 15C), is also observed. In the more distinctly trough-bedded sets, deeper erosion occurred prior to deposition of individual troughs. The depth of erosion varies from trough to trough.

Based on outcrop studies in the Powder River, Wind River, and Hanna Basins, trough cross bedding (Figure 14C and D) commonly forms more than 80% of the beds in this facies; planar-tangential sets comprise less than 20% of the outcrops of this facies studied. Planar-tabular cross bedding occurs only in trace amounts.

Where planar-tabular beds are observed (Figure 16), the lateral changes may be abrupt. As has been reported in numerous fluvial examples, the down current lower portion of the sand wave that forms the planar tabular beds commonly shows ripples and climbing-ripples that flow obliquely up the face which dips obliquely toward the major current flow direction. These associated ripples are rare in Shannon outcrops, but where they are present, they always occur in the thinning down-current portion of the planar-tabular sand waves. As can be seen in Figure 16D, the ripples form thin sets which climb in directions nearly the reverse of the currents forming the large scale cross bedding. These features are observed in measured section W2B-77 at 110 feet.

At several locations the co-sets of beds between two obvious erosional (truncation) surfaces were observed to thin very gradually laterally. In general the thinning of the co-sets appears to be in a direction oblique to the transport direction. Commonly the co-sets between truncations, which are well defined by weathering of the outcrop, are 6 to 10 feet in thickness. Within the co-sets, more commonly than not, individual beds or sets were observed to thin in a down current direction, but the package as a whole seems to thin oblique to the down current flow. More detailed studies will be required to substantiate the processes by which co-sets or facies thin laterally.

Using a hand-held grain size comparitor, the average grain size for this facies was determined to be 250 microns (medium-grained sandstone); the percent glauconite ranges from 5-20%. Based on visual examination the facies is generally moderately well sorted. The vertical sequence of sedimentary features, lateral continuation of individual beds and other factors seem to vary similarly in all the sub-types of cross bedding.

In the Salt Creek area, individual beds have been traced laterally up to 100 yards. Whether lack of outcrop continuity or actual cessation of individual beds is responsible for the lateral dimensional limits was not determined.

Lithologically, this facies is similar to that of the *Bar Margin Facies* except that it is generally 25 to 50 microns coarser, contains less glauconite, (15%) and fewer shale clasts (4%)(Table 2). Shale clasts range

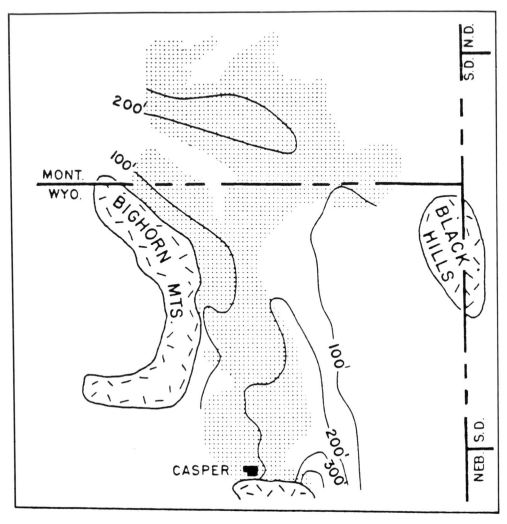

FIG. 6.—*Left*, Generalized Shannon Sandstone isopach map for the Powder River Basin. *Right*, Detailed Shannon interval isopach in the area north of the Salt Creek study area. The total Shannon thickness is comparable to either the upper *or* lower Shannon described in the study of the Salt Creek area. Stippled areas are areas of "relatively clean resistive sandstone facies" which include most of the fields discovered prior to 1976. (Modified from Crews *et al.*, 1976).

from 0 to 8% of the units studied as is shown in summary Table 3.

As will be discussed more thoroughly in the section on the *Bar Margin Facies*, rip-up-clasts and lenses of limonitic clay (after sideritic clay) are almost always limited to the Central Bar and Bar Margin Facies. Shale rip-up clasts are almost entirely limited to the same facies. Where limonitic clay and shale clasts and lenses combined exceed 8% by volume of a unit, it is designated a Bar Margin Facies (Type 1). Commonly the percentage of glauconite increases proportional to the increase in shale and siderite clasts.

Burrowing is almost totally absent in this facies; a few bedding plane traces and an occasional trace of burrowing within beds were observed. Burrowing lo-

cally ranges up to 10% by volume, and averages 4% in all units studied (Table 3).

A simplified "Markov-chain"-type analysis was used to describe facies successions (Figure 17 and Table 4) in the Shannon shelf-ridge complexes. This analysis was made to determine the most common vertical sequence of facies types. In vertical sequence the Central Bar Facies is most commonly (38% of contacts) underlain by Interbar Sandstones (Figure 17). This contrasts with the vertical sequence of Shannon facies at Hartzog Draw Field where Central Bar Facies most commonly is underlain by Bar Margin Facies (Tillman and Martinsen, 1979). Tables 4 and 5 summarize the percentage and unit number occurrences of vertical facies sequences for all facies observed in out-

SHANNON ISOPACH

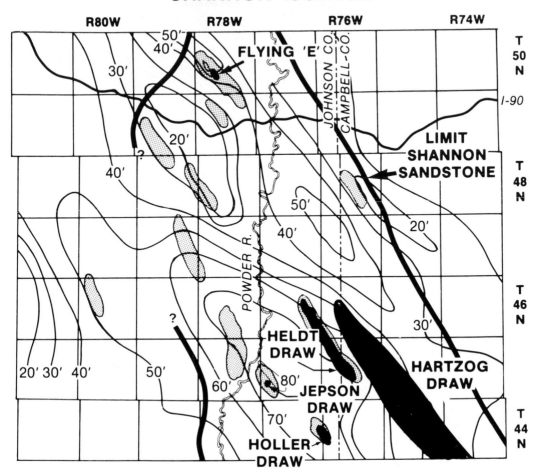

crop. The Central Bar Facies is succeeded by a variety of facies; most commonly (66% of the occurrences) the outcrop of the overlying unit is "covered". More often than not the "covered" intervals are shale. Shelf Sandstones and Bar Margin Facies (Type 1) also are observed above this facies.

Bar Margin Facies

The Bar Margin Facies is very similar to the Central Bar Facies in that they both are composed predominantly of high angle cross-bedded sandstones (Table 2). The depositional processes that were responsible for the formation of most of the Bar Margin Facies are similar to those that formed the Central Bar Facies. Two Bar Margin Facies types have been defined, each with certain aspects which differ from the Central Bar Facies.

Bar Margin Facies (Type 1).—Bar Margin Facies (Type 1) is differentiated from the Central Bar Facies by its greater amount of clasts and/or beds of shale and limonitic clay (after sideritic clay)(18% vs 4%); it generally has a finer mean grain size (225 vs 250 microns) and a higher percentage of glauconite (20 vs 12%, Table 3). Physical structures form more than 95% of the facies. Cross bedding is predominantly of the trough and planar-tangential type; planar-tabular bedding is rare (Figures 18 and 19). Individual sets are locally thicker (up to 24 inches) than in the Central Bar Facies; however, the maximum continuous thickness of beds of Bar Margin Facies (Type 1) is less than that for Central Bar Facies; this is especially true where observed in subsurface cores from Hartzog Draw Field. Where trough cross bedding forms most of the beds and clasts and lenticular beds of shale and/or limonite exceed 8% by volume, the Bar margin Facies (Type 1) designation is used. Lenses of limonitic clay (after sideritic clay) up to 3 inches thick are common in this facies (Figure 20B). Limonitic clay clasts in the facies result from reworking and redeposition of these lenses. Locally the clasts of limonitic clay and/or shale may form up to 30% (by volume) of the lithologic unit (Figure 20A and B). Accompanying the increase in

	CENTRAL MARINE BAR FACIES	CENTRAL BAR (PLANAR LAMINATED) FACIES	BAR MARGIN FACIES (TYPE 1)	BAR MARGIN FACIES (TYPE 2)	INTERBAR FACIES
LITHOLOGY	Predominantly medium grained quartzose sandstone, moderately glauconitic; local siderite clasts.	Fine to medium grained quartzose sandstone.	Fine-medium grained sandstone, shale and limonite rip-up clasts and lenses, very glauconitic.	Fine-medium grained sandstone with only rare shale interbeds. Fewer clasts and lenses and less glauconitic than (Type 1).	Thinnly interbedded fine to very fine-grained silty sandstone and silty shale, slightly glauconitic.
SEDIMENTARY STRUCTURES	Predominantly moderate angle trough and planar-tangential cross bedding. Trough sets commonly horizontally truncated.	Mostly sub-horizontal plane-parallel laminated sandstone, 0.5'-thick laminasets. Minor shale and sandstone ripples.	Mostly moderate angle troughs, some current ripples, shale clasts rarely show preferred orientations.	Interbedded sequences of several beds of troughs overlain by several beds of ripples.	Predominantly horizontal ripple-form bedding surfaces marked by interbedded shales. Trace of wave ripples; current ripples predominate
BURROWING	Sparse	Sparse	Sparse	Sparse	Moderate to locally high
RESERVOIR POTENTIAL	Excellent	Limited?	Good	Moderate to Good	Limited
OUTCROP OCCURENCES	Moderately Common	Uncommon	Common	Common	Uncommon

TABLE 1.—FACIES OF THE SHANNON SHELF-RIDGE COMPLEXES IN THE AREA OF SALT CREEK ANTICCLINE, WYOMING

clasts is almost always a substantial increase in the amount of glauconite. This facies commonly appears to be green to dark green in outcrop and individual laminae may contain up to 75% glauconite. It is common for more than 15% of the total volume of the facies to be glauconite.

Burrowing is easier to recognize in this facies than in the other cross-bedded facies; burrowing in this facies averages about 5%. Vertical, horizontal, and oblique burrows are sparse to locally common. Limonite lenses and beds may be vertically burrowed and filled with coarse and sometimes glauconitic sand (Figure 20C and D). Horizontal ⅛ inch diameter "white" burrows are also common in this facies.

This facies has the widest variety of vertically associated facies of any facies. Underlying facies types are Central Bar Facies, Bar Margin (Type 1), Interbar Sandstone, Interbar, Bioturbated Shelf Sandstones and Shelf Sandstone Facies (Tables 4 and 5). The facies that overlie this facies are even more diverse than those underlying it. (Table 4).

Bar Margin Facies (Type 2).—Bar Margin Facies (Type 2) is transitional between the Central Bar Facies and the Interbar Sandstone Facies (Table 2). It may be defined as a predominantly cross-bedded sandstone interbedded with numerous (more than 25%) thin rippled sandstone beds. It may form very thick sequences (e.g., top of measured Section W2A; Figures 10 and 21A).

Typically the cross-bed sets range from troughs to planar-tangential (Figure 22A and B) and to planar-tabular. Thicknesses of these cross-bedded sets are less than for the other types of cross-bedded facies. Sets generally range from 3 to 12 inches in thickness and commonly have truncated horizontal bedding boundaries (Figure 21B). The interbedded rippled beds are almost always sandstone and seldom exceed 3 inches in thickness; however, the ripple and ripple-form beds seldom exceed 40% (by volume) of the facies. Ripple-form beds are those which have a rippled upper surface and internally are indistinctly rippled to "massive appearing".

Burrowing in this facies averages 15% by volume and ranges from 5-30% where burrowing is observed (Table 3). Thallasinoides and a variety of ⅛-¼ inch diameter burrows are observed (Table 3). Concentric burrows flaring upward to a maximum diameter of 4 inches have been observed (Figure 22C and D).

This facies may be glauconitic, but the maximum concentration is never as great as the amount observed in the Bar Margin Facies (Type 1). The average percentage of glauconite in Bar Margin (Type 1) and Bar Margin Facies (Type 2) is similar (20%) (Table 3), however. Mean grain size in the cross-bedded part of the units is similar in both Bar Margin Facies types; however, when the mean size of the rippled facies is "averaged in", the Bar Margin Facies (Type 2) is slightly finer (215 microns) than the Bar Margin Facies

INTERBAR SANDSTONE FACIES	SHELF SANDSTONE FACIES	BIOTURBATED SHELF SANDSTONE FACIES	SHELF SILTSTONE FACIES	BIOTURBATED SHELF SILTSTONE FACIES	SHELF SILTY SHALE FACIES
Fine grained sandstone. Virtual absence of silty shale.	Very fine grained sandstone. Trace of laminated shale.	Shaly, slightly sandy dark gray siltstone, traces to moderate amounts of glauconite.	Siltstone and very fine grained sandstone. Some shale.	Siltstone, very fine grained sandstone and shale. Some glauconite and limonite.	Silty shale; rare thin (⅛") silty sandstone lenses.
Predominantly horizontal ripple-form bedding surfaces. Bedding commonly indistinct. Trace of wave ripples; current ripples predominate.	Subhorizontally to low angle laminated. Rare troughs.	Few physical structures preserved. Scattered thin rippled sand and horizontal laminasets. Bedding commonly destroyed.	Somewhat mottled to massive-appearing. Some low-angle bedding and lamination.	Highly mottled. Few physical structures preserved. Trace of horizontal diffuse bedding.	Current ripples and sub-horizontal laminae. Bedding surfaces indistinct, horizontal.
Low to Moderate	Sparse	More than 75% Burrowed	Moderate	Abundant, more than 75% burrowed.	Low to Moderate
Limited	Limited?	None	None	None	None
Very Common	Uncommon	Common	Uncommon	Common	Poorly preserved

(Type 1). Limonitic clay clasts may be observed in trace amounts and clay clasts and/or beds may be present (especially associated with ripples) in amount of up to 5%.

Bar Margin (Type 2) Facies are underlain by many of the same facies as the Bar Margin (Type 1) Facies (Tables 4 and 5). However, the facies overlying are more limited and include Central Bar Sandstones, Bioturbated Shelf Sandstones and Bioturbated Shelf Siltstones.

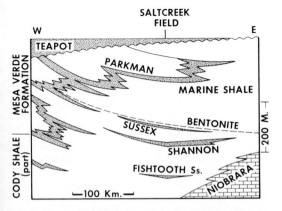

FIG. 7.—Stratigraphic reconstruction of the Upper Cretaceous in central Wyoming. Bentonite marker above the Sussex Sandstone is assumed to be time-synchronous. (After Gill and Cobban, 1966).

Interbar Facies and Interbar Sandstone Facies

The Interbar Facies is composed predominantly of rippled beds (Figure 23). In the subsurface, particularly in cores at Hartzog Draw Field, the Interbar Facies is composed of rippled interbedded shale and sandstone (in nearly equal amounts) along with some siltstone (Martinsen and Tillman, 1978, 1979; Tillman and Martinsen, 1979; and in press). The interbedded shales and siltstones, both in the subsurface and in the Salt Creek area, are commonly up to 60% burrowed and most often the burrows are sand or silt filled.

Very little interbedded shale was observed in the predominantly rippled facies in outcrops of the shelf-ridge complexes at Salt Creek. Only Unit 3 in Measured Section W4D was classified as Interbar Facies (Figure 23A). Interbar Facies, as defined, contain at least 25% "shale". The term Interbar Sandstone Facies is introduced in order to differentiate the rippled sandstone units which are abundant at Salt Creek from the interbedded shale and sandstone rippled facies typical of the Shannon at Hartzog Draw Field.

The Interbar Sandstone Facies is composed almost entirely of rippled sandstone beds. The rippled beds form closely spaced, nearly horizontal beds (Figure 24A and B). Asymmetrical (current) ripples form from 40 to 95% of the beds. Scattered symmetrical ripple-form (wave ripples) may be observed, particularly on bedding surfaces. Physical structures in this facies range from 55 to 95% by volume and average 75%. In some areas rippled sandstone beds up to 3 inches thick

Measured Section W1A-77; Shannon Sandstone Outcrop,
NE SW Section 3 T39N R79W, Natrona Co., Wyo.

Jacob Staff Section
Measured by: R. Wolff and R. Scott
Described by: R. Tillman and R. Martinsen
October, 1977

Midwest Quadrangle

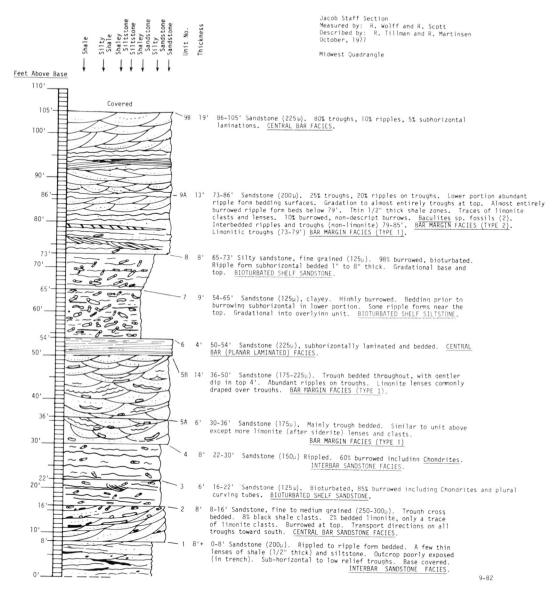

9B 19' 86-105' Sandstone (225μ). 80% troughs, 10% ripples, 5% subhorizontal
         laminations. CENTRAL BAR FACIES.

9A 13' 73-86' Sandstone (200μ). 25% troughs, 20% ripples on troughs. Lower portion abundant
         ripple form bedding surfaces. Gradation to almost entirely troughs at top. Almost entirely
         burrowed ripple form beds below 79'. Thin 1/2" thick shale zones. Traces of limonite
         clasts and lenses. 10% burrowed, non-descript burrows. Baculites sp. fossils (2).
         Interbedded ripples and troughs (non-limonite) 79-85'. BAR MARGIN FACIES (TYPE 2).
         Limonitic troughs (73-79') BAR MARGIN FACIES (TYPE 1).

8   8'   65-73' Silty sandstone, fine grained (125μ). 98% burrowed, bioturbated.
         Ripple form subhorizontal bedded 1" to 8" thick. Gradational base and
         top. BIOTURBATED SHELF SANDSTONE.

7   9'   54-65' Sandstone (125μ), clayey. Highly burrowed. Bedding prior to
         burrowing subhorizontal in lower portion. Some ripple forms near the
         top. Gradational into overlying unit. BIOTURBATED SHELF SILTSTONE.

6   4'   50-54' Sandstone (225μ), subhorizontally laminated and bedded. CENTRAL
         BAR (PLANAR LAMINATED) FACIES.

5B 14' 36-50' Sandstone (175-225μ). Trough bedded throughout, with gentler
         dip in top 4'. Abundant ripples on troughs. Limonite lenses commonly
         draped over troughs. BAR MARGIN FACIES (TYPE 1).

5A 6'   30-36' Sandstone (175μ). Mainly trough bedded. Similar to unit above
         except more limonite (after siderite) lenses and clasts.
         BAR MARGIN FACIES (TYPE 1)

4   8'   22-30' Sandstone (150μ) Rippled. 60% burrowed including Chondrites.
         INTERBAR SANDSTONE FACIES.

3   6'   16-22' Sandstone (125μ). Bioturbated, 85% burrowed including Chondrites and plural
         curving tubes. BIOTURBATED SHELF SANDSTONE.

2   8'   8-16' Sandstone, fine to medium grained (250-300μ). Trough cross
         bedded. 8% black shale clasts. 2% bedded limonite, only a trace
         of limonite clasts. Burrowed at top. Transport directions on all
         troughs toward south. CENTRAL BAR SANDSTONE FACIES.

1   8'+ 0-8' Sandstone (200μ). Rippled to ripple form bedded. A few thin
         lenses of shale (1/2" thick) and siltstone. Outcrop poorly exposed
         (in trench). Sub-horizontal to low relief troughs. Base covered.
         INTERBAR SANDSTONE FACIES.

9-82

FIG. 8.—Measured section W1A; located on the west flank of the breached Salt Creek Anticline (Fig. 3). The lower
Shannon sandstone extends up to 54 feet. The upper Shannon sandstone (32 feet thick) includes units 9A and 9B.

alternate with sandy to silty more burrowed beds.
When this occurs the sandstones are more resistant and
the siltier beds weather back slightly (Figure 25A and
B). Because of the uniform grain size and uniform
cementation within most beds, internal ripple laminations are difficult to observe (Figure 25C). More commonly the shape and orientation of the ripple can be
better judged by the outline of the top of the ripple on

the bedding surface (Figure 23C and D). Where little
internal lamination can be observed, but where well-
exposed bedding contacts showing the outline of the
upper surface of the ripple are observed, the term
"ripple-form" is used to describe the bedding surface
(Figure 26A). Thin ripple-form bedding is most often
observed in vertical sections, but it may also be observed on bedding surfaces. Where symmetrical rip-

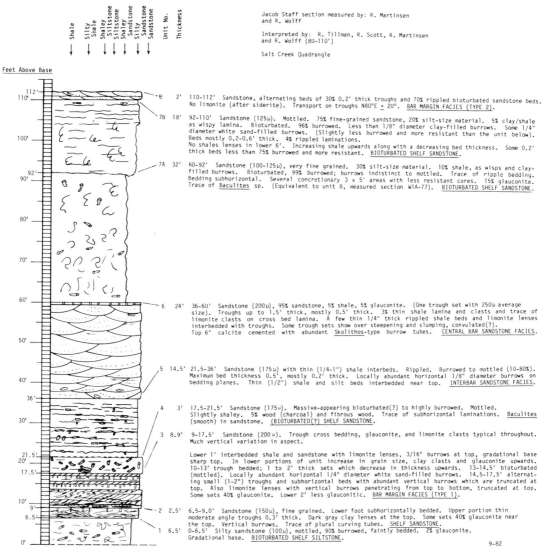

Fig. 9.—Measured section W1B; located on the west flank of the breached Salt Creek Anticline (Fig. 3). The lower Shannon sandstone includes units 2 through 6. The upper Shannon sandstone (20 feet thick) includes units 7B and 8.

ples are observed on bedding surfaces, crest line orientations are variable as is discussed in more detail elsewhere. The average bed thickness observed is about 5 inches. A few shallow erosional channels were observed (Figure 24C).

Glauconite is present in percentages ranging from 2 to 15%, which is considerably lower than the amount observed in the commonly superjacent and coarser-grained Bar Margin Facies. Concentrations of

glauconite in some laminae may be observed occasionally (Figure 26C).

Burrowing in this facies is not particularly conspicuous since most of the burrows are sand filled and do not contrast with the unburrowed portion of the rippled beds. Where silty or shaley interbeds are observed, burrows may be more easily recognized (Figures 23B and 26C). A variety of burrows have been recognized including *Thallasinoides* on bedding plane surfaces

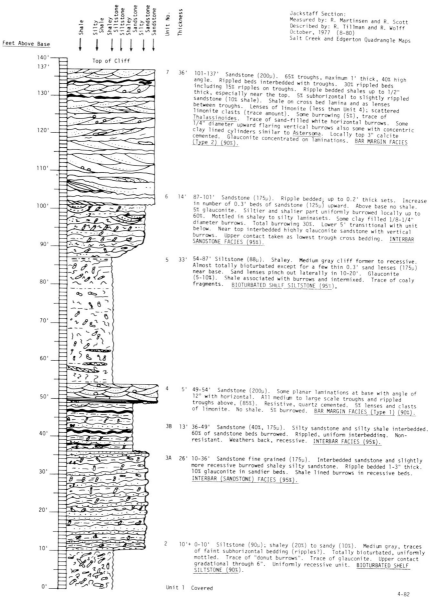

Measured Section W2A-77, Shannon Sandstone
SWNW Section 8 T39N R78W; Strike N30°E Dip 6°SE
Castle Rocks, Natrona Co., Wyoming

Jackstaff Section:
Measured by: R. Martinsen and R. Scott
Described by: R. Tillman and R. Wolff
October, 1977 (8-80)
Salt Creek and Edgerton Quadrangle Maps

7   36'   101-137' Sandstone (200μ). 65% troughs, maximum 1' thick, 40% high
angle. Rippled beds interbedded with troughs. 30% rippled beds
including 15% ripples on troughs. Ripple bedded shales up to 1/2"
thick, especially near the top. 5% subhorizontal to slightly rippled
sandstone (10% shale). Shale on cross bed lamina and as lenses
between troughs. Lenses of limonite (less than Unit 4); scattered
limonite clasts (trace amount). Some burrowing (5%), trace of
Thalassinoides. Trace of sand-filled white horizontal burrows. Some
1/4" diameter upward flaring vertical burrows also some with concentric
clay lined cylinders similar to Astersoma. Locally top 3" calcite
cemented. Glauconite concentrated on laminations. BAR MARGIN FACIES
(Type 2) (90%).

6   14'   87-101' Sandstone (175μ). Ripple bedded; up to 0.2' thick sets. Increase
in number of 0.3' beds of sandstone (125μ) upward. Above base no shale.
5% glauconite. Siltier and shalier part uniformly burrowed locally up to
60%. Mottled in shaly to silty laminasets. Some clay filled 1/8-1/4"
diameter burrows. Total burrowing 30%. Lower 5' transitional with unit
below. Near top interbedded highly glauconite sandstone with vertical
burrows. Upper contact taken as lowest trough cross bedding. INTERBAR
SANDSTONE FACIES (95%).

5   33'   54-87' Siltstone (88μ). Shaley. Medium gray cliff former to recessive.
Almost totally bioturbated except for a few thin 0.3' sand lenses (175μ)
near base. Sand lenses pinch out laterally in 10-20'. Glauconite
(5-10%). Shale associated with burrows and intermixed. Trace of coaly
fragments. BIOTURBATED SHELF SILTSTONE (95%).

4   5'   49-54' Sandstone (200μ). Some planar laminations at base with angle of
12° with horizontal. All medium to large scale troughs and rippled
troughs above, (85%). Resistive, quartz cemented. 5% lenses and clasts
of limonite. No shale. 5% burrowed. BAR MARGIN FACIES (Type 1) (90%).

3B   13'   36-49' Sandstone (40%, 175μ). Silty sandstone and silty shale interbedded.
60% of sandstone beds burrowed. Rippled, uniform interbedding. Non-
resistant. Weathers back, recessive. INTERBAR FACIES (95%).

3A   26'   10-36' Sandstone fine grained (175μ). Interbedded sandstone and slightly
more recessive burrowed shaley silty sandstone. Ripple bedded 1-3" thick.
10% glauconite in sandier beds. Shale lined burrows in recessive beds.
INTERBAR (SANDSTONE) FACIES (95%).

2   10'+ 0-10' Siltstone (90μ); shaley (20%) to sandy (10%). Medium gray, traces
of faint subhorizontal bedding (ripples?). Totally bioturbated, uniformly
mottled. Trace of "donut burrows". Trace of glauconite. Upper contact
gradational through 6". Uniformly recessive unit. BIOTURBATED SHELF
SILTSTONE (90%).

Unit 1 Covered

4-82

FIG. 10.—*Left*, Measured section W2A; located on Castle Rocks buttes on southeast flank of the Salt Creek Anticline (Fig. 3). The lower Shannon sandstone includes units 3A, 3B, and 4. The upper Shannon sandstone includes units 6 and 7. *Right*, Measured section W2B; located on the east flank of the Salt Creek Anticline (Fig. 3). The lower Shannon sandstone is units 2, 3, and 4. The upper Shannon sandstone is units 5B, 6, and 7.

and a few plural curving tubes (Figure 27). Burrowing in this facies ranges from 0% to 45%.

The lower contact of this facies, where it is observed in outcrop, ranges from transitional to sharp (Figure 25A). The upper contact is usually transitional through 3 feet or more, but it is occasionally sharp (Figure 24B).

Interbar Sandstone Facies are most commonly un-

derlain by Bioturbated Shelf Sandstone (50% of the occurrences) and Bioturbated Shelf Siltstones (25%) (Tables 4 and 5). The most common facies overlying Interbar Sandstones are Central Bar Facies (38%) (Figure 17). Bar Margin (Types 1 and 2) (21%) and Interbar and Interbar Sandstone Facies also overlie this facies.

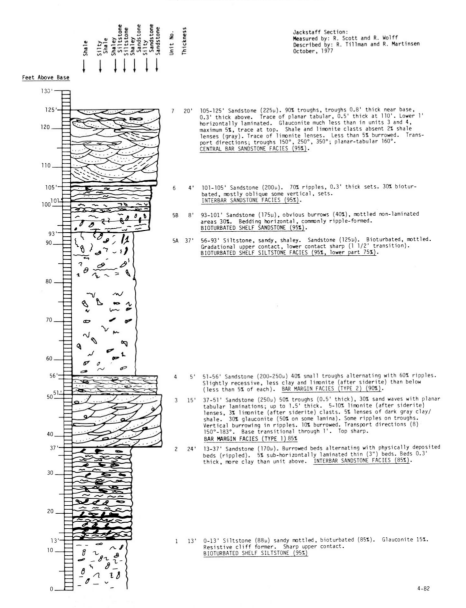

Measured Section W2B-77
Shannon Sandstone, "Salt Creek, North"
NW1/4SE1/4 Section 12-T40N-R79W, Natrona Co., Wyo.

Jackstaff Section:
Measured by: R. Scott and R. Wolff
Described by: R. Tillman and R. Martinsen
October, 1977

Shale
Silty Shale
Shaley Siltstone
Siltstone
Shaley Sandstone
Silty Sandstone
Sandstone
Unit No.
Thickness

Feet Above Base

7   20'   105-125' Sandstone (225µ). 90% troughs, troughs 0.8' thick near base, 0.3' thick above. Trace of planar tabular, 0.5' thick at 110'. Lower 1' horizontally laminated. Glauconite much less than in units 3 and 4, maximum 5%, trace at top. Shale and limonite clasts absent 2% shale lenses (gray). Trace of limonite lenses. Less than 5% burrowed. Transport directions; troughs 150°, 250°, 350°; planar-tabular 160°. CENTRAL BAR SANDSTONE FACIES (95%).

6   4'   101-105' Sandstone (200µ). 70% ripples, 0.3' thick sets. 30% bioturbated, mostly oblique some vertical, sets. INTERBAR SANDSTONE FACIES (95%).

5B   8'   93-101' Sandstone (175µ), obvious burrows (40%), mottled non-laminated areas 30%. Bedding horizontal, commonly ripple-formed. BIOTURBATED SHELF SANDSTONE (95%).

5A   37'   56-93' Siltstone, sandy, shaley. Sandstone (125µ). Bioturbated, mottled. Gradational upper contact, lower contact sharp (1 1/2' transition). BIOTURBATED SHELF SILTSTONE FACIES (95%, lower part 75%).

4   5'   51-56' Sandstone (200-250µ) 40% small troughs alternating with 60% ripples. Slightly recessive, less clay and limonite (after siderite) than below (less than 5% of each). BAR MARGIN FACIES (TYPE 2) (90%).

3   15'   37-51' Sandstone (250µ) 50% troughs (0.5' thick), 30% sand waves with planar tabular laminations; up to 1.5' thick. 5-10% limonite (after siderite) lenses, 3% limonite (after siderite) clasts. 5% lenses of dark gray clay/shale. 30% glauconite (50% on some lamina). Some ripples on troughs. Vertical burrowing in ripples. 10% burrowed. Transport directions (8) 150°-183°. Base transitional through 1'. Top sharp. BAR MARGIN FACIES (TYPE 1) 85%

2   24'   13-37' Sandstone (170µ). Burrowed beds alternating with physically deposited beds (rippled). 5% sub-horizontally laminated thin (3") beds. Beds 0.3' thick, more clay than unit above. INTERBAR SANDSTONE FACIES (85%).

1   13'   0-13' Siltstone (88µ) sandy mottled, bioturbated (85%). Glauconite 15%. Resistive cliff former. Sharp upper contact. BIOTURBATED SHELF SILTSTONE (95%)

4-82

Bioturbated Shelf Sandstone

This facies is volumetrically very important in outcrops in the Salt Creek area; however, it is almost totally absent in the subsurface at Hartzog Draw Field (Martinsen and Tillman, 1978, 1979). The increase in the occurrence of this facies corresponds with the greater percentages of sand observed in the Salt Creek area (when compared to the subsurface at Hartzog Draw Field).

The Bioturbated Shelf Sandstone contains from 70-95% burrows (Table 3); physical sedimentary structures range from a trace to 30% within the facies. A variety of burrow types may be recognized in this fac-

ies. *Asterosoma,* commonly oriented at 45°, may be recognized by its nested concentric clay linings (Figure 28A). Burrows with an upward increasing diameter (*Asterosoma?*) are also observed (Figure 28B). Vertical burrows in siderite lenses (Figure 28C) are seen in some units along with *Chondrites* and plural curving tubes. The most common physical structures observed are ripples which average about 10%. Horizontal laminations and low angle cross bedding were also observed.

Lithologically, where this facies was observed in 19 locations, it has an average mean grain size of 150 microns; the range of mean grain sizes is from 125-175

MEASURED SECTION W3-77, SHANNON SANDSTONE
NE SEC. 24 T40N R80W; 1/4 MILE WEST OF INTERSTATE,NORTH OF SMOKEY GAP ROAD.
NATRONA CO., POWDER RIVER BASIN, WYOMING

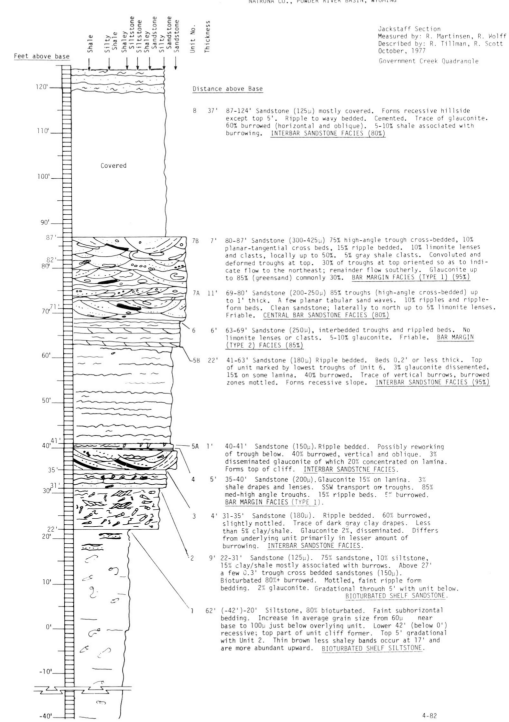

Jackstaff Section
Measured by: R. Martinsen, R. Wolff
Described by: R. Tillman, R. Scott
October, 1977

Government Creek Quadrangle

Distance above Base

8   37'   87-124' Sandstone (125μ) mostly covered. Forms recessive hillside
          except top 5'. Ripple to wavy bedded. Cemented. Trace of glauconite.
          60% burrowed (horizontal and oblique). 5-10% shale associated with
          burrowing. INTERBAR SANDSTONE FACIES (80%)

7B 7'   80-87' Sandstone (300-425μ) 75% high-angle trough cross-bedded, 10%
          planar-tangential cross beds, 15% ripple bedded. 10% limonite lenses
          and clasts, locally up to 50%. 5% gray shale clasts. Convoluted and
          deformed troughs at top. 30% of troughs at top oriented so as to indi-
          cate flow to the northeast; remainder flow southerly. Glauconite up
          to 85% (greensand) commonly 30%. BAR MARGIN FACIES (TYPE 1) (95%)

7A 11'   69-80' Sandstone (200-250μ) 85% troughs (high-angle cross-bedded) up
          to 1' thick. A few planar tabular sand waves. 10% ripples and ripple-
          form beds. Clean sandstone; laterally to north up to 5% limonite lenses.
          Friable. CENTRAL BAR SANDSTONE FACIES (80%)

6   6'   63-69' Sandstone (250μ), interbedded troughs and rippled beds. No
          limonite lenses or clasts. 5-10% glauconite. Friable. BAR MARGIN
          (TYPE 2) FACIES (85%)

5B 22'   41-63' Sandstone (180μ) Ripple bedded. Beds 0.2' or less thick. Top
          of unit marked by lowest troughs of Unit 6. 3% glauconite dissemented,
          15% on some lamina. 40% burrowed. Trace of vertical burrows, burrowed
          zones mottled. Forms recessive slope. INTERBAR SANDSTONE FACIES (95%)

5A 1'   40-41' Sandstone (150μ). Ripple bedded. Possibly reworking
          of trough below. 40% burrowed, vertical and oblique. 3%
          disseminated glauconite of which 20% concentrated on lamina.
          Forms top of cliff. INTERBAR SANDSTCNE FACIES.

4   5'   35-40' Sandstone (200μ). Glauconite 15% on lamina. 3%
          shale drapes and lenses. SSW transport on troughs. 85%
          med-high angle troughs. 15% ripple beds. 5% burrowed.
          BAR MARGIN FACIES (TYPE 1).

3   4'   31-35' Sandstone (180μ). Ripple bedded. 60% burrowed,
          slightly mottled. Trace of dark gray clay drapes. Less
          than 5% clay/shale. Glauconite 2%, disseminated. Differs
          from underlying unit primarily in lesser amount of
          burrowing. INTERBAR SANDSTONE FACIES.

2   9'   22-31' Sandstone (125μ). 75% sandstone, 10% siltstone,
          15% clay/shale mostly associated with burrows. Above 27'
          a few 0.3' trough cross bedded sandstones (150μ).
          Bioturbated 80%+ burrowed. Mottled, faint ripple form
          bedding. 2% glauconite. Gradational through 5' with unit below.
          BIOTURBATED SHELF SANDSTONE.

1   62'   (-42')-20' Siltstone, 80% bioturbated. Faint subhorizontal
          bedding. Increase in average grain size from 60μ   near
          base to 100μ just below overlying unit. Lower 42' (below 0')
          recessive; top part of unit cliff former. Top 5' gradational
          with Unit 2. Thin brown less shaley bands occur at 17' and
          are more abundant upward. BIOTURBATED SHELF SILTSTONE.

4-82

FIG. 11.—Measured section W3. The lower Shannon sandstone is here designated to include only units 2 through 5A. The
upper sandstone includes 5B through 7B. Location of section shown in Figure 3.

FIG. 12.—Shannon outcrops, Salt Creek, Wyoming. *A*, Stacked sequences of Shannon Sandstone shelf-ridge complexes at Castle Rocks, Salt Creek Anticline, Natrona County, Wyoming. General location of measured section W2A-77 (Fig. 10). *B*, Stacked sequences of shelf-ridge complexes which combine to form a 137-feet thick Shannon section at Castle Rocks; measured section W2A-77. Units 5-7 form upper cliff; units 2-4 are below on hillside. Telephone poles (upper left) for scale. Mesa Verde sandstones in middle distance (to right). *C*, Typical vertical sequence of Shannon shelf-ridge facies. Diffusely horizontally laminated *Bioturbated Shelf Siltstone* (Unit 5; overlain by *Interbar Sandstone Facies* (Unit 6). Top few beds (above arrow) are troughs in *Bar Margin Facies* (Unit 7); measured section W2A-77. *D*, Diffusely horizontally bedded *Bioturbated Shelf Sandstone*, Unit 5 (below arrow), overlain by *Interbar Sandstone Facies*, Unit 6. Above upper arrow is trough-bedded *Bar Margin Facies*, Unit 7: measured section W2A-77, 54–110 feet.

microns (Table 3). Where siltstone occurs, it may form up to 30% of the unit, while shale averages about 10% by volume. Glauconite may range up to 15% by volume, but commonly it is present in only trace amounts.

Most commonly, in 52% of the sequences observed (Figure 17), this facies overlies a Bioturbated Shelf Siltstone Facies. In 83% of the occurrences observed this facies is overlain by Bar Margin, Interbar, or Interbar Sandstone Facies. Both upper and lower contacts, where observed, are gradational through more than 3 feet.

Bioturbated Shelf "Siltstone"

This facies commonly forms the lowest unit in the shelf-ridge sequence. It consists almost entirely of siltstone and very fine-grained sandstone; the average mean size is about 100 microns. It appears to be shaley

but averages only about 20% shale and less than 10% fine to medium sand (Table 3). Bedding within the unit is usually diffuse and horizontal (Figure 29D). Glauconite is present in amounts ranging from a trace to 4%.

Burrowing averages about 87% and ranges from 75-90% by volume. The number of burrow types is limited and the predominantly oblique to horizontal burrows are silt filled and contrast little with the material adjacent to the burrows. Physical structures average about 13% and range from 5-40% in volume. Small ripples and low angle to subhorizontal laminations form most of the physical structures. Burrowing has destroyed most of the physical structures. The basal contact is usually covered (75% of the occurrences); however, it is locally exposed (Figure 29A). The upper contact ranges from gradational (through

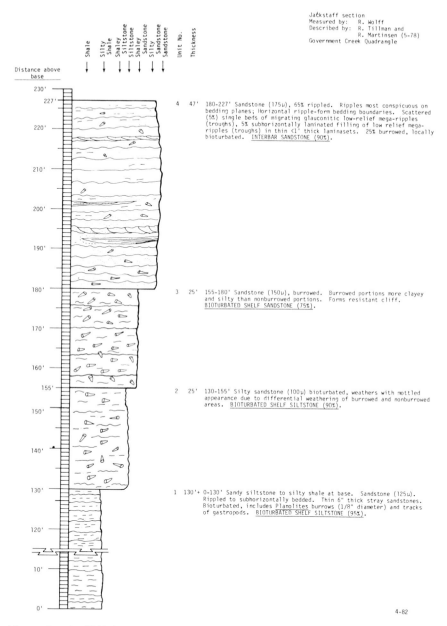

MEASURED SECTION W4A-78, SHANNON SANDSTONE
S 1/2 NE 1/4 NE 1/4 SEC. 2 T39N R80W
NATRONA CO., POWDER RIVER BASIN, WYOMING

Jackstaff section
Measured by: R. Wolff
Described by: R. Tillman and
                       R. Martinsen (5-78)
Government Creek Quadrangle

4   47' 180-227' Sandstone (175μ), 65% rippled. Ripples most conspicuous on
bedding planes; Horizontal ripple-form bedding boundaries. Scattered
(5%) single beds of migrating glauconitic low-relief mega-ripples
(troughs), 5% subhorizontally laminated filling of low relief mega-
ripples (troughs) in thin <1' thick laminasets. 25% burrowed, locally
bioturbated. INTERBAR SANDSTONE (90%).

3   25' 155-180' Sandstone (150μ), burrowed. Burrowed portions more clayey
and silty than nonburrowed portions. Forms resistant cliff.
BIOTURBATED SHELF SANDSTONE (75%).

2   25' 130-155' Silty sandstone (100μ) bioturbated, weathers with mottled
appearance due to differential weathering of burrowed and nonburrowed
areas. BIOTURBATED SHELF SILTSTONE (90%).

1 130'+ 0-130' Sandy siltstone to silty shale at base. Sandstone (125μ).
Rippled to subhorizontally bedded. Thin 6" thick stray sandstones.
Bioturbated, includes Planolites burrows (1/8" diameter) and tracks
of gastropods. BIOTURBATED SHELF SILTSTONE (95%).

4-82

Fig. 13.—Measured section W4A; located on west-facing excarpments west of the Salt Creek Anticline (Fig. 3). This section is interpreted to include only the lower Shannon sandstone.

more than 3 feet) to sharp (Figure 29B and C). This facies is overlain by Bioturbated Shelf Sandstones and Bioturbated Shelf Siltstones in nearly 66% of the occurrences observed (Figure 17).

This unit has no reservoir potential and forms a characteristic silt kick on Gamma-Ray and SP logs. Although not easily observed in outcrop, it is found in the subsurface to grade into a Silty Shelf Shale Facies.

Shelf Sandstone Facies

This facies is characterized by subhorizontal to low angle laminations. Up to 3-inch thick trough cross beds may also be observed (Table 3). Burrowing averages about 10% and ranges from 0-20%. Plural curving tubes and *Skolithos* type burrows were observed. The *Skolithos* locally are closely spaced (1 inch).

TABLE 2.—DETAILED TABULATION. *BAR MARGIN FACIES (TYPE 2)*[1]

MEASURED SECTION W2A-77, SHANNON SANDSTONE "CASTLE ROCKS",
SW NW SEC. 8 T39N R78W, NATRONA COUNTY, POWDER RIVER BASIN, WYOMING

BAR MARGIN FACIES (TYPE 2) (90%)

Described by: R. Tillman 1977,
1979, 1980, and
R. Martinsen, 1977

Columnized Data Format:
%/Max. set thickness/Additional notes

Unit 7; 101-137' (36')

Lithology
1. Sandstsone (%); Glauconite on laminations (40% on lamina)   92%/1.0'   (220μ)
2. Siltstone (%)   —
3. Shale (%); a) Laminated; b) Clasts   5%/0.5'/a
4. Siderite (%); a) Lenses; b) Clasts   3%/0.05'/a,b
5. Glauconite (%); a) Laminated; b) Clasts   40%/b

Physical Structures   95%/1.0'
1. High angle cross-bedding (20° +); a) Troughs;
   b) Planar   20%/1.0'/a
2. Moderate angle cross-bedding (10-20°) a) Troughs;
   b) Planar   25%/0.5'/a
3. Subhorizontal to low angle bedding; a) Trough;
   b) Planar   5%/0.1'/a,b
4. Horizontal laminations   —
5. Rippled; a) Sandstone; b) Shale   30%
6. Rippled interbedded sandstone and shale   —
7. Ripples superimposed on troughs   15%/0.2'/moderate
8. Reworked: a) By waves & currents; b) Bedding destroyed, massive; c) By bioturbation

Biogenic Sedimentary Structures   5%
1. Identified burrows; d) "donut burrows"; t) *Teichichnus;* a) *Asterosoma;* c) *Chrondrites*   Tr *Thallasinoides*
2. Distinct burrows; aa) <⅛"; a) ⅛"-¼";   tr- a,f,l
   b) ¼" -½"; c) >½"; d) Silt filled; e) Clay   tr- b,f,j
   filled; f) Sand filled; g) Silt lined; h) Clay   tr- a,k,h
   lined; i) Spreiten; j) Vertical; k) Oblique;   tr- aa,e,k
   1) Horizontal
3. Bioturbated   tr - patches
4. Total interval burrowed (%, footage)   10%/3.6'
5. Diversity (Number of burrow types); 3-low, 7-moderate; 10-high.   5-moderate

Bed Thickness (inches)   12"

Contact Relations
1. Upper; a) Very sharp (<0.5' transition);
   b) Sharp (0.5-1' transition); c) Transitional (1-3' transition); d) Gradational (>3' transition);
   e) Contact, erosional (truncated) angular; f) Contact erosional parallel;
   g) Covered   a or b
2. Lower; a), b), c), d), e), f), g)   d

[1]Summary form for tabulation of detailed descriptors of Shannon Sandstone shelf-facies. This form is subdivided into five major subdivisions; the first three are used to tabulate percentages of the various components observed in cores and outcrops. When using this type of form, the percentages for sandstone, siltstone, and shale listed under "lithology" total to 100%. Set thicknesses should be indicated, and further description may be indicated by a, b, or c. Commonly the mean size in microns is listed opposite sandstone percentage. The percentage of siderite and glauconite are estimated. Percentage values for "physical structures" and "biogenic sedimentary structures" total to 100% Individual percentages under each heading are totaled to give the total percentages. Percentages for burrow types that can be identified are listed in section one. Burrow types that cannot be identified by name are shown in section two using descriptors. In this section the volume percent of burrowing is followed by descriptions of size, type of filling, and orientation. Bed thickness is recorded and upper and lower contact relationships for each unit may be described by indicating which predetermined definition is appropriate.

This facies name is used where neither ripples nor burrowing are prominent; if ripples or burrowing are present, "Interbar Sandstone" and "Bioturbated Shelf Sandstone" respectively are used (Table 2). The Shelf Sandstone Facies is observed in outcrop at three loca- tions; at one location the facies is near the base of a shelf bar sequence and is overlain by a Bar Margin Facies (Type 2). The other occurrences are at the top of a shelf bar sequence and are overlain by a recessive covered unit.

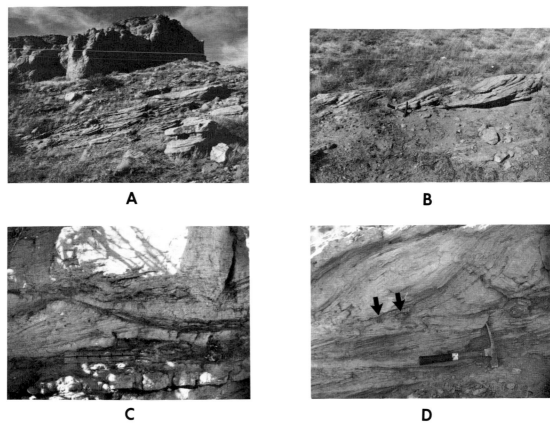

A B C D

FIG. 14.—*Central Bar Facies*. *A*, Hillside composed predominantly of only moderately resistive trough-bedded *Central Bar Facies*, Shannon sandstone. Note that the Central Bar Facies is somewhat recessive due to the friable nature of the sandstone. Overlying cliff face consists of coarsening-up cycle with *Bioturbated Shelf Sandstone* at base and thin *Bar Margin Facies (Type 2)* at top. Unit 6, measured section W1B-77, located near west side of Salt Creek Field, Wyoming. *B*, Planar tangential cross-bedding in generally recessive (friable) sandstone. Transport direction S50°W, toward upper left. Unit 9B, 90 feet in measured section W1A-77; *Central Bar Facies*. *C*, Low relief troughs with thin "silty shale drapes" (dark gray), and a few shale clasts. Shale also drapes over ripples (near base). Scale is six inches long. Unit 2, 16 feet; measured section W1A-77; *Central Bar Facies*. *D*, Sigmoidal trough cross-bedding and low relief truncated troughs (below point of pick). Rounded limonite clasts (arrows). Lower part of Unit 6, measured section W1B-77; *Central Bar Facies*.

Shelf Siltstone Facies and Shelf Silty Shale Facies

The Shelf Siltstone Facies ranges from mottled and massive-appearing to subhorizontally and low angle laminated (Table 3). Burrowing in this facies is variable; although it averages less than 30%, it ranges from 0 to 75%. This facies name is used "by default" where neither ripples nor burrowing are prominent (Interbar and Bioturbated Siltstone, respectively) (Table 2). The Shelf Siltstone Facies is observed in outcrop only twice; at both locations it is overlain by a Bioturbated Shelf Sandstone Facies; the upper contact is gradational through nearly 6 feet.

The Shelf Silty Shale Facies has been observed only below the lowest shelf-ridge complex in outcrop and in cores. No shales outcrop above the uppermost sandstone of the shelf-ridge complexes. The shale facies has a massive appearance; silt-size material is gener-

ally disseminated. Scattered silty rippled laminations may be observed. The unit is partially burrowed by very small (⅛-inch diameter) clay-filled burrows which give the unit a slightly mottled texture. Foraminifera from outcrop samples are tabulated in Table 6; several of the samples were from the Shelf Silty Shale Facies.

Evidence of the presence of relatively high energy conditions during deposition of the Shelf Silty Shale Facies is inferred, at least for the subsurface, by the state of preservation of many of the foraminifera tests. In cores from Hartzog Draw Field, the foraminifera, especially the larger ones, exhibit damage or loss of the final one or two chambers, or an abraded periphery may be developed in planispiral tests (Dailey, personal communication). This facies is assumed to be the normal facies for this part of the shelf and would be expected in areas of the shelf where no shelf-ridge complexes were being deposited.

CROSS SECTIONS

A series of subsurface and surface stratigraphic cross sections were constructed in the area of study.

Subsurface Cross Sections

Four primarily subsurface cross sections were constructed; sections A-A', B-B', and C-C' (Figures 30, 31, and 32) are east-west sections, and are oriented basically transverse to the north-south striking Shannon Sandstone ridges (Figure 3). Measured section D-D' (Figures 33, 34) is a north-south section. Locations of the sections are shown in Figure 3. Because they are the most commonly available logs in the greater Salt Creek Field area, SP and resistivity logs were used for constructing both the subsurface cross sections and Shannon isopach maps, to be discussed later. Sections B-B' and C-C' include both surface sections and subsurface logs. The surface sections were converted to "simulated logs" based on the known SP and resistivity (and GR) responses in Shannon wells which have been cored (Tillman and Martinsen, 1979, and in press). Included in the cross sections are wells from three fields in the area which produce from the Shannon. These fields are Shannon field, Teapot field, and Meadow Creek field.

In both surface outcrops and in the subsurface of the Salt Creek area two Shannon sandstones are recognized. Based on the position within the Shannon interval where the sandstones develop they have been termed upper and lower. The detailed facies recognized in outcrop are difficult or impossible to recognize in the types of subsurface logs generally available in the Salt Creek area. In addition to having mostly old logs, the entire Shannon Sandstone interval in the greater Salt Creek field area is relatively sandy. Distinct inter- and intra-unit shales and shaley-siltstones characteristic of the Shannon interval in areas such as the greater Hartzog Draw field (GHDF) area are absent. An accurate and consistent determination of what is "sandstone" is difficult due to a lack of well-defined resistivity responses, even within some productive sandstones. Therefore "sandstone" was not quantitatively defined in terms of amount of SP or resistivity responses. Instead, SP and resistivity responses were "qualitatively normalized". Sandstone identified on the cross sections probably consists mostly of the cross-bedded sandstones of the Central Bar and Bar Margin facies, and the cleaner, less burrowed, ripple-bedded sandstones of the Interbar Sandstone Facies. The more silty and burrowed portions of the Interbar Facies, as well as the Bioturbated Sandstone and Siltstone Facies comprise the remainder of the Shannon interval with the section above and below the Shannon interval consisting predominantly of Shelf Silty Shale

A

B

C

FIG. 15.—*Central Bar Sandstone, sand waves. A,* Sigmoidal base tangential mega-ripples interbedded with sub-horizontal laminations. Sussex Sandstone, east side of Salt Creek Anticline; Measured section W2C-81; *Central Bar Facies. B,* Bidirectional flow recorded in subsequent layers of trough cross-bedding at 87 feet in Unit 6B, *Central Bar Sandstone.* In lower bed, flow is diagonally out and to left. This type of variation in flow direction is typical of top bed at this location only. Flow in the top bed was toward back right. Measured section W3B-77. *C,* Recumbent (to right) planar tangential trough beds at top of *Central Bar Sandstone.* Recumbency is interpreted to be the result of soft sediment deformation caused by subsequent currents flowing over bar. Unit 6B at 87 feet, measured section W3-77.

TABLE 3.—QUANTITATIVE SUMMARY OF SEDIMENTARY FEATURES, SHANNON FACIES, SALT CREEK[1]

	CENTRAL BAR FACIES (N = 12)		CENTRAL BAR FACIES PLANAR LAMINATED (N = 4)		BAR MARGIN FACIES (TYPE 1) (N = 12)	
COLUMIZED DATA FORMAT: %/Max. set thickness/Additional notes *% Sandstone = 100% - (% Silt + % Shale)	Mean Value (Percent)	Range of Values	Mean Value (Percent)	Range of Values	Mean Value (Percent)	Range of Values
Lithology						
1. Sandstone* Mean grain size, microns	245	(200-300)	205	(125-225)	220	(200-300)
2. Siltstone (%)						
3. Shale (%); a) Laminated; b) Clasts	4	(1-8)			3%a,3%b	(0-5,0-5)
4. Limonite (%); a) Laminated; b) Clasts	2%a,2%b	(1-5,1-5)			6%a,6%b	(1-10,3-10)
5. Glauconite (%)	12	(5-20)			20	(5-30)
Physical Structures						
1. High angle cross-bedding (20°+); a) Troughs; b) Planar-tabular; c) Planar-tangential; d) Curved-tangential	90 14	(20-100) (10-20)	100	100	95 40	(80-100) (20-75)
2. Moderate angle cross-bedding (10-20°) a) Troughs; b) Planar-tabular; c) Planar tangential; d) Planar	70	(5-95)			56	(25-45)
3. Subhorizontal to low angle bedding (<10°); a) Trough; b) Planar-tabular; c) Planar tangential; d) Planar	5	5	Present		23	(10-50)
4. Horizontal laminations	7	(2-10)	98	(95-100%)		
5. Rippled; a) Sandstone; b) Shale	7	(5-10)	3	(2-5%)	26	(15-50)
6. Rippled interbedded sandstone and shale					5	(5)
7. Ripples superimposed on troughs	10	10			20	(20)
8. Reworked: a) By waves & currents; b) Massive appearing; c) By bioturbation						
Bed Thickness (inches)	12"		12"	(6-18")	10"	(3-18")
Contacts						
1. Upper: a) very sharp (<0.5' transition); b) Sharp (0.5-1' transition); c) Transitional (1-3' transition); d) Gradational (>3' transition); e) Contact, erosional (truncated) angular; f) Contact erosional parallel; g) covered			a			
2. Lower: a), b), c), d), e), f), g)						

[1]Refer to caption for Table 2 for description of procedures utilized for tabulating data. For each Facies, the mean percentage and range of values for each descriptor are tabulated; "N" refers to the number of outcrops of a Facies used in tabulations. This table differs from Table 2 in that, instead of recording the % sandstone under "lithology" descriptor 1, the percent sandstone is assumed to be equal to 100 − (% silt + % shale). The mean size (in microns) is recorded opposite "sandstone."

Facies. Because the subsurface cross sections do not show the individual facies, they do not display the complexities of the Shannon that are apparent from the surface cross sections.

The subsurface cross sections are meant to show gross sandstone build-up trends. Each sandstone build-up within both the upper and lower sandstone sequences consists of many individual sandstone bed sets which have only local reservoir continuity. This is most apparent from the fact that there is Shannon production (Shannon Field) within approximately one mile of the Shannon outcrop and is consistent with outcrop observations and the proposed mechanism for Shannon deposition.

In terms of gross trends, however, several important features of the Shannon are apparent from the subsurface cross sections; some of these trends are subtle. All the subsurface cross sections indicate that there is much more variability in the thickness of the lower sandstone than in the thickness of the upper sandstone. Also, the variability in thickness is greater in the east-west cross sections (Figures 30, 31, 32) than in the north-south cross section (Figures 33, 34). This variability is to be expected in that stratigraphic variability

in shelf sandstones is usually greatest perpendicular to depositional strike, and the Shannon sandstones strike approximately north-south.

Another important observation apparent from the subsurface cross sections is that although only two sandstones may be developed within a single well, these sandstones may develop through slightly different stratigraphic intervals in relation to markers 1 and 2. This difference is especially apparent on the east (right) side of cross section B-B' (Figure 31). In the subsurface in areas such as the greater Hartzog Draw area there are consistent patterns observed in the Shannon sandstones in relation to their stratigraphic location between markers 1 and 2 (or equivalent bentonite markers). The base of the Shannon sandstone in the greater Hartzog Draw Field area (GHDF area) rises from south to north, and either east to west or west to east within an individual ridge (Martinsen and Tillman, 1978, 1979; Tillman and Martinsen, 1979). Cross sections A-A', B-B', C-C', and D-D' do not show the same clear cut patterns as would similar cross sections in the GHDF area, but some subtle patterns can be discerned, especially in the lower sandstone. To help interpret the cross sections and to be better able to

BAR MARGIN FACIES (TYPE 2) N = 8		INTERBAR SANDSTONE FACIES N = 22		INTERBAR FACIES N = 1	BIOTURBATED SHELF SANDSTONE N = 19		BIOTURBATED SHELF SILTSTONE N = 18		SHELF SANDSTONE N = 2		SHELF SILTSTONE N = 2	
Mean Value (Percent)	Range of Values	Mean Value (Percent)	Range of Values		Mean Value (Percent)	Range of Values	Mean Value (Percent)	Range of Values	Mean Value (Percent)	Range of Values	Mean Value (Percent)	Range of Values
215	(200-300)	190	(125-200)	175	150	(125-175)	105	(90-175)	163	(150-175)	113	(100-125)
		12	(5-30)		32	(10-60)	75	(60-95)			75	(70-80)
3	(0-10)	8	(5-20)		10	(5-20)	20	(10-30)			25	(20-30)
2%a,2%b	(0-5,1-5)				9	(tr-20)	8	(1-15)	5%a	(5)		
20	(10-30)	7	(2-15)	10	12	(2-15)	4	(2-5)				
90	(70-100)	75	(55-95)	95	10	(1-30)	13	(5-40)	90	(80-100)	40	(40)
17	(5-40)											
54	(30-60)	5	(5)									
		10	(3-20)		7	(5-10)	9	(5-15)				
		9	(5-15)		5	(5)	7	(5-10)				
34	(5-60)	70	(40-95)	95	11	(1-30)	10	(5-15)				
6"	(3-12")	5"	(2-12")		5"	(1-8")			3"	3"		
b		b			d	b,d	b					d
c		d (a,d)			d	(d,g)						

recognize sandstone patterns that may be present, "ridge and swale" areas (of the lower sand only) as defined on the isopachous maps, are noted on cross sections. In the east-west cross sections, within a single ridge of the lower sandstone the base of the sandstone gently rises to the west. In some ridges the base appears to remain level, and in one area rises slightly to the east. No consistent rising or falling pattern is visible in strike section D-D', either within a single ridge or along the general line of section.

Spearing (1976), however, published a north-south cross section through the Salt Creek area (Figure 33A) which clearly shows that the base of the lower sandstone rises from north to south (opposite to the trend observed in the GHDF area). We do not fully understand why Spearing's cross section shows a rising pattern while ours do not, but it may be due to the fact that his cross section covers a much larger area, and the complexities of the Shannon sandstone in the Salt Creek Field area are such that the patterns are either modified there or are present and obscured. Possibly section D-D', which is only 16 miles in length as compared to Spearing's which is approximately 34 miles (55 km), indicates the "trees" but obscures the "forest". We believe the "rise and fall" of the development of the cleanest sandstones within the Shannon interval are typical of Shannon sandstone type deposi-

tion and probably represent deposition in response to one or more of the following: positive paleo-sea floor topography, asymmetry of the ridge geometry during formation, and hydrodynamic flow such that the high energy sands are limited to certain areas which migrate as the ridge grows.

Cross Sections of Outcrop Measured Sections

Cross sections E-E' and F-F' are north-south cross sections subparalleling strike cross section D-D', but are constructed using outcrop measured sections (Figure 3). These outcrop sections are subdivided in much more detail than are the subsurface sections. On the scale of the cross section, both vertical and lateral facies changes within the Shannon are abrupt. Upper and lower Shannon sandstones are identified and correspond with similar sand distributions in nearby subsurface wells.

Section E-E' (Figure 35) is located along the eastern margin of mapped thicks in both the upper and lower sandstones (Figures 38 and 39). The upper part of the lower Shannon sandstone in this section varies laterally over a distance of 0.7 miles from a thick Central Bar Facies on the southeast to a similar thickness of sandstone consisting of a thin Central Bar (planar laminated) Facies overlying a thick Bar Margin (Type 1) Facies on the northwest. Most of the Central Bar sec-

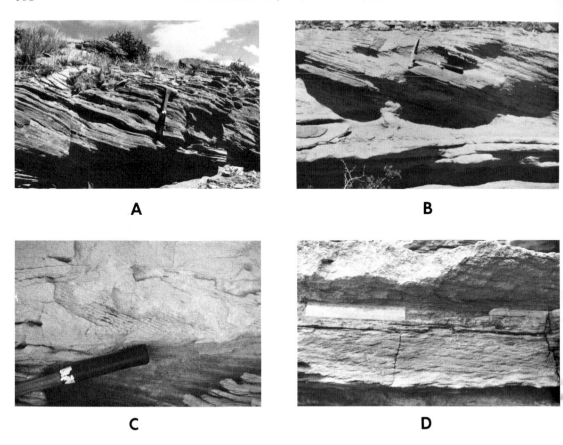

Fig. 16.—*Planar-tabular sand wave beds. A,* Planar-tabular laminations in sand wave. Note the very uniform thickness of laminasets which contrasts with systematic thickening of laminasets in trough beds. Transport to the right (south). One of only several occurrences of this type of bedding observed in outcrops in the Salt Creek area. Unit 7 at 110 feet, *Central Bar Sandstone;* measured section W2B-77. *B,* Planar-tabular sand wave; transport to right (south). Set thickness is 3 feet. This type of sand wave is very rare in this and other facies of the Shannon. In Central Bar Sandstones, troughs are most abundant and planar-tangential beds are second in abundance. Unit 7 at 110 feet, *Central Bar Sandstone;* measured section W2B-77. *C,* Planar-tabular laminations in sand wave. Note the very uniform thickness of laminasets which contrasts to systematic thickening of laminasets in trough beds. Transport to the right (south). One of only several occurrences of this type of bedding observed in outcrops in the Salt Creek area. Unit 7 at 110 feet; measured section W2B-77; *Central Bar Sandstone. D,* Reverse-climbing-ripples at toe of sand wave. Flow which produced sand wave, part of which overlies the ripples, is to the right (south). Reverse-climbing-ripplies flow to left (north) oblique to foresets on sand wave. Note ripple set boundaries climb to left, obliquely upstream; individual ripple laminae dip downstream. Unit 7 at 110 feet, measured section W2B-77; *Central Bar Sandstone.*

tion on the southeast changes facies to the northwest by addition of siderite, shale rip-up clasts, and glauconite. The lower portion of the lower Shannon sandstone also shows facies changes from southeast to northwest; the vertical sequence of Interbar Sandstone overlain by thin Central Bar Facies in section W1A changes to a Bar Margin (Type 1) overlying Shelf Sandstone Facies to the south in W1B. These fairly abrupt facies changes, both vertically and laterally, are believed to be typical of the Shannon as a whole.

Similar abrupt lateral and vertical changes in facies are also observed in cross section F-F' (Figure 36)

which was constructed along the western margin of Salt Creek Field (Figure 3); it also lies along the western margin of mapped thicks in both the upper and lower sandstones (Figures 38 and 39). In Section F-F' the Central Bar Facies is the highest energy deposit and the Bioturbated Shelf Siltstones are the lowest energy deposits. Bar Margin Facies (Type 1 and 2) are also high to moderate energy deposits. Interbar Sandstone Facies are relatively low energy sandstone deposits. In this cross section Bioturbated Siltstone or Shelf Siltstone is always overlain by Interbar Sandstones or Bioturbated Sandstones. At both ends of the section in

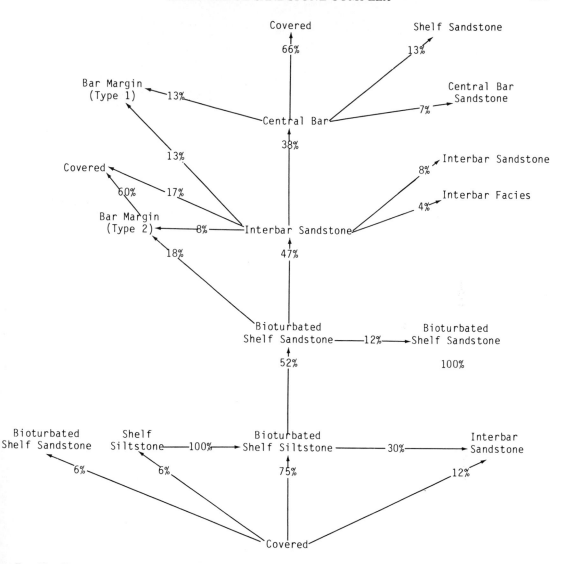

FIG. 17.—Vertical sequence of facies probabilities in order of deposition (bottom to top). Measured sections W1A to W5. Individual percentage values may be considered as the probability of finding the overlying facies indicated. The percentages are compiled from 127 facies successions.

the lower Shannon sandstone the siltstone facies thicken at the base at the expense of the suprajacent Interbar Sandstone. These changes correspond with mapped thickenings and thinnings of the bar in a longitudinal direction and indicate lateral variability in the sandstones along strike, as well as perpendicular to strike.

Some of the thickening of the basal sandstones may have resulted from very minor sub-regional erosion at the base of the Interbar Facies. In the lower Shannon sandstone in F-F', the Interbar Facies is overlain by Central Bar Facies at W2C and W2D and Bar Margin (Type 1) Facies in W2B. The uppermost facies in the

lower Shannon is quite variable; the uppermost outcropping unit of the lower Shannon sandstone includes nearly all the sandstone facies, from north to south, Bar Margin (Type 2), Central Bar(?) (W2C), Shelf Sandstone, Interbar and Bar Margin (Type 1) Facies.

On the southern end of the thick where sections W2E and W2A are located, high energy Central Bar Facies are absent even though total sandstone thicknesses correspond with the three more northerly sections. These sections, however, are located downcurrent from a mapped thick on ridge #3, and thus probably were in areas not favorable to deposition of the higher energy facies. Interbar Sandstone Facies

TABLE 4.—PERCENTAGES OF OVERLYING SHANNON FACIES, MEASURED SECTIONS (W1A-W5), SALT CREEK AREA, WYOMING[1]

FACIES	% of Total Number of Facies	Central Bar Sandstone	Central Bar (Planar)	Bar Margin (Type 1)	Bar Margin (Type 2)	Interbar Sandstone	Interbar	Bioturbated Shelf Sandstone	Shelf Sandstone	Bioturbated Shelf Siltstone	Shelf Siltstone	Covered Interval
Central Bar Sandstone	9	7%		13					13			66
Central Bar (planar laminated)	3	25			25	25			25			
Bar Margin (Type 1)	10	8	8	8	17	8		17	8	8		17
Bar Margin (Type 2)	8	20						10		10		60
Interbar Sandstone	20	38	8	13	8	8	4		4			17
Interbar Facies	6		33						33			33
Bioturbated Shelf Sandstone	14		12	18	47	6		12				6
Shelf Sandstone	2			33								66
Bioturbated Shelf Siltstone	14	6				30		52	6	6		
Shelf Siltstone	2								100			
Covered Interval	13					13		6			75	6

[1]This table is designed to show the probabilities of occurrence of one facies succeeding another. Reference Facies (underlying facies) are listed to the left. The overlying facies are listed across the top. An example of how the table is to be read: the *Interbar Sandstone Facies* (left column) is overlain by the *Central Bar Sandstone Facies* in 38% of the occurences; the *Bar Margin Facies* (Type 1) in 13% of the occurences, etc.

predominate in these two locations. Even lower energy interbedded rippled shale and sandstone Interbar Facies (as contrasted to Interbar Sandstone Facies) occur in W2A. The only high energy facies in the lower Shannon sandstone at the south end of the ridge thick is a 5-foot thick Bar Margin Facies (Type 1) at the top of the sandstone. The relative lack of high energy facies in the southern portion of the thick suggests a strong asymmetry of depositional process. High energy sandstones appear to be absent on the south or down-current end of the thick. They are replaced in this area by low energy sandstones. Outcrops south of and lateral to the upper portion of the high energy sandstone facies on the west flank of the ridge are relatively thick (25′ +) Bioturbated Shelf Siltstones which are even lower energy deposits than the subjacent sandstones.

The upper Shannon sandstone in F-F' also suggests a north to south decrease in energy. Although grass cover at the top of most of the sections suggests that facies overlying the uppermost resistant sandstone is siltstone or shale, there are areas of friable sandstones, such as unit 9 in measured section W1A (Figure 8), which weathers back to form a grassy slope. It is improbable, however, that facies coarser than siltstones lie above the tops of the measured sections in F-F'.

Central Bar Facies are observed in the upper Shannon sandstone in the three northerly sections (W2B, W2C, W2D), but not in the southerly sections. The small thick area centered on W2C and W2D (Fig-

ure 38) as well as the more northerly, slightly thinner section, all contain Central Bar Facies overlying Interbar Sandstones or Bar Margin (Type 2) Facies.

The vertical sequence associated with the upper Shannon higher energy facies in section F-F' is typical for that seen throughout the Salt Creek area. Interbar Sandstone and Interbar Facies underlie the higher energy Central Bar and Bar Margin at each section. Central Bar Facies and Bar Margin (Type 2) Facies are the uppermost units in all the upper Shannon sandstones in F-F'. The lowest energy sandstone facies (interbedded rippled shale and sandstone) observed is the Interbar Facies. The Interbar Facies occurs only in the southernmost section, and it occurs in both the upper and lower Shannon. Also, in the upper Shannon sandstone in the southern area, Bar Margin Facies (Type 2) sandstones are the highest energy sandstones observed; no Central Bar Facies are observed. This distribution of facies is further evidence for the asymmetry of processes of ridge formation in which depositional energy appears to decrease in a downcurrent direction.

Cross section G-G' (Figure 37) uses the base of the lower Shannon sandstones as a datum for correlation. As can be seen on the map (Figure 3), five of the measured sections are closely spaced near the south end of the cross section. In the westerly part of the study area where this section traverses from north to south, the lower sandstone is relatively thick (Figure 39), while the upper sandstone is relatively thin to ab-

TABLE 5.—OCCURRENCES OF OVERLYING SHANNON FACIES,
MEASURED SECTIONS (W1A-W5), SALT CREEK AREA, WYOMING[1]

	Central Bar Sandstone	Central Bar (planar laminated)	Bar Margin (Type 1)	Bar Margin (Type 2)	Interbar Sandstone	Interbar Facies	Bioturbated Shelf Sandstone	Shelf Sandstone	Bioturbated Shelf Siltstone	Shelf Siltstone	Covered	Total Number of Units Observed
Central Bar	2C		2C, 2D				1B 1A				1A,2B,2C, 2D,3,4D, 4D,5,5	15
Central Bar (Planar)	4D			4C	2D				1A			4
Bar Margin (Type 1)	3	1A	1A	28, 4CC	3		4CC 4C	2D	2A		2C, 3	12
Bar Margin (Type 2)	1A, 2C						1B		2B		4C,4CC,2E, 1B,2A	10
Interbar Sandstone	2D,2D, 2B,2C, 1A,1B, 4D,5	4D 4C	3 1A, 2B	2A, 2C	2C, 3	2A		5			4A 4C 2E, 3	24
Interbar Facies			2A					5			5	3
Bioturbated Shelf Sandstsone			4CC 4CC	1B 1A, 2E	4D,4A, 4D,1B, 4B,2C, 4C,1A, 2E,2B, 3	2D	1B 4CC				5	17
Shelf Sandstone				1B					2D		5	3
Bioturbated Shelf Siltstone		2D			2C 2A, 2B, 1B, 2A		4C,4D, 2E,4A, 2C, 3,4C, 1A,2B	1B	4A			17
Shelf Siltstone							4B 2D,2E					3
Covered					1A, 6		5		2D,2E, 2B,2C, 1B,2A, 3,4A, 4C,4CC	4B	Total	15 127

[1]This table is designed to show the distribution of recurrences of facies overlying particular facies. Reference facies (underlying facies) are listed to the left. The overlying facies are listed across the top. An example of how to read the table is: the *Central Bar Sandstone* (left column) is overlain by another *Central Bar Facies* at one location in Measured Secton W2C and is overlain by *Bar Margin Facies* (Type 1) at one location in W2C and W2D, etc.

sent in outcrop sections. The lower Shannon sandstone in this section exemplifies the highly variable lateral nature of the various facies within the complex. The W4D to W4CC series of sections covers a distance of 2.5 miles. The higher energy Bar Margin Facies at the base of two of the three closely spaced sections near the middle of the cross section extend less than 0.5 miles and in adjacent sections the upper Bar Margin Facies cannot be correlated between wells.

The lateral sequence in the upper portion of the lower Shannon sandstone from W4D on the south to W5 on the north is suggestive of relatively abrupt lateral facies changes; abrupt at least on the scale of the cross section. The Central Bar Facies in the lower Shannon in W4D grades northward to Interbar Facies, which in turn grades into Bar Margin (Type 1 and Type 2) Facies, which grade back into Central Bar Facies in the northernmost measured section (W5). For nearly

A

B

C

FIG. 18.—*Bar Margin Facies.* A, General view of vertical sequence of shelf-ridge complex facies. At the base is a diffusely bedded *Bioturbated Shelf Siltstone,* Unit 5 (33 feet thick), overlain by a rippled *Interbar sandstone Facies,* Unit 6 (more resistant; 14 feet thick). At the top of the photograph (above arrow) are low-relief troughs interbedded with ripples, Unit 7; measured section W2A-77; *Bar Margin Facies (Type 2).* B, Classic *Bar Margin Facies (Type 1),* consisting of abundant limonite and shale clasts scattered within a series of planar-tangential beds up to 2 feet thick. The planar-tangential beds are thicker than is usual in most of the Shannon. Transport direction is to right (south). Unit contains more than 10% limonite (after siderite) in lenses and clasts. Unit 6B, measured section W3-77. C, High energy *Bar Margin Facies*

three miles in the middle portion of the cross section, Central Bar Facies are absent and Bar Margin Facies are thin, indicating that only low to moderate energy sandstones were deposited in this area, even though there were thick deposits (Figure 38) of sand accumulated in the area. Bioturbated Sandstone Facies and Interbar Sandstone Facies form over two-thirds of the sandstones along the line of section.

Log responses in nearby subsurface wells located near the line of cross section (G-G') appear to show a silty character which probably is a reflection of the absence of Central Bar Facies. This silty character is typical of many wells within the thickest areas of lower Shannon sandstone development and may indicate that the cleanest (cross bedded) sandstones were not always deposited in the thickest portions of the ridges. The upper Shannon sandstone in this area is very thin (Figure 39) and is assumed to pinch out a short distance to the west. It is interesting to note that even in this area of thin sands, several of the higher energy sand facies occur; Central Bar Facies (to the north and south), Bar Margin (Type 2) Facies, and Interbar Sandstone Facies. The finer-grained deposits form covered intervals along this line of outcrop section.

ISOPACH MAPS

Isopach maps were constructed for the upper and lower Shannon sandstones using outcrop and well data in the Salt Creek area. As discussed previously, "sandstone" as indicated on the maps probably consists of Central Bar, Bar Margin and some Interbar Sandstone Facies sandstones, or in other words, the "cleaner" sandstones within the Shannon interval.

Lower Shannon Sandstone.—The isopach map of the lower sandstone shows a distinctly linear, north-south depositional pattern suggestive of ridge and swale depositional patterns. The most prominent ridge (ridge #2) occurs in R78W (Figure 38). This ridge exceeds 18 miles in length, is about 7 miles wide at its widest point, and is up to 75 feet thick. It appears to be a coalescence of several small ridges and is multi-crested. The dimensions of this total ridge complex are comparable to those of Hartzog Draw Field (Martinsen and Tillman, 1978, 1979; Tillman and Martinsen, 1979 and in press), while the dimensions of the smaller ridges, which are components of the complex, are comparable to those of several of the smaller Shannon producing fields such as Heldt Draw Field (Davis, 1976; Seeling, 1978), all of which are located about 25 miles northeast of Salt Creek. In addition to this large ridge complex, two other areas of thick north-south trending sandstones are observed. One is in Ranges 79 and 80 West (ridge #3), and another is indicated

(Type 1) composed of high energy subhorizontal planar laminations at base, overlain by small trough and rippled sandstone. "Ripples-on-troughs" are above arrow. Resistant, quartz-cemented. Clasts and lenses of limonite comprised in unit. Unit 4 (49–54 feet); measured section W2A-77.

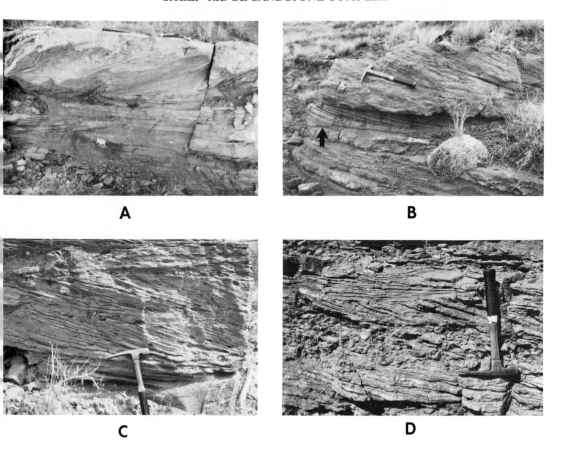

Fig. 19.—*Bar Margin Facies* (Type 1). *A*, Large trough deposited by current flowing out of photo. Trough filled asymmetrically from right. Smaller, horizontally truncated troughs below show similar transport direction. Scale in centimeters. Unit 5 (33 feet), measured section W1A-77. *Bar Margin Facies (Type 1). B*, Glauconitic laminated trough-bedded *Bar Margin Facies (Type 1)*, Unit 4 (above arrow), overlying rippled *Interbar Sandstone Facies*, Unit 3 (below arrow). Transport direction in trough of Unit 4 is to right. Measured section W3-77. *C*, Low relief planar-tangential troughs in *Bar Margin Facies (Type 1)*. Cut-and-fill nature seen here (arrow) is generally not observed in the Shannon; instead most troughs are horizontally truncated. Unit 4, measured section W2A-77. *D*, Trough cross-bedded sandstone containing abundant sub-rounded limonite rip-up clasts (ovoid "blobs" in zone between hammer-head and white stripe). Unit consists of 20% limonite (after siderite) clasts. Note subhorizontal truncation of troughs that is typical of the Shannon. Transport direction is to south (right). Unit 6B at 82 feet, measured section W3B-77. *Bar Margin Facies (Type 1)*.

within the east half of Range 77 West (ridge #1).

The north-south pattern of the ridges at Salt Creek Field contrasts somewhat with the strongly northwest-southeast pattern typical of the Shannon sandstones in the GHDF area. If sand body orientation is compared with the orientation of the Shannon age shoreline as reconstructed by Gill and Cobban (1973; Figure 45 herein), it can be seen that it also changes orientation from north to south and may have influenced the orientation of the Shannon sandstone ridges. The sandstone ridges are not oriented parallel to the paleo-shoreline, but at an angle to it. The angle between the shorelines and the shelf ridge complexes is commonly oblique such that in areas to the south (such as at Salt Creek) where the shoreline is more southwest-northeast, the ridges formed north-south. Sediment transport directions as collected by us, and others reported by Spearing (1976), are predominantly to the southwest, and thus indicate southwestward (as opposed to due south) current directions in the Salt Creek area during Shannon time. Likewise, in the GHDF area, where Shannon sandstone bodies are oriented northwest-southeast, sediment transport directions were almost due south (Tillman and Martinsen, in press). Similar orientations of sand ridges at approximately a 30° angle to current flow has frequently been observed on the Atlantic shelf of the eastern United States (Stubblefield *et al.*, this volume).

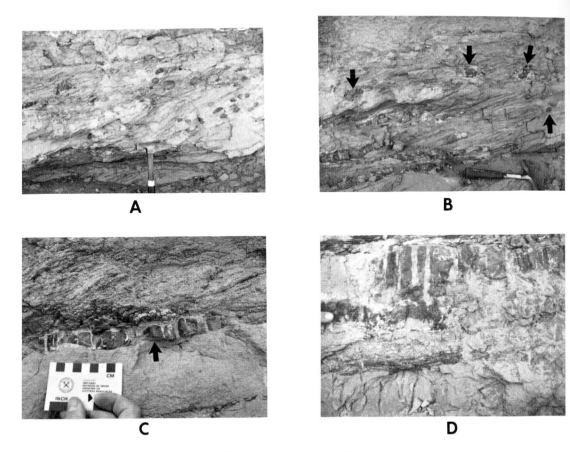

FIG. 20.—*Limonite lenses and clasts, Bar Margin Facies (Type 1). A*, Rounded to oblate limonite clasts (after siderite) in horizontally truncated troughs of *Bar Margin Facies (Type 1)*. Unit 3 at 13 feet, measured section W1B-77. *B*, Lenses of limonite (after siderite; left of hammer) and reddish limonite clasts (arrows) in highly glauconitic Shannon Sandstone composed of low-relief troughs. Unit 3 at 12.5 feet, measured section W1B-77; *Bar Margin Sandstone (Type 1). C*, Limonite lens (arrow) with closely spaced vertical sand-filled burrows (white). Scale in centimeters. Measured section W1B-77; *Bar Margin Facies (Type 1). D*, Vertical white glauconitic sand-filled burrows in 2-inch-thick limonite lens (finger for scale). Measured section W1B-77 at 10 feet; *Bar Margin Facies (Type 1)*.

Upper Shannon Sandstone

The pattern of sandstone deposition of the upper sandstone is quite different from that of the lower sandstone (Figure 39). While the lower sandstone has a north-south component in its depositional pattern, the upper sandstone is more oblate than linear, and this geometry sets the upper sandstone apart from both the lower sandstone, and from most other Shannon sandstones which display strongly linear geometries. Spearing (1976) also published isopach maps of the upper and lower sandstones (Figure 40). His isopach maps cover a broader area than do ours, and in a general way his maps resemble ours. The differences may be attributed in part to the inclusions of the silty and burrowed sandstone units as well as the cleaner sandstones within his isopach thicknesses. Both Spearing's and our maps show an oblate geometry for the upper sandstones and an elongate geometry for the lower sandstones. The average thickness of the upper sandstone is less than for the lower sandstones, reaching a maximum of 65 feet in thickness.

In outcrop the upper sandstone appears to be composed of the same types of facies as the lower sandstone, and, as such, must be a product of the same or similar depositional processes as the lower sandstone. In both the upper and the lower sandstones, sediment transport was to the southwest.

DIRECTIONAL DATA

Excellent trough cross bedding was commonly observable in three dimensions in the Central Bar and Bar Margin Facies outcrops in the Salt Creek area. Depositional current transport directions were measured in

FIG. 21.—*Bar Margin Facies (Type 2)*. *A*, Looking southeast from measured section W2A-77. Basal unit (below arrow) is a *Bioturbated Shelf Siltstone*, Unit 5. Between arrows is an *Interbar Sandstone Facies*, Unit 6. At the top is *Bar Margin Facies (Type 2)*, Unit 7; measured section W2A-77 at Castle Rocks. *B*, Typical *Bar Margin Facies (Type 2)* consisting of interbedded planar-tangential troughs to rare planar-tabular beds and rippled beds with ripple-form horizontal bed geometry. Unit 7 at 120 feet, measured section W2A-77. *C*, Typical *Bar Margin Facies (Type 2)* consisting of interbedded planar-tangential troughs to planar-tabular beds (one bed down from hammer) and rippled beds. Upper ripple-formed surfce typical of rippled beds. Unit 7 at 120 feet, measured section W2A-77. *D*, Typical low-relief trough (at hammer) displaying tangential contact with underlying rippled sandstone in *Bar Margin Facies (Type 2)*. Unit 7 at 125 feet, measured section W2A-77.

outcrop sections W1A, W1B, W2A, W2B, and and W3; readings were taken only on those troughs in which good to excellent three-dimensional reconstruction was possible. Regional dip and strike were taken at most of the localities for regional correction of local dip and strike by use of a "dip-azimuth program." The mean transport direction for individual units is plotted, and the boundaries of one standard deviation are indicated by dashed lines (Figure 41). Outcrops in which directional data were obtained are discussed in three groupings based on geographic location.

Measured Sections W1A and W1B are in the south-central part of the study area (Figures 3, 35) and are located about 0.7 miles apart. Thirteen readings of trough bed transport directions were taken in two units in section W1A (Figure 41); the transport directions

ranged from 145° to 230°. Mean transport direction in Unit 6, a Central Bar (planar laminated) Facies is 204°; a similar mean transport direction, 210°, was obtained from troughs in the Central Bar and Bar Margin Facies (Units 9A and 9B). In measured section W1B three readings were taken in the Central Bar Facies (Unit 6) and a mean transport direction of 197° was obtained (Figure 41).

The easternmost outcrops in the Salt Creek area were studied in measured sections W2A in the south and W2B in the north (Figure 3). In measured section W2A directional measurements were taken from two units, both of which were at the top of bar complexes. In Unit 4, a Bar Margin Facies (Type 1), the mean transport direction is to the SSE at 157°; in the upper-most unit of the upper bar, a Bar Margin Facies (Type

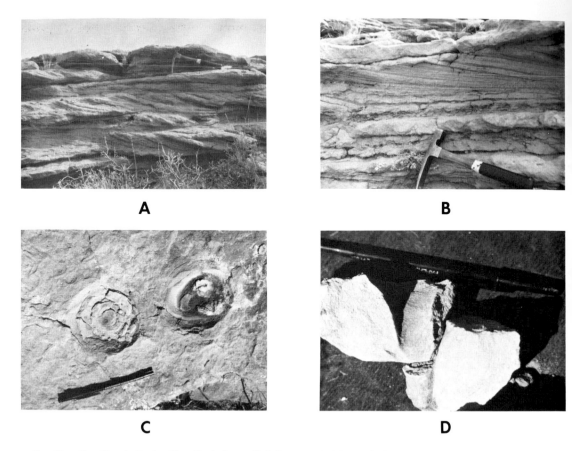

A **B**

C **D**

FIG. 22.—*Bar Margin Facies (Type 2)*. A, Low relief planar-tangential troughs interbedded with thin rippled beds in *Bar Margin Facies (Type 2)*. Note that tops of cross-bedded sets all are eroded and form a sub-horizontal bed. Current transport direction is to right. Unit 7 at 130 feet, measured section W2A-77. B, Low relief planar-tangential troughs interbedded with thin rippled beds in *Bar Margin Facies (Type 2)*. Excellent definition into light and dark laminae is due to the high concentration of glauconite in the dark (green) layers). Current transport direction to right. Unit 7, measured section W2A-77. C, Four-inch-diameter concentric burrows on top of bedding surface; burrows narrow downward as shown in Figure 22D. Unit 7, measured section W2A-77; *Bar Margin Facies (Type 2)*. D, Relatively rare upward-flaring, smooth-walled burrow. Photo about 80% of natural size. Sand-filled, downward-narrowing portion of burrow shown on left portion of lower block. Burrow more than triples in width at top. Unit 7, measured section W2A-77; *Bar Margin Facies (Type 2)*.

2), the mean transport direction is 186° (Figure 41). The range for all transport directions in this section is relatively small (100°) ranging between 110-210°. This relatively narrow range of transport directions is especially remarkable when you consider that these features were formed, as will be discussed later, on or near the middle shelf, a location where one might expect that a wide variety of transport directions might be encountered.

Two units in measured sections W2B contain large-scale trough cross beds which are well exposed. Planar-tabular beds are more common in this section than in other measured sections. As was observed in section W2A, the trough-bedded units here are also at the top of stacked bar complexes. The mean transport directon for Bar Margin Facies (Type 2), Unit 4, is 182°; the mean transport direction for Unit 7, a Central Bar Facies, has a mean transport direction of 154°.

As can be seen in Figure 40, the transport directions in measured section W3, which is one of the more westerly sections, are multi-modal. In this section relatively abundant contorted and convoluted troughs are observed at the top of the lower Shannon sandstone bar sequence (Figure 37). The upper three feet of Unit 6B (Figures 11 and 16) contains interbedded northeast flowing troughs and south to southwest flowing troughs, suggesting that during the final phases of bar formation, northeast flowing currents alternated with

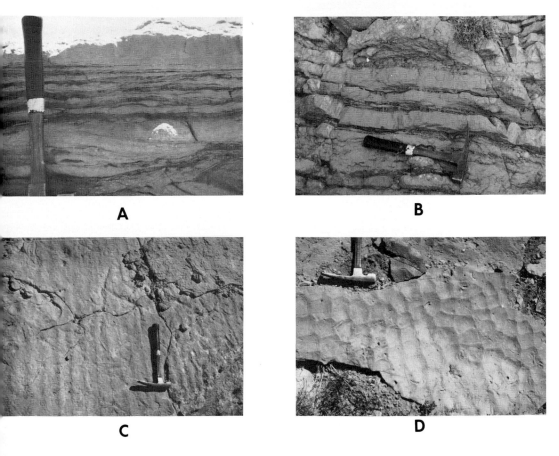

Fig. 23.—*Interbar and Interbar Sandstone Facies*. A, Ripples defined by alternating sand ripples and silty shale ripples (dark) interbedded with low-relief trough beds, *InterbarFacies*. *Interbar Facies* contain interbedded shales while the *Interbar Sandstone Facies* is all sandstone. Unit 7 at 136 feet; measured section W2A-77. B, Rippled sandstone near top of Unit 9A, which is designated as a *Bar Margin Facies (Type 2)*. This sequence of several successive beds of ripples is not typical of this facies; however, the presence of ripples interbedded with troughs is important in defining this facies. Note vertical burrows are in silty shale interbeds. Measured section W1A-77. C, Long-crested, parallel symmetrical-wave ripples on bedding surface, orientation is N30°W. At 30 feet in Unit 4, measured section W1A-77; *Interbar Sandstone*. D, Interference ripples on bedding surface in *Interbar SandstoneFacies*. Formed by two nearly equal currents at 90° to each other. Unit 6, measured section W2A-77.

southerly and southwesterly flowing currents. This is the only place where abundant contorted beds and the anomolous NE transport direction were observed in outcrop. In addition, a substantial number of WSW oriented troughs were observed; these are also somewhat anomalous. The capping facies of the lower bar sequence at W3, Unit 4, is a Bar Margin Facies (Type 1) and has a mean transport bearing of 193° which is "normal".

Wave ripple crests were observed on bedding surfaces at several locations. Their crest orientations range from 60° to 185°, a range of 125°. This variation in ripple crest orientations suggests that small-scale features, which form parts of the Bar Margin Facies (Type 2) and most of the Interbar Sandstone Facies, are formed by currents which are variable in their flow directions. A fairly large variation in orientations of ripple crests was also indicated by Spearing (1975, 1976) who tallied 171 readings of crest orientations (Figure 42). Further measurements on ripple crest orientations should be taken to determine if ripple crest variations are really as variable as the limited number of measurements taken so far in this study would suggest.

Seventy-two trough bed transport directions were measured in outcrop sections during the course of this study. It is significant that two-thirds of the transport directions (one standard deviation) fall within an arc of

FIG. 24.—*Interbar Sandstone Facies. A*, Lower portion of measured section W2B-77 exhibiting typical vertical Shannon shelf-ridge sequence. Unit 1 (below arrow) is *Bioturbated Shelf Sandstone*. Middle unit (between arrows) is 24-feet thick *Interbar Sandstone Facies*, Unit 2. Less resistant beds above arrow at top are trough-bedded *Bar Margin Facies (Type 1)*. *B*, Ripple-formed, horizontally-bedded, highly-rippled *Interbar Sandstone Facies* (below arrow) overlain by trough-bedded *Bar Margin Facies (Type 1)*. Measured section W2B-77, Units 2 and 3, on east side of Salt Creek Anticline. *C*, Shallow 15 feet-wide erosional channel in otherwise sub-horizontally bedded, rippled *Interbar Sandstone*. Note that channel is filled with rippled beds. Measured section W1A-77, Unit 4.

FIG. 25.—*Interbar Sandstone Facies. A, Shelf Siltstone (Bioturbated)*, Unit 2, at base, overlain by 26-feet thick *Interbar Sandstone (Unit 3A)* consisting of 1-3 inch thick ripple-form horizontal beds interbedded with some highly burrowed beds. Measured section W2A-77. *B*, Interbedded sandstone and slightly more recessive, burrowed (some shale lined) shaley, silty sandstone. Beds 1-3 inches thick. Sandier beds include 10% glauconite. Contact of units 3A and 3B just above head to right. Unit 2 is *Shelf Siltstone (Bioturbated)* at lower left. Unit 3A is an *Interbar Sandstone*. Unit 3B, which is siltier and more recessive, is also designated as an *Interbar Sandstone Facies*. Measured section W2A-77. *C*, Ripple-form bedded *Interbar Sandstone Facies*. Ripple-form shape on horizointal bedding boundaries is especially prominent

60° centering on 188° (Figure 43). This average transport direction is very close to that recorded for the same general area in Figures 6-8 by Spearing (1976); 200° ± 15°.

PALENTOLOGY

Micropaleontology

The earliest reported micropaleontology work on the Shannon, cited by Parker (1958), included results of studies by N. and D. Curtis of several thousand individuals of planktonic and benthic foraminifera from two wells in southeastern Montana. Their work indicated that the depth of water for Shannon deposition in that area ranged from 90 to 400± feet. They believed they could recognize a general decrease in depth of water toward the top of the cored section in one of the wells. Genera which were identified included: *Textularia* sp. 2, *Robulus* sp., *Pullenia* sp. 1, *Bulimina* sp. 1, *Bulimina* sp. 2, *Anomalina* sp. 2, *Spiroplectammina* sp. 1. Spearing (1975, 1976) reported on the work of C.H. Ellis (personal communication) who found that "the marine shales beneath the Shannon sandstone....contain an arenaceous shelf-type foraminiferal assemblage in outcrops at Salt Creek."

Recent work by Dailey (personal communication, 1978) shows that samples from four Salt Creek outcrop locations include a relatively diverse suite of foraminifera (Table 6). Twenty-six species were identified from shales and silty shales ranging from 250 to 20 feet below the basal Shannon sandstones. In addition to foraminifera, Inoceramus prisms, probable *Baculites* fragments, and rare ostracods were identified in the micropaleontology preparations.

Dailey described the faunal assemblages recovered from the measured sections as largely of agglutinated character, low to medium-low diversity, and devoid of planktonic organisms. *Haplophragmoides* dominates these assemblages, while primary accessory types are *Bathysiphon* and *Spiroplectinata* in sections W1A, W2 and W3 and *Gaudryina* in section W4A. Modern species of *Haplophragmoides* and *Bathysiphon* are abundantly represented in two environments: (1) transition and nearshore waters, and (2) deeper depths of the slope and abyssal depths. The absence of typical deep-water elements rules out a slope-abyssal interpretation. Of the remaining conspicuous morphotypes, *Gaudryina* is not indicative of any particular bathymetric zone and *Spiroplectinata* is without recent analogs. Also present are a few *Lenticulina* and a single specimen of *Epistomina;* these two genera are known today only from middle shelf and deeper depths, although *Lenticulina* probably extended into shallower depths prior to middle Tertiary time.

Special weathering characteristics are required to see ripples within beds. Below hammer is *Bioturbated Shelf Sandstone.* Interval shown in photograph is 18-22 feet in measured section W1A-77.

A

B

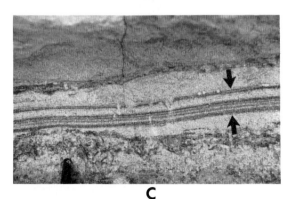

C

Fig. 26.—*Interbar Sandstone Facies. A,* Typical thin-bedded (1-3 inches thick) internally ripple-laminated, ripple-form bedded *Interbar Sandstone Facies,* Unit 2, measured section W2B-77. *B,* Interbedded 1-inch thick fine-grained rippled sandstones and 1/8-1/4-inch sandy shales typical of some parts of *Interbar Sandstone Facies.* Note that the ratio of sand to "shale" is about 10:1. Unit 6, measured section W2A-77. *C,* Subhorizontal laminations containing high concentrations of glauconite (between arrows). White sand-filled horizontal burrows (1/8 inch diameter) cut upper glauconite laminae. High degree of bioturbation in sandstone shown in lower part of photograph. Dark color in bioturbated zone is glauconite. Pen tip for scale. *Interbar Sandstone Facies,* Unit 6, measured section W2A-77.

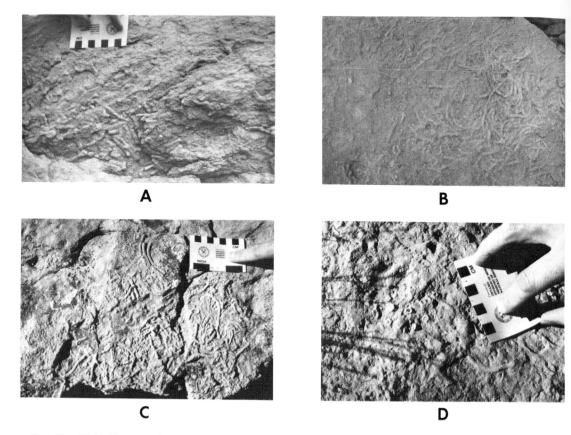

FIG. 27.—*Shelf-ridge trace fossils. A,* Concentration of *Thallasinoides* on base of bedding surface in *Interbar Sandstone.* Scale in centimeters. Unit 4, measured section W1A-77. *B,* Two sizes of horizontal curving non-branching burrows on bedding surface in *Interbar Sandstone Facies.* Unit 6, measured section W2A-77. *C, Thallasinoides* parallel to bedding surface. Unit 5 at 28 feet, measured section W1B-77; *Interbar Sandstone Facies. D,* Plural curving tubes. View of upper surface of bedding plane. Unit 5, measured section W1B-77; *Interbar Sandstone.*

Considered together, the general features of this fauna indicate shelf depths. The fauna suggest deposition either on the inner-shelf or some shallow part of the middle-shelf or outer-shelf. Distinctive foraminifera are suggestive of shallow (innermost shelf) deposits as deeper (than middle) shelf depth fauna are absent. One or more additional factors, such as turbulence, subnormal salinity, or low pH, strongly influenced the character of this fauna as well as many other interior seaway faunas. Depth alone cannot account for the impoverished nature of these assemblages. Samples collected in the shaley portions interbedded with Shannon sandstones were barren.

According to Dailey, the assemblages consist almost entirely of taxa whose ranges are either indeterminate or too broad for precise age dating. Only a single sample from a measured section (W9; Sec. 8 T42N R80W), which is north of the area discussed in detail herein, included elements of age importance. At 80 feet in Measured Section W9, "a single planktonic

individual assignable to *Rugotruncana subcircumnodifera* was recognized. This species has been reported from the late Campanian to late Maastrichtian....therefore this sample is probably not older than late Campanian age."

As can be noted by comparing Table 6 with the fauna discussed by Parker (1958), only one genus was found to be common to both the Curtis' and Dailey's faunal lists. The reason for these apparently differing faunas is not known.

Macropaleontology

The Shannon sandstone was deposited between the *Scaphites hyppocripis* and *Baculites* sp. (smooth) zones in the Salt Creek area (Table 7). The *Scaphites* is a coiled form while the *Baculites* sp. is elongate and straight. The *Baculites* sp. (smooth) (80.5 m.y.) is the oldest of a single lineage of endemic forms which continued to *Baculites eliesi* (71 m.y.) (Gill and Cobban, 1973). Potassium-argon dates were used by Gill and

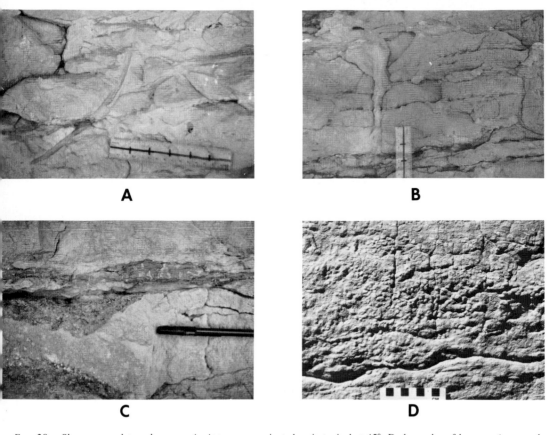

FIG. 28.—Shannon sandstone burrows. A, Asterosoma oriented as is typical at 45°. Darker color of burrows (some sub-horizontal) is due to concentric clay "linings." Six inch scale. Unit 3 at 19 feet, measured section W1A-77; Bioturbated Shelf Sandstone. B, Vertically oriented upward-swelling burrow. Ripple-form beds with bioturbated (75% or more burrowed) ripple-form sets that are 2-3 inches thick. Unit 3 at 20 feet; measured section W1A-77; Bioturbated Shelf Sandstone. C, White sand-filled vertical burrows of 1/8 inch diameter in limonite lens which is interbedded with partially obscured ripple-form sandstone beds that are 1-3 inches thick. Dark colored glauconite grains on left. Bioturbated Shelf Sandstone, Unit 3 at 20 feet; measured section W1A-77. D, Highly horizontally burrowed sandstone showing symmetrical ripple-form bed surfaces (wave ripples). Scale in centimeters. Unit 3 at 20 feet; measured section W1A-77. Bioturbated Shelf Sandstone.

Cobban (1973) to determine ages and to indicate that the ammonite zones each span about a half million years.

Baculites sp. (smooth) which is commonly found in shales associated with the Shannon Sandstone at Salt Creek is of moderate size (2-inch diameter), has a low degree of taper, an ovate cross section, and contains no ribs on the flanks. The centers may be smooth or weakly ribbed (Figure 44A). The distinct suture pattern for this form is illustrated by Gill and Cobban (1973, p. 8, Fig. 3A). This form of *Baculites* sp. (smooth) is smaller than the *Baculites* sp. (smooth) of approximately 78 m.y. of age.

The *Baculites* sp. (smooth) of Shannon age gives way stratigraphically to the *Baculites* sp. (weak flank ribs) (80 m.y.) which resembles the smooth form in most ways except for the presence of arcuate ribs on

the flanks (Figure 43B). This ammonite zone is associated with post Sussex shales at Salt Creek. The Ardmore bentonite is in the upper part of this ammonite zone (Table 7).

Trace Fossils

A wide diversity in size, orientation, and type of fill material suggests a relatively hospitable environment for burrowers in portions of the shelf-ridge complex. Rare *Teichichnus*, *Thallasinoides*, *Chrondrites*, and plural curving tubes were identified (Figures 27 and 28). Common Cretaceous shoreline trace fossils such as *Ophiomorpha*, *Asterosoma*, and *Rhizocorallium* were not observed.

The amount of burrowing in individual facies falls into two distinctly different modes. The Bioturbated Shelf Sandstone and Bioturbated Shelf Siltstone Facies

A **B** **C** **D**

FIG. 29.—*Fine-grained Shannon Shelf-Ridge Facies. A, Shelf Silty Shale* overlain by *Bioturbated Shelf Siltstone* (above arrow). Location of paleontologic samples at measured section W2AA-77 located just south of, and stratigraphically below, measured section W2A-77. *B, Bioturbated Shelf Siltstone Facies,* Unit 1 (below arrow) 13 feet thick. Above arrow is *Interbar Sandstone Facies,* Unit 2, which is 24 feet thick. Measured section W2B-77, Salt Creek Field, Wyoming. *C,* Cliff formed by *Bioturbated Shelf Siltstone,* Unit 1 (0-22 feet) overlain by *Bioturbated Shelf Sandstone Facies* (Unit 2, above lower arrow) and thin *Interbar* (Units 3 and 5A) and *Bar Margin Facies (Type 1),* Unit 4 (note person for scale, arrow). Measured section W3-77. *D,* Middle portion of photo displays two successive *Bioturbated Shelf Sandstones,* both of which are more than 75% burrowed. Unit 7A below arrow has slightly finer grain size (100 μ) than overlying Unit 7B (125 μ). Unit 7A has fewer "shale" interbeds and has generally subhorizontal bedding where visible. Measuredsection W1B-77.

averaged 87% burrowing; they range from 75-95% (by volume) burrowed (Table 3).

The maximum percentage of burrowing recorded for all other facies was 45% for several outcrops of Interbar Sandstone Facies. The highest average percentage of burrowing among the non-bioturbated facies was 27%, also in the Interbar Sandstone Facies. Bar Margin and Central Bar Facies average 4-5% burrowed with maximum values of 10 and 30%, respectively (Table 3).

SHANNON SHELF-RIDGE COMPLEX MODEL

The Shannon sandstones in the Powder River Basin of Wyoming were deposited as shelf-ridge complexes on a wide shelf during early Campanian time in the interior seaway (Figure 45) stretching from Alaska to Mexico. The wide shelf model proposed by Asquith

(1970), in which the shelf-slope boundary can be recognized, is believed to be applicable for deposition during Shannon time.

The Shannon sandstones in the Salt Creek area were deposited 70-100 miles east of the contemporaneous shoreline (Figures 45, 46) during the last stages of the Telegraph Creek-Eagle regression about 80.5 million years ago (Santonian to Campanian). During the regression, the shoreline moved east 240 miles in 5.5 million years at a rate of about 50-70 miles per million years (Gill and Cobban, 1973). Based on interpretations of Gill and Cobban (1973), this regression began as early as 83.5 million years ago and probably paused briefly about 80 million years ago (Table 7). The Shannon was deposited not only during this period of regression, but also possibly during a minor stillstand.

TABLE 6.—FORAMINIFERS IN SALT CREEK MEASURED SECTIONS[1]

SPECIES	W1A		W2C			W3		W4A			
	-250'	-260'	-25'	-60'	-100'	-20'	-40'	0'	19'	54'	97'
AMMODISCUS SP.			F		F	R					
BATHYSIPHON BROSGEI TAPPAN			R						F		
BATHYSIPHON VITTA NAUSS	A	F	F	R	R	F	R				
CITHARINA N. SP.									R		
DOROTHIA SMOKYENSIS WALL										F	C
EPISTOMINA FAX NAUSS											R
FLABELLAMMINA SP.					R						
GAUDRYINA KANSASENSIS (MORROW)								F	C	A	
GAVELINELLA KANSASENSIS (MORROW)					R						
GLOMOSPIRELLA SP.								R			
HAPLOPHRAGMOIDES BONANZAENSE STELCK & WALL					C						
HAPLOPHRAGMOIDES COLLYRA NAUSS	C	C									
HAPLOPHRAGMOIDES EXCAVATUS CUSHMAN & WATERS	A	A	C	A	A			F	F	F	A
?HAPLOPHRAGMOIDES KIRKI WICKENDEN				R							
HAPLOPHRAGMOIDES SP. INDET.						C	F				
LENTICULINA MUENSTERI (ROEMER)									F		
LENTICULINA ROTULATA (LAMARCK)								F	F		
LENTICULINA TAYLORENSIS (PLUMMER)									R	R	
POLYMORPHINID SP. INDET.									R		
REOPHAX CONSTRICTUS (ROEMER)											F
REOPHAX RECTA (BEISSEL)					C						
SACCAMMINA ALEXANDERI LOEBLICH & TAPPAN		R									
SPIROPLECTINATA BENTONENSIS (CARMAN)	C	A	F	A	A						R
TROCHAMMINA DIAGONIS (CARSEY)										R	
?TROCHAMMINA RUTHERFORDI STELCK & WALL		C									
TROCHAMMINA SP. INDET.				R						F	

[1]Identification by Dailey (personal communications, 1978). Agglutinated forms predominate; calcareous perforate forms are rare. Abundances indicated are as follows: 41 or more = abundant (A); 10-40 = common (C); 4-9 = few (F); 1-3 = rare (R).

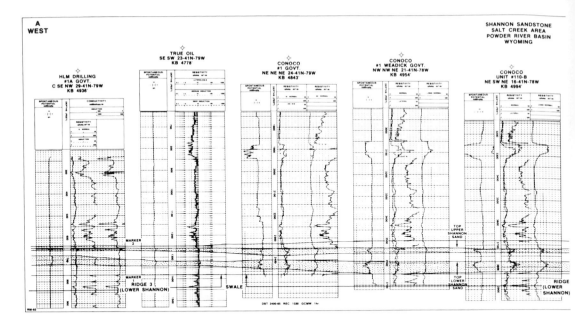

FIG. 30.—East-west cross-section A-A' across northern subsurface expression of Salt Creek Anticline. Location is shown in Figure 3. Includes Meadow Creek Oil Field which produces from the Shannon. Wells in Meadow Creek Field are Conoco Unit #138 and the D&D #4 Irvine. The upper and lower Shannon sandstones are designated. Ridge and swale topography, as interpreted from the isopach map, is delineated for the lower Shannon sandstone.

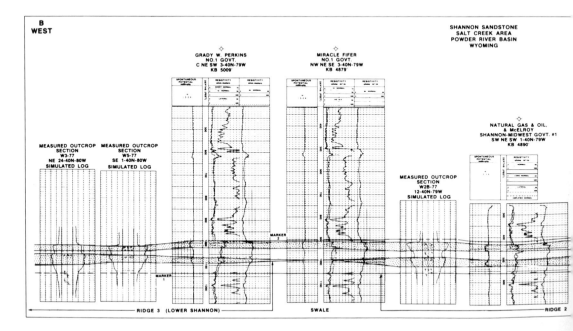

FIG. 31.—East-west cross-section B-B' across Salt Creek Anticline. Three surface sections are correlated with subsurface logs. Location is shown on Figure 3. The Pan American No. 37 Government well is located in Shannon Field, the first commercial production from the Shannon. Ridge and swale topography, as interpreted from the isopach map, is delineated for the lower Shannon sandstone.

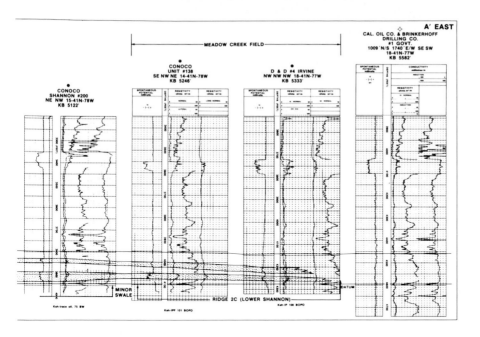

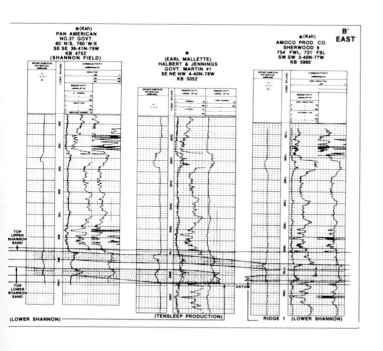

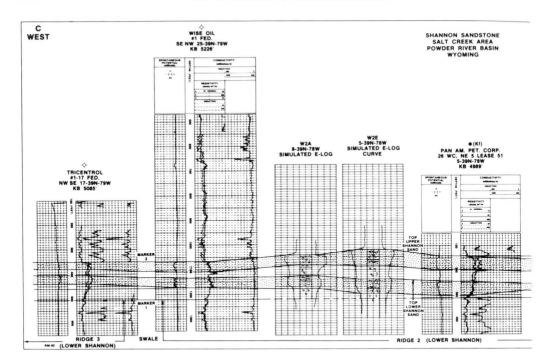

FIG. 32.—East-west cross-section C-C' across south end of Salt Creek Anticline. Location is shown in Figure 3. Two outcrop sections are correlated with subsurface logs. The Tricentral 1-17 Federal, on the west end of the cross-section, shows curve reversal due to fresh-water invasion which is typical for many of the very shallow wells in the area west of the anticline. Ridge and swale topography, as interpreted from the isopach map, is delineated for the lower Shannon sandstone.

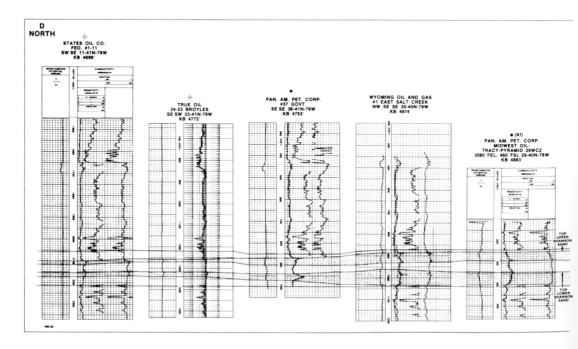

FIG. 33.—North-south cross-section D-D' located just east of Shannon outcrop cross-section F-F' (Fig. 35). Location is given in Figure 3. Included are wells that produced from the Shannon at Teapot and Shannon Fields.

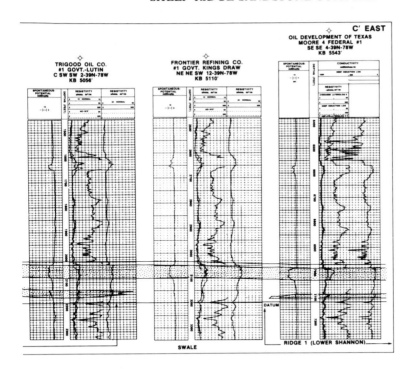

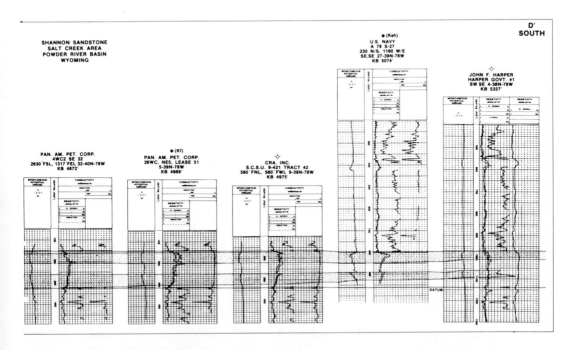

TABLE 7.—AMMONITE, TIME AND FORMATION RELATIONSHIPS[1]

	AGE (million years)
Baculites obtusus — — — — — — — — — Regression	79.5
Ardmore bentonite • • • • • • • •	
Baculites sp. (weak flank ribs) — — — — Transgression	80.0
Sussex Sandstone	
Baculites sp. (smooth) — — — — — — —	80.5
Shannon Sandstone Regression	
Fish Tooth Sandstone	
Scaphites hyppocripis — — — — — — —	81.0
Niobrara Shale	

[1]Relationship of ammonite zones to formations in Salt Creek area, Wyoming. Ages of ammonite zones shown along with indications of periods of transgression and regression as interpreted by Gill and Cobban (1973, p. 18). Contrast periods of transgression and regression with Weimer (1983, p. 362).

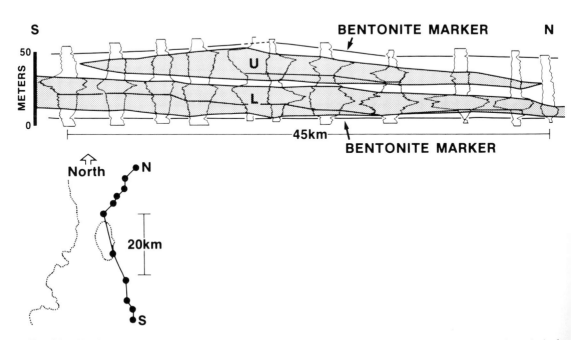

FIG. 34.—North-south Shannon cross-section through Salt Creek area. Subsurface logs are SP and resistivity logs. A single outcrop section with letters U and L on it is included. (Spearing, 1976). Compare with section D-D' (Figure 33) which is 18 miles long. Length of this section is 29 miles (45 km). A rise from the south to north of the base of the lower sandstone is observed. See text for discussion of implied genetic significance of this rise.

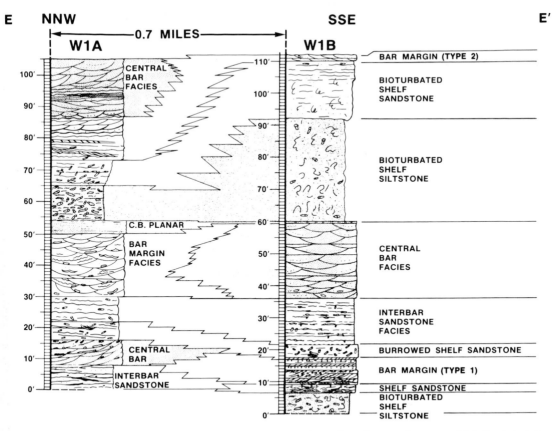

FIG. 35.—Stratigraphic section (E-E'); relatively closely spaced sections (Figure 3) on southwest part of Salt Creek Anticline Shannon outcrop. Lateral changes from *Central Bar Facies* to lower energy facies and doubling of thickness of the upper Shannon sandstone between sections are shown. Detailed descriptions of the measured sections are Figures 8 and 9. The lower Shannon sandstone extends up to 54-60 feet on the section and is capped by *Bioturbated Shelf Siltstones*. The upper Shannon is the upper sandstone at this location.

The end of the "Shannon regression" was marked by a minor, but rapid, transgression which may have lasted less than 1 million years prior to continued regression. The Sussex sandstones at Salt Creek (Brenner, 1979) were deposited during this minor transgression probably during a sequence of minor regression-stillstand and regression.

Combining interpretations related to stratigraphic sequence, lithology, sedimentary structures, micropaleontology (foraminifera) and trace fossils, it is suggested that the Shannon sandstones cropping out in the Salt Creek area were deposited as offshore sand ridges on the mid-shelf of the western Pierre-Cody Seaway below daily effective wave base at a water depth probably greater than 60' and less than 300'. The shelf-ridge sandstone complexes are completely enveloped on all sides by fine-grained marine shale deposits.

This interpretation of environment of deposition coincides with prior interpretations made by several previous workers, among them Parker (1958) and Spearing (1975). Spearing indicated that sand transport was to the south with the source probably somewhere to the north-northwest. Although specific evidence is lacking, it may be possible that a delta system fed large amounts of sand into the Western Powder River Basin at a location in southern Montana (Figure 45). In the Billings, Montana area thick shoreline sandstones have been observed, and erosion of these may have served as a source of sediment for the Shannon. The shoreline sandstones of the Eagle Sandstone in the Billings area are described by Shelton (1965).

By integration of the vertical sequence relationships observed in measured sections with the lateral variations observed in surface and subsurface stratigraphic cross sections, a three-dimensional model for Shannon sand-ridge complexes was developed. Similarities between the Shannon Sandstone model and the model developed by Porter (1976) for the Upper Cretaceous Hygiene Sandstone in Colorado are relatively great. The three-dimensional Shannon shelf-ridge model is shown in Figure 47. The processes of deposition are described as follows.

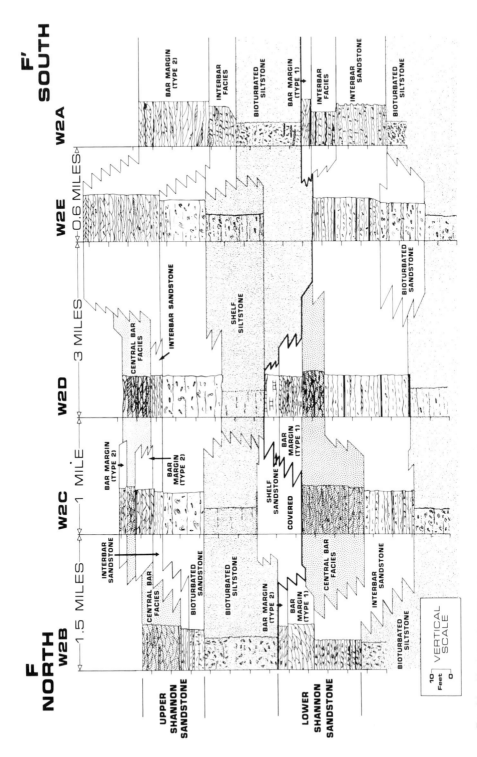

FIG. 36.—North-South stratigraphic cross-section of outcrops along section F-F'. Location (Figure 3) is just west of subsurface cross-section D-D' (Figure 33). *Central Bar Facies* are absent in the southerly measured sections where *Bar Margin Facies* are the highest energy facies. Vertical sequences of *Interbar Sandstone Facies* to *Central Bar Facies* are prominent. Detailed descriptions of measured sections W2A and W2B are given in Figures 10A, B.

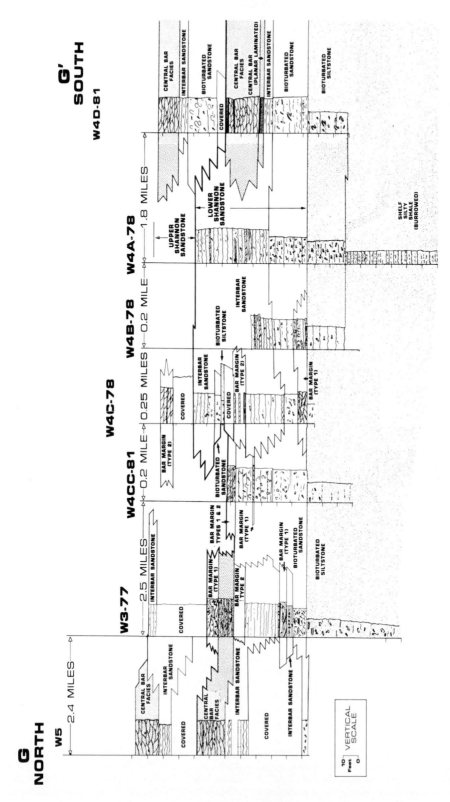

FIG. 37.—North-south stratigraphic cross section of outcrops along G-G' on west-facing escarpment west of Salt Creek Anticline. Location is in Figure 3. The upper and lower Shannon sandstones are designated. Rapid lateral facies changes are the rule. Detailed descriptions of measured sections W3 and W4A are Figures 11 and 13.

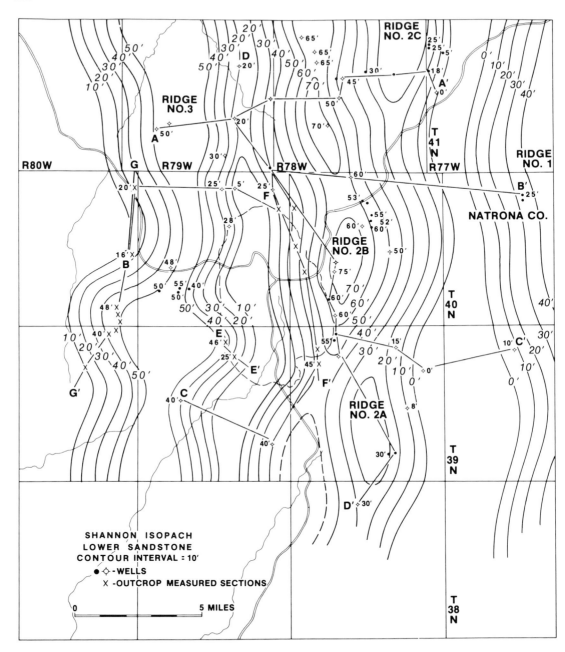

FIG. 38.—Lower Shannon sandstone isopach map, Salt Creek area, Natrona County, Wyoming. Note grain of map is north-south. Ridges are numbered from right to left. Thin areas between ridges are designated as swales. The maximum thickness of sandstone encountered in the lower Shannon interval is 75 feet. All control points and interpreted thicknesses are shown.

The bases of Shannon shelf-ridge complexes form gently sloping surfaces at the top of the Shelf Silty Shale Facies. Coarser-grained deposits built upward from that surface. The highest energy depositional currents produced Central Bar and Bar Margin (Type 1) Facies. Prior to and possibly concurrent with deposi-tion of these high energy sand facies, slightly less com-petent currents flowed between the bars, producing the rippled sandstones and interbedded siltstones and shales of the Interbar Sandstone and Interbar Facies. These less competent currents may also have deposited sediments on the crests of the ridges which were later

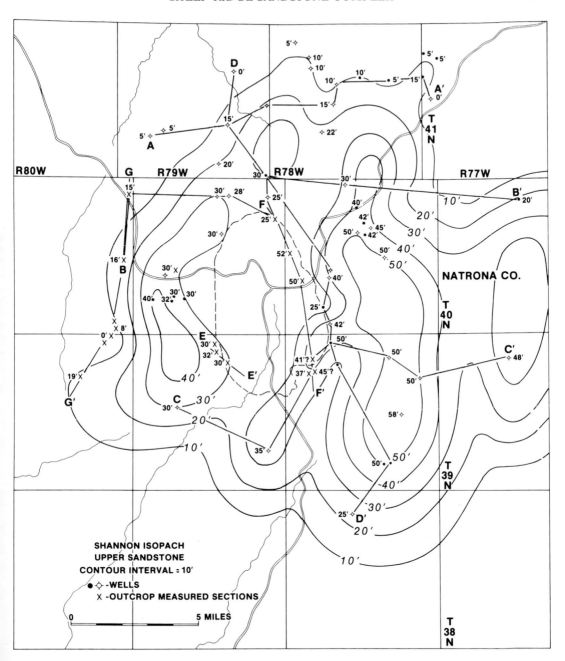

FIG. 39.—Upper Shannon sandstone isopach map, Salt Creek area, Natrona County, Wyoming. Note that the "grain" is much more subdued than in the lower Shannon sandstone. Only in some areas where the upper Shannon sandstone is 40 feet or more thick is there locally a suggestion of a NNW-SSE lineation. A more oblate pattern for the sandstone as a whole and a more oblate pattern for individual ridges is observed here than in the lower sandstone.

eroded. Short- and long-term fluctuations of nearly unidirectional currents active in the area created an interfingering of each of these facies. During periods of very slow deposition, shale was deposited as drapes over the whole shelf-ridge complex. In the area of the shelf-ridge complex, during periods of slow deposition, burrowing organisms were able to move in and establish themselves well enough so as to thoroughly burrow portions of the previously deposited facies producing the Bioturbated Shelf Siltstone Facies. Intro-

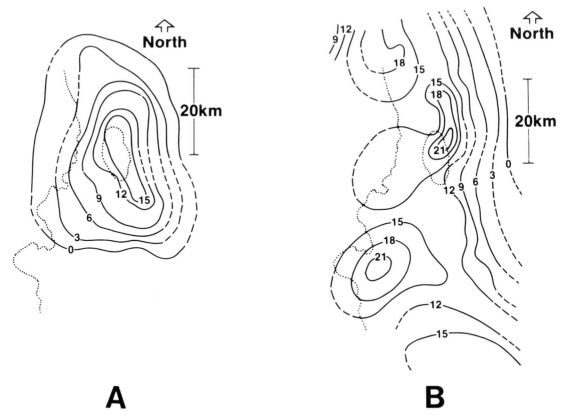

A   **B**

Fig. 40.—*A,* Isopach map of upper Shannon sandstone. Area of eroded Shannon is indicated by dotted line and is centered over the Salt Creek Anticline; the Shannon has been removed to the west. Isopach thicknesses are in meters (Spearing, 1976). The north-south dimension of the map is 31 miles. The contrast with the lower Shannon configuration (Fig. 40B) is considerable. *B,* Isopach map of lower Shannon sandstone. Area of erosion of Shannon is indicated by dotted lines. Isopach thicknesses are in meters (Spearing, 1976). The contrast with the upper Shannon configuration (Fig. 40A) is considerable.

duction of strong unidirectional currents into the area, after these periods of relatively slow deposition and burrowing, resulted in the ripping-up of thin shale and siderite beds and redeposition of shale and siderite rip-up clasts within the subsequently deposited or, in some areas, contemporaneously deposited Bar Margin Facies. Central Bar Facies were deposited during periods of high current flow and abundant sand supply.

Although the Shannon sandstones at Salt Creek Field are very similar to the Shannon sandstones in the greater Hartzog Draw Field area, and the same basic processes are responsible for deposition in both areas (Martinsen and Tillman, 1978, 1979; Tillman and Martinsen, 1979), there are some important differences. The Shannon interval at Salt Creek Field is considerably more sandy overall than the Shannon interval in the GHDF area, and distinct inter- and intrabedded shaley siltstones and shales are absent. The closer spacing and more oblate form of the upper Shannon ridges at Salt Creek Field appear to be the result in part of lateral coalescence of ridges.

It is probable that controls on sand deposition were different at Salt Creek than in the GHDF area. We believe these differences are due to the fact that a paleo-high existed in the Salt Creek Field area during Shannon time, which acted as a focus for sand deposition. Furthermore, this paleo-high was actively growing during Shannon time, as evidenced by the presence of stacked bars and by the change in geometry from the lower sandstone to the upper sandstone. During upper Shannon sandstone deposition this high was significant enough to be the dominant control on sand deposition, and overrode the "normal" shelf depositional processes which tend to form separate linear sand bodies on surfaces with less relief and with a smaller length to width ratio. At no time, however, did the paleo Salt Creek structure cause the Shannon sands to be subaerially exposed. It is suggested that the deposition of a second (upper Shannnon sandstone) ridge complex directly over the lower ridge complex is related to recurrent structural movement immediately prior to the deposition of the upper sandstone. The Salt Creek

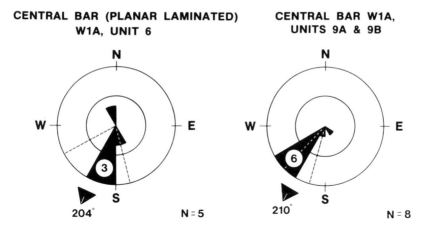

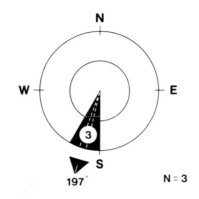

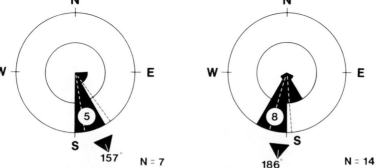

FIG. 41.—Directional data for Shannon outcrops in measured sections W1A, W1B, and W2A. Facies type and unit number are indicated for each data set. More than 90% of the directional readings are from trough cross-beds that were observable in three dimensions. N = the number of readings obtained from individual units. The mean transport direction is indicated by an arrow. One standard deviation from the mean transport direction is indicated by dashed radiating lines. The number in the white circle is the maximum number of data points included within the 15° arc which is the longest radius of the rose diagram.

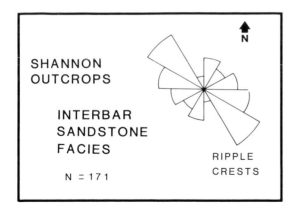

SHANNON
OUTCROPS

INTERBAR
SANDSTONE
FACIES

N = 171

RIPPLE
CRESTS

FIG. 42.—Summary of directional data of ripples from Shannon outcrops. Measured by Spearing (1976). Orientation is for ripple crests; flow direction perpendicular to ripple crests.

structure, which has a closure of 1500 feet in the prolific Second Frontier Sandstone, is one of the most prominent productive anticlines in Wyoming. Successive periods of local structural movement such as suggested by Weimer (1983) for portions of the Cretaceous interior seaways are believed to have formed local positive topography in the Salt Creek area localizing the oblately distributed upper Shannon sandstone over the anticline. The occurrence of stacked Shannon shelf ridge complexes is unusual and may occur only in areas of successive local uplifts of the sea bottom.

Recurrent movement on Laramide structures has been well documented in other areas of the Rocky Mountains. A direct analogy can be made by Reynolds' (1976) study of the Lost Soldier Anticline area. Reynolds studied depositional patterns in relation to timing of the formation of the Lost Soldier anticline, and concluded that the anticline was a positive area throughout the Late Cretaceous. Specifically, he inter-

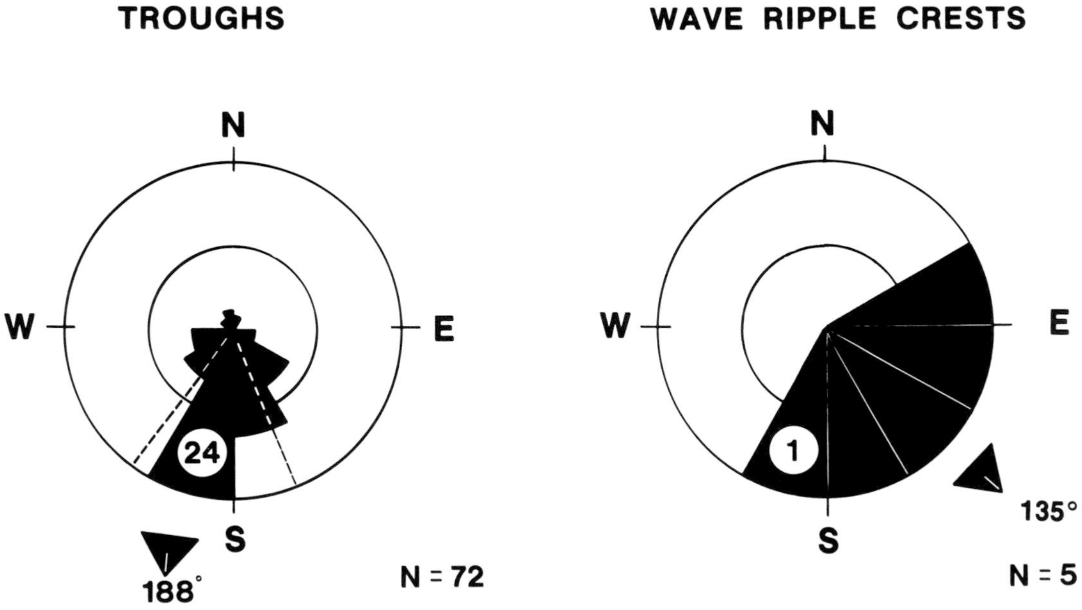

TROUGHS

N

W — — E

(24)

S

▼
188°

N = 72

WAVE RIPPLE CRESTS

N

W — — E

(1)

S

▲
135°

N = 5

FIG. 43.—Summary of directional data for Shannon measured sections W1A, W1B, and W2A. Directional data are subdivided into transport directions determined from troughs and wave-ripple crests. The mean transport direction is indicated by an arrow. The number in the white circle is the maximum number of data points represented by the longest radius of the rose diagram. The mean transport direction of the troughs is about 40° west of the general trend of the well-documented Shannon and Sussex bars (N32°W) in the Powder River Basin (Tillman and Martinsen, in press; Berg, 1975). The average orientation of crests of the symmetrical (wave) ripples indicates probable transport to the southwest perpendicular to their crest direction. Similar directions of transport, 188° and 225° respectively, are suggested by the troughs and wave-ripple crests. One standard deviation from the mean transport direction is indicated by dashed radiating lines.

A

B

FIG. 44.—*Baculites;* Upper Cretaceous time markers. *A*, An assortment of *Baculites* (ribbed and smooth?) collected between the Shannon and Sussex Sandstones at measured section W2C-81. Note slight crenulation on uppermost edge of fossil in upper left (*Baculites*, ribbed). Borings (branching) parallel to edge of shell are seen in *Baculites* on lower right. *B, Baculites* (ribbed) time marker commonly utilized for accurate zonation in the Upper Cretaceous of the Rocky Mountains. This is from shales just below Sussex Sandstone at top of measured section W2C-81, Johnson County, Wyoming.

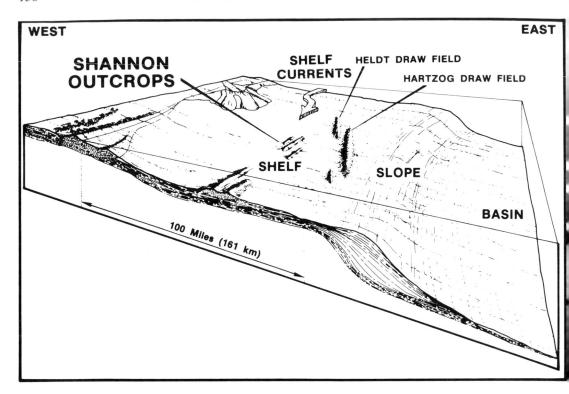

FIG. 45.—Shelf model for the Shannon Sandstone on the western flank of the Powder River Basin. NNW-SSE trending shelf-ridge sandstones of the Shannon are superimposed on the model developed by Asquith (1970). South-flowing shelf currents are believed to have redistributed sands supplied from a large shore line complex in southern Montana.

preted anomalously thick deposits of Sussex and Shan-non sandstones as indicators of the existence of that paleo-high during their deposition. Unlike fluvial sys-tems, which gravitate towards lows, shelf sand deposi-tion may be most prominent on and be enhanced by sea bottom bathymetric highs. The total Shannon sand-stone interval in the Lost Soldier area and in the Salt Creek Field area are relatively thick, especially when compared to the average Shannon Sandstone interval in areas such as the greater Hartzog Draw Field area, which in addition to being thinner contains a greater percentage of interbedded shales. The depositional pat-terns of the Sussex and · Shannon sandstones in the greater Lost Soldier area are oblate and similar to the upper Shannon Sandstones in the Salt Creek Field area.

Exactly how sands were carried eastward from the western shoreline, which presumably included Eagle shoreline and delta sandstones in southern Montana, is a mystery. Exactly what the processes were, in time and space, that caused the currents that moved the sand south is also not known. Storm enhancement of shelf subparallel currents is the most likely mechanism of sand movement. However, the processes involved in moving and depositing the sand must account for the formation of large shelf-ridge complexes containing

linear trending areas of high energy cross-bedded sand-stones as well as oblate areas of coalescing bars. Both the lower energy and the high energy current deposits which form the complexes generally show small ($\pm 30°$) variations in transport directions. The average transport direction in the cross-bedded facies at Salt Creek is SSW. Elongate features forming the thicker portions of the shelf-ridge complex trend nearly north-south at Salt Creek.

Based on subsurface studies (Seeling, 1978; Martinsen and Tillman, 1978; Tillman and Martinsen, 1979, and in press) and the outcrop data presented here, it appears that once begun the shelf ridges build upwards (rather than laterally) in areas from 0.5 to 3 miles wide. Any assignment of processes to the Shan-non shelf ridge complexes must account for a very strong preference for vertical rather than lateral ac-cumulation of thick sections of cross-bedded sandstones.

Tidal currents have been suggested by several au-thors as the mechanism for formation of strong uni-directional bottom currents; however, in the senior authors' experience, micro- or the low end of the meso-tidal range are the highest tides indicated by Upper Cretaceous sandstone outcrops, and offshore subtidal currents generated by tidal changes are interpreted to

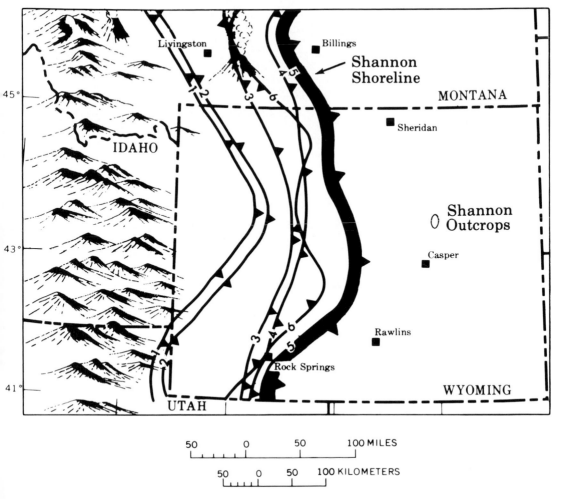

Fig. 46.—Interpreted "shoreline locations" for Lower Campanian-Upper Santonian. Commonly referred to as the Telegraph Creek-Eagle Regression (and minor transgression). Barbs indicate direction of each shoreline movement. Shoreline 1 is 83 mybp (*Desmoscaphites erdamanni* ammonite zone) at the base of the Montana Group (Santonian). Shoreline 5 is *Baculites* sp. (smooth) which includes the Shannon. Shoreline 6 (transgressive) is in *Baculites* (weak flank ribs) and includes the Sussex (Gill and Cobban, 1973).

be of relatively low velocity.

Storm currents, superimposed on other bottom currents, are believed by us to be the major cause of nearly shelf parallel currents which seem to have been responsible for the major shelf-ridge buildups of the Shannon. Swift and Rice (this volume) examine in more detail the various aspects of storm and storm-related current deposition in the interior seaway during the Upper Cretaceous.

Shelf-shoreline comparison.—In any depositional systems analysis, incomplete or limited data from a single outcrop, log, or core may indicate a general marine environment for sandstones like the Shannon. However, with limited data it may be difficult to differentiate shoreline-shoreface deposits from shelf-ridge sandstone deposits. Spearing (1976) compared the

stratification sequences and sedimentary structures of the offshore, shallow marine Shannon Sandstone with those of shoreline deposits. He found some similarities, but also very significant differences. The Shannon is not interpreted to be a shoreline deposit for the following reasons:

"The Shannon intertongues both landward and seaward with marine shales. No evidence is present for any associated fluvial, flood plain, or lagoonal facies, and there are no consistent landward-seaward facies changes as one would expect in a shoreline sequence.

In a shoreline sand, one would expect capping remnants of beach deposits composed of very low-angle, swash-generated beds sloping gently

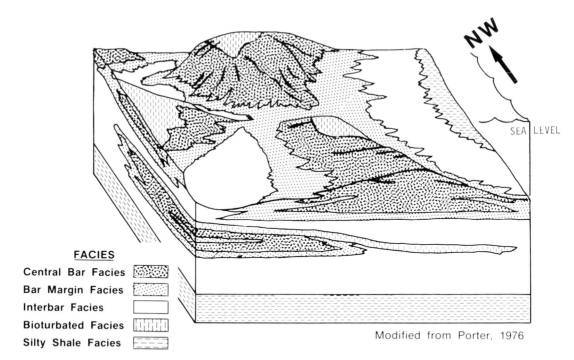

FACIES

Central Bar Facies

Bar Margin Facies

Interbar Facies

Bioturbated Facies

Silty Shale Facies

Modified from Porter, 1976

FIG. 47.—Model of facies distribution of mid-shelf Shannon Sandstone shelf-ridge complex. This diagram has extreme vertical exaggeration, but the abrupt lateral changes indicated here are substantiated by outcrops (Figures 35-37).

seaward. No beach stratification has been observed on the Shannon outcrops, nor could any evidence of emergence, such as root traces or soil zones be found. The (relatively) sharp base and phosphate pebbles beneath the lowermost sand sequence suggest slow sand migration over a sea bottom and not shoreline progradation. Many shoreline progradational units have sharp transgressive bases, but they overlie shallower marine or nonmarine facies.

Shannon stratification closely resembles only the lower parts of a shoreline sequence; but, as both shoreline and offshore marine sand bodies are shoaling sequences, they should have some similar features. It is their differences, sometimes subtle, that are significant in this case.

Suspension clay laminae occur (in) Shannon sand sequences, indicating that alternating traction and suspension deposition persisted (during portions of) Shannon deposition. It is unlikely that suspended clays or clay laminae would remain for long in the energetic wave breaker zone on a shoreline. Indeed, clay laminae are rare in the foreshore or upper shoreface facies of most shoreline sequences. Furthermore, abundant clay clasts of the type found in the cross-bedded facies of the Shannon sequences are not common in shoreline deposits. And where clay clasts do occur on beaches, they are almost always riddled with holes bored by organisms.

No bored clay clasts were observed anywhere in the Shannon. Glauconite is abundant throughout the Shannon, ranging from 1% to (locally more than) 17%; the highest percent galuconite is found in the cross-bedded facies. Soft clay grains such as glauconite are normally removed by attrition in the shoreline setting. Other sand bodies indentified as shoreline deposits in the Cretaceous of the Rocky Mountains generally have very little glauconite."

CONCLUSIONS

The Shannon sandstones in the Powder River Basin of Wyoming were deposited as shelf-ridge complexes on a wide shelf during early Campanian time in the interior seaway stretching from Alaska to Mexico. The shelf-ridge complexes probably were deposited during a period of regression and still stand. The wide shelf model proposed by Asquith (1970), in which the shelf-slope boundary can be recognized, is believed to be applicable for deposition during Shannon time.

The Shannon sandstone shelf-ridge (bar) complexes in the Salt Creek area are composed of a series of interfingering facies with abrupt vertical transitions and surprisingly variable lateral facies. Cross-bedded potential reservoir facies are the Central Bar Facies, Bar Margin Facies (Types 1 and 2), and rare Central Bar (planar laminated) Facies. These facies are differentiated on the basis of percentage of clay clasts, percentage of limonite clasts and lenses, and the percentage of interbedded rippled beds within the cross-bedded units.

Sandstone facies that formed under relatively lower energy conditions include Interbar Sandstone Facies and Bioturbated shelf Sandstones. Rare Interbar Facies (interlaminated shale and sandstone) and Shelf Sandstone Facies are also encountered. The low energy facies form substantially more inferior reservoirs than do the cross-bedded facies; more commonly they are not reservoir facies. Additional facies include Bioturbated Shelf Siltstone, Shelf Siltstone, and, below the base of a few of the outcrop sections, Shelf Silty Shale Facies.

Cross sections constructed using a series of closely spaced outcrop sections indicate lateral variation of facies types are common over relatively short distances. Central Bar Facies are not present in all measured sections; instead in some areas such as the Castle Rocks area (outcrop section W2A), the highest energy facies are Bar Margin Facies (Types 1 and 2). Most commonly Central Bar Facies grade laterally into Bar Margin Facies or Interbar Sandstone Facies. The most common vertical sequence of sandstone facies are coarsening-upward sequences where Central Bar or Bar Margin Facies overlie Interbar Sandstones.

Average grain sizes for sandstone facies are: Bioturbated Shelf Sandstone and Shelf Sandstone 150 microns, Interbar sandstones 175 microns, Bar Margin (Type 2) Facies 200 microns, Bar Margin (Type 1) Facies 225 microns, and Central Bar Facies 250 microns.

Two stacked shelf-ridge complexes are recognized over most of the Salt Creek area. Both sequences contain generally coarsening-upward sequences of facies. Subsurface cross sections and isopach maps indicate that the complexes contain north-south trending "thicks". The mapped "thicks" contain both cross-bedded and ripple-bedded facies; however, the cross-bedded reservoir sandstones (Central Bar and Bar Margin Facies) are difficult to differentiate from ripple-bedded Interbar Sandstones on the older subsurface logs available in the study area. It was not determined whether the increase in total subsurface sand thickness corresponds to a similar increase in the thickness of the cross-bedded sandstone facies.

Evidence from fossils, both foraminifera and trace fossils, suggests that the shelf-ridge (bar) complex was deposited at middle shelf to outer part of the inner shelf *water depths*. The foraminiferal assemblage was dominated by *Haplophragmoides* and consisted largely of agglutinated types of low to medium diversity. *Baculites* ammonite zonation may be used to subdivide the Lower Campanian into approximately half million year intervals (Gill and Cobban, 1973). The presence of *Baculites* (smooth) in outcrops at Salt Creek identifies it as having been deposited about 80.5 million years ago. Bentonite markers in the subsurface generally parallel the *Baculites* zones and are useful in subdividing the Lower Campanian in subsurface cross sections.

Trace fossils are abundant in the Bioturbated Shelf Sandstone Facies and Bioturbated Shelf Siltstone Facies; burrowing in these facies ranges from 75-95% (by volume). The highest average percentage of burrowing in the other facies in the shelf-ridge complex is 27%. The cross-bedded facies average less than 5% burrowing. Identifiable shoreline-type trace fossils were absent.

Excellent directional data, obtained from high-angle trough cross bedding in the Central Bar Facies and Bar Margin Facies, fall within a fairly narrow range of transport directions. The average transport direction is SSW (188°) and two-thirds of the transport directions are within an arc of 60°. Most current ripples indicate transport in a similar direction; however, long crested "symmetrical" ripple transport directions are more variable. A strong preference for vertical accumulation of cross-bedded sandstones is indicated.

A plausible depositional site for the Shannon outcrops at Salt Creek is one in which shelf-ridge complexes were deposited below daily effective wave base at water depths generally between 60 and 300 feet. These ridges were deposited approximately 70-100 miles east of the Shannon age shoreline, as reconstructed by Gill and Cobban (1973), on a shelf that was approximately 100 miles wide. Sediment transport in the Salt Creek area was accomplished predominantly by unidirectional southwest flowing shelf currents. Periodically storms intensified current velocities such that medium-grained sand was sorted out from the bulk of shelf sediment and was transported and deposited in the form of mega-ripples and sand waves. Stacking of these mega-ripples and sand waves developed north-south trending sand ridges up to 75-feet thick. A paleo-high existed, and was actively growing, during Shannon time in the Salt Creek Field area and acted as a focus for ridge formation. The Salt Creek high never achieved enough relief so as to subaerially expose the Shannon sands, but was significant enough to influence ridge geometry such that the sandstones ridges at Salt Creek are more oblate and less linear than Shannon sandstone ridges in other areas of Wyoming.

ACKNOWLEDGEMENTS

Field work for this study was supported by Cities Service and was done mainly by Tillman and Martinsen. Assisting in measuring and describing sections were Dick Scott, Roger Wolff, Jim Rine, Doug Jordan, Tom Moslow, J.J. Morris, Dave Watson, and Ann Kramer.

Don Dailey of Cities Service Research analyzed the foraminifera and was responsible for interpretations of the micropaleontologic data. W. A. Cobban aided with discussions of and identification of *Baculites* sp.

Dar Spearing was very helpful in the initial stages of the study in providing outcrop locations. Roger Slatt,

Jim Ebanks, J.J. Morris and Mary Walker aided in editing. Drafting was provided by Cities Service.

Melinda Everley and Mary Walker typed the manuscript.

REFERENCES

ASQUITH, D. O., 1970, Depositional topography and major marine environments, Late Cretaceous, Wyoming: Am. Assoc. Petroleum Geologists Bull., v. 54, p. 1184–122 4.

_____, 1974, Sedimentary models, cycles, and deltas, Upper Cretaceous, Wyoming: Am. Assoc. Petroleum Geologists Bull., v. 58, p. 2274–2283.

BAKER, F. E., 1957, History of Salt Creek Oil Field: Wyoming Geol. Assoc., Wyoming Oil and Gas Fields, p. 388–390.

BARLOW, J. A., AND J. D. HAUN, 1966, Regional stratigraphy of Frontier Formation and relation to Salt Creek Field, Wyoming: Am. Assoc. Petroleum Geologists Bull., v. 50, p. 2185–2196.

BARRATT, J. C., AND A. J. SCOTT, 1982, Phayles Sandstone (Upper Cretaceous) deltaic and shelf bar complex, central Wyoming [Abs.]: Am. Assoc. Petroleum Geologists Bull., v. 66, p. 545.

BERG, R. R., 1975, Depositional environments of the Upper Cretaceous Sussex Sandstone, House Creek Field, Wyoming: Am. Assoc. Petroleum Geologists Bull., v. 59, p. 2099–2110.

BRENNER, R. L., 1979, A sedimentologic analysis of the Sussex Sandstone, Powder River Basin, Wyoming: Wyoming Geol. Assoc. Earth Sci. Bull., v. 12, p. 37–47.

COSTELLO, W. R., AND J. B. SOUTHARD, 1981, Flume experiments on lower–flow regime bed forms in coarse sand: Jour. Sed. Petrology, v. 51, p. 849–864.

CREWS, G. C., J. A. BARLOW, JR., AND J. D. HAUN, 1976, Upper Cretaceous Gammon, Shannon and Sussex Sandstones, central Powder River Basin, Wyoming: Wyoming Geol. Assoc., Guidebook 28th Ann. Field Conf., p. 9–20.

DAVIS, M. J. T., 1976, An environmental interpretation of the Upper Cretaceous Shannon Sandstone, Heldt Draw Field, Wyoming: Wyoming Geol. Assoc., Guidebook 28th Ann. Field Conf., p. 125–138.

ELDRIDGE, G. H., 1889, Some suggestions upon the methods of grouping the formations of the middle Cretaceous and the employment of an additional term in the nomenclature: Am. Jour. Sci., v. 38, p. 313–321.

GILL, J. R., AND W. A. COBBAN, 1966, The Red Bird section of the Upper Cretaceous Pierre Shale in Wyoming: U.S. Geol. Surv. Prof. Paper 393–A, 73 p.

_____ AND _____, 1973, Stratigraphy and geologic history of the Montana Group and equivalent rocks, Montana, Wyoming, and North Dakota: U.S. Geol. Surv. Prof. Paper 776, 37 p.

MARTINSEN, R. S., AND R. W. TILLMAN, 1978, Hartzog Draw, new giant oil field [Abs.]: Am. Assoc. Petroleum Geologists Bull., v. 62, p. 540.

_____ AND _____, 1979, Facies and reservoir characteristics of shelf sandstones, Hartzog Draw Field, Powder River Basin, Wyoming [Abs.]: Am. Assoc. Petroleum Geologists Bull., v. 63, p. 491.

MEIJER DREES, N. C., AND D. W. MHYR, 1981, The Upper Cretaceous Milk River and Lea Park Formations in southeastern Alberta: Bull. Canadian Petroleum Geol., v. 29, p. 42–74.

PARKER, J. M., 1958, Stratigraphy of the Shannon Member of the Eagle Formation and its relationship to other units in the Montana Group of the Powder River Basin, Wyoming and Montana: Wyoming Geol. Assoc., Guidebook 13th Ann. Field Conf., p. 90–102.

PORTER, K. W., 1958, Marine shelf model, Hygiene Member of the Pierre Shale, Upper Cretaceous Denver Basin, Colorado, in R. Epis and R. Weimer (eds.), Studies of Colorado Field Geology: Prof. Contrib. Colorado School of Mines, p. 251–263.

REYNOLDS, M. W., 1976, Influence of recurrent Laramide structural growth on sedimentation and petroleum accumulation, Lost Soldier area, Wyoming: Am. Assoc. Petroleum Geologists Bull., v. 60, p. 12–33.

RICE, D. D., AND W. A. COBBAN, 1977, Cretaceous stratigraphy of the Glacier National Park area, northwestern Montana: Bull. Canadian Petroleum Geol., v. 25, p. 828–841.

SEELING, A., 1978, The Shannon Sandstone, a further look at the environment of deposition at Heldt Draw Field, Wyoming: The Mountain Geologist, v. 15, p. 133–144.

SHELTON, J. W., 1965, Trend and genesis of lowermost sandstone unit of Eagle Sandstone at Billings, Montana: Am. Assoc. Petroleum Geologists Bull., v. 49, p. 1385–1397.

SPEARING, D. R., 1975, Shannon Sandstone, Wyoming, in Depositional Environments as Interpreted from Primary Sedimentary Structures and Stratification Sequences: Soc. Econ. Paleontologists and Mineralogists Short Course No. 2, p. 104—114.

_____, 1976, Upper Cretaceous Shannon Sandstone: An offshore, shallow marine sand body: Wyoming Geol. Assoc., Guidebook 28th Ann. Field Conf., p. 65–72.

TILLMAN, R. W., AND R. S. MARTINSEN, 1979, Hartzog Draw Field, Powder River Basin, Wyoming, in R. W. Flory (ed.), Rocky Mountain High: Wyoming Geol. Assoc., 28th Ann. Meeting, Core Seminar Core Book, p. 1–38.

WEIMER, R. J., 1983, Relation of unconformities, tectonics, and sea level changes, Cretaceous of the Denver Basin and adjacent areas, in M.W. Reynolds and E.D. Dolly (eds.), Mesozoic Paleogeography of the West–Central United States: Rocky Mountain Sec. Soc. Econ. Paleontologists and Mineralogists, Rocky Mountain Paleogeog. Symp. 2, p. 359–376.

WILMARTH, M. G., 1938, Lexicon of geologic names of the United States (including Alaska): U.S. Geol. Survey Bull. 896, 2396 p.

WIDESPREAD, SHALLOW-MARINE, STORM-GENERATED SANDSTONE UNITS IN THE UPPER CRETACEOUS MOSBY SANDSTONE, CENTRAL MONTANA

DUDLEY D. RICE
U.S. Geological Survey, Denver, Colorado 80225

ABSTRACT

The Upper Cretaceous (Cenomanian) Mosby Sandstone Member of the Belle Fourche Shale and correlative units form a large (greater than 300,000 sq km) lobe of sandstones that extend southeast from the Dunvegan delta in northeastern British Columbia and northwestern Alberta. The sandstones occupy the shallow, flat, western shelf of a north-south trending epicontinental seaway in Alberta, southwestern Saskatchewan, and central Montana.

Sandstone bodies are concentrated on ancestral highs such as Bow Island Arch in southeastern Alberta, Bowdoin Dome in north-central Montana, and Central Montana Uplift that were episodically uplifted and that affected sedimentation. The sandstone bodies contain coarsening-upward cycles that indicate shoaling. The bodies occur as elongate northwest-southwest trending sand ridges that are tens of kilometers long. On the Central Montana Uplift, the postulated sand ridges have been redistributed around the shoals into still smaller bodies that are a few to several kilometers long.

On the Central Montana Uplift, the Mosby is composed of thin (average 1.5 m) very fine-grained sandstone units that consist of individual beds intercalated with shale or amalgamated. The base of each bed is a planar to undulating erosional surface, and the top is gradational with the overlying shale. The dominant sedimentary structure is hummocky cross-stratification. The upward sequence of sedimentary structures suggests decreasing energy levels. Ripple cross-lamination and wave ripples are commonly developed near the top of beds or units where the beds are amalgamated. Trace fossils are restricted to the rippled surfaces and consist mainly of horizontal *Ophiomorpha* and *Thalassinoides* burrows. Unoriented shells, including gastropods, bivalves, and ammonites, are concentrated in lenses or concretions near the base of units and where the units grade laterally into shale.

Sediments were transported by the interaction of waves and southward-flowing, geostrophic currents enhanced by wind forcing, as much as 1,100 km, from the Dunvegan delta and the sand was concentrated on positive features as coarsening-upward cycles. Regional and local characteristics suggest that the sediments forming the upper part of coarsening-upward cycles were subsequently redistributed by storm flows on the Central Montana Uplift. The sediments were suspended on the shoals and the bottom was irregularly scoured by passing storm flows. As the currents decelerated, the sediments were quickly dropped parallel to the undulating surface producing hummocky cross-stratification. Shells were concentrated by pressure gradients created by passing wave surges. The upward sequence of sedimentary structures within a single bed suggests deposition during the waning stages of a single storm event. Reworking of sediment by oscillatory waves, burrowing along rippled surfaces, and deposition of interbeds of finer-grained silt and mud indicate the return to fair-weather conditions. The morphology and lateral facies of individual bodies are controlled by bottom topography and by water depth.

INTRODUCTION

The Upper Cretacious Mosby Sandstone Member of the Belle Fourche Shale occurs within a dominantly fine-grained sequence in central Montana. The Mosby is interpreted to have been deposited on an open marine shelf during episodes of progradation. Other examples of shelf sandstones in the northern Great Plains are the Medicine Hat Sandstone of Alberta and Shannon Sandstone Member of the Gammon Shale in eastern Montana (Fig. 1).

Shelf sandstones reported in previous studies are as much as 20 m thick, are enclosed in marine shale, and can be demonstrated to have been deposited some distance from the shoreline. Most of the shelf sandstones form coarsening-upward sequences indicative of shoaling and or upward increase in energy level, occur as elongate bodies, and have directional features that indicate southward transport. Characteristics and interpretation of depositional regimes for this type of shelf sandstone are given by Walker (1979) and Brenner (1980). In contrast, the Mosby Sandstone Member in central Montana consists of thin (average 1.5 m) sand-

stone units that have sharp bases and gradational tops. The units commonly display sedimentary features that indicate an overall upward decrease in energy level. The purpose of this paper is to describe these sandstone units and to interpret their origin in terms of processes operating in shallow marine environments.

The Mosby was named as a member of the Colorado Shale by Lupton and Lee (1921) for exposures near the Mosby Post Office on the Musselshell River in central Montana (Fig. 2). Reeves (1927) used the Mosby as a valuable marker in mapping the geology of the Cat Creek and Devil's Basin oil fields and adjacent areas. Cobban (1952) formally established the type section in the area in which it was named by Lupton and Lee (1921).

The Mosby Sandstone Member is extensively exposed and was studied in detail along more than 30 km of outcrop exposed on subsidiary structures of the Central Montana Uplift, namely the Cat Creek, Flatwillow, and Pike Creek anticlines, and Devil's Basin. These structures and part of studied outcrops are shown and discussed in Johnson and Smith (1964).

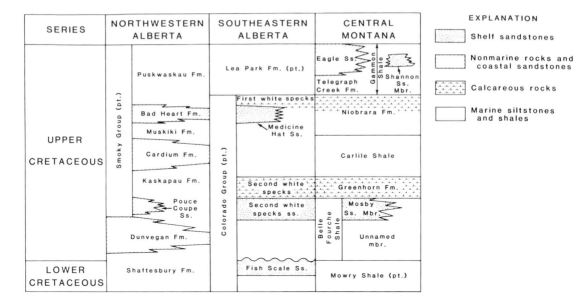

SERIES	NORTHWESTERN ALBERTA	SOUTHEASTERN ALBERTA	CENTRAL MONTANA
UPPER CRETACEOUS	Smoky Group (pt.) — Puskwaskau Fm.	Lea Park Fm. (pt.)	Eagle Ss — Telegraph Creek Fm. / Gammon Shale / Shannon Ss. Mbr.
	Bad Heart Fm.	First white specks — Medicine Hat Ss.	Niobrara Fm.
	Muskiki Fm.		Carlile Shale
	Cardium Fm.	Colorado Group (pt.)	
	Kaskapau Fm.	Second white specks	Greenhorn Fm.
	Pouce Coupe Ss.	Second white specks ss.	Mosby Ss. Mbr.
	Dunvegan Fm.		Belle Fourche Shale / Unnamed mbr.
LOWER CRETACEOUS	Shaftesbury Fm.	Fish Scale Ss.	Mowry Shale (pt.)

EXPLANATION

- Shelf sandstones
- Nonmarine rocks and coastal sandstones
- Calcareous rocks
- Marine siltstones and shales

FIG. 1.—Correlation chart of selected Cretaceous rocks in central Montana and Alberta. Northwestern Alberta column modified from Williams and Burk (1964).

The Mosby was also examined at Porcupine Dome, a subsidiary structure of the Central Montana Uplift, and the Highwood, Little Rocky, and North Moccasin Mountains, which are Tertiary intrusive features. The general location of studied outcrops is marked on Figure 2, but the lateral extent of the outcrops is not shown.

The Central Montana Uplift represents an area of complex structural and sedimentation events during much of Paleozoic and Mesozoic time (Norwood, 1965). During the 'Laramide orogeny in Late Cretaceous and Paleocene time, the complex was uplifted to its final configuration and has many subsidiary fields. The entire Central Montana Uplift and local areas of Tertiary igneous activity may have had precursor elements that occupied the same relative positions and that affected earlier Cretaceous sedimentation.

STRATIGRAPHY

The Mosby Sandstone Member is in the upper part of the Belle Fourche Shale (Fig. 1). The Belle Fourche is about 90 m thick in the study area and is composed of dark-gray silty to sandy shale, in part bentonitic, of Cenomanian Age. The lower 18 m contain abundant ironstone concretions. The rest of the formation below the Mosby includes numerous laminae and lenses of siltstone and very fine-grained sandstone that locally are a few meters thick. The Mosby consists of one or more ledge-forming sandstone units, as much as 2.7 m thick, that are interbedded with, and are overlain by, noncalcareous shale that has lenses of siltstone and sandstone. The lower contact of Mosby is placed at the base of the lowest sandstone unit. Invertebrate marine

fossils are abundant in the Mosby and are representative of the late Cenomanian *Dunveganoceras albertense* zone (Cobban, 1952).

The Mosby Sandstone Member is sharply, but conformably, overlain by the Greenhorn Formation of Cenomanian and Turonian Age (Fig. 1). The Greenhorn is made up of dark-gray calcareous shale that weathers very light gray, and maintains a thickness of about 8 to 9 m throughout the study area. The basal contact of the Greenhorn is an abrupt change from noncalcareous to calcareous shale and is marked by a thin, persistent bed of septarian limestone concretions. About 1 m above the base is a widespread bentonite bed about 1 m thick. This bentonite bed can be correlated northwestward across the Sweetgrass Arch to Glacier National Park (Cobban, 1951). The limestone concretions and the bentonite bed make excellent markers for measuring sections.

The Mosby Sandstone Member is equivalent to the shallow, gas-bearing Phillips sandstone, an informal subsurface unit, in the Bowdoin Dome area of north-central Montana. On the Bow Island Arch in southeastern Alberta, the Mosby correlates with the Second white specks sandstone that is overlain by the Second white specks zone (Fig. 1). The Second white specks zone consists of a calcareous shale that is speckled with white fecal pellets (Hattin, 1975) and is equivalent to the Greenhorn Formation.

GEOLOGIC SETTING

Fine-grained rocks of early Late Cretaceous age were deposited on the western shelf of a north-south trending epicontinental seaway that extended from the

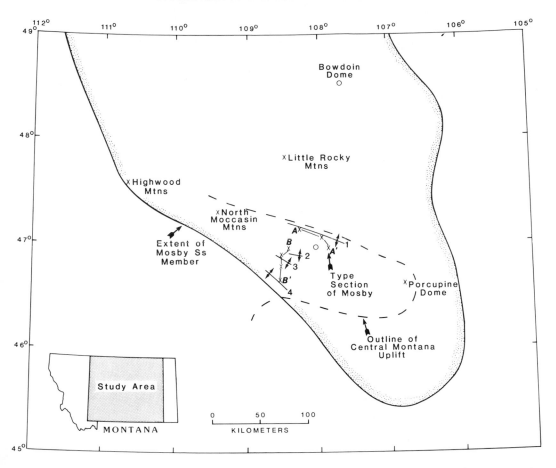

FIG. 2.—Map of study area in Montana showing the lateral extent of Mosby, location of outcrops (x), lines of cross section, wells with electric logs (o), and structural features (1, Cat Creek anticline; 2, Flat Willow anticline; 3, Pike Creek anticline; 4, Devil's Basin).

Arctic Ocean to the Gulf of Mexico during most of Late Cretaceous time (Williams and Stelck, 1975). The Mosby and correlative units form an isolated south-ward-projecting lobe of shelf sandstones in central Montana, Alberta, and southwestern Saskatchewan (Fig. 3). The Mosby grades laterally in three directions (west, east, and south) into offshore marine siltstones and shales. The western and eastern shorelines at this time can only be approximated because rocks repre-senting coastal facies are not preserved.

On the Sweetgrass Arch and the Disturbed belt to the west, the uppermost part of the Belle Fourche Shale, including the Mosby Sandstone Member, is rep-resented by the Floweree Member of the Marias River Shale (Mudge, 1972; Cobban et al., 1976; Rice and Cobban, 1977) (Fig. 4). The Floweree attains a thick-ness as much as 18 m and is composed of sandy shale that has thin beds (less than 30 cm) of mostly siltstone and some sandstone, and layers of chert granules and pebbles. The Floweree unconformably overlies Lower Cretaceous rocks, and equivalent strata of the lower

part of the Belle Fourche are missing because of ero-sion and (or) nondeposition. The missing section can probably be explained by paleotectonic movement on the ancestral Sweetgrass Arch (Peterson, 1966).

In the southern Alberta Foothills, the entire Belle Fourche Shale correlates with the upper part of the Sunkay Member of the Blackstone Formation (Stott, 1963). The Sunkay consists of several cycles that grade from shale upward to siltstone and that have local thin layers of coarse sandstone at the top.

The Mosby Sandstone Member grades eastward and southward into noncalcareous shales included in the Belle Fourche and Cody Shales. Further to the east, the noncalcareous shales pass into calcareous shales in-cluded in the Greenhorn Formation (Fig. 3). Where exposed in the Black Hills area, the contact between the Belle Fourche and Greenhorn is difficult to deter-mine because of the interfingering of noncalcareous and calcareous shales (Robinson et. al., 1964). How-ever, rocks assigned to the *Dunveganoceras albertense* zone are generally calcareous shales of the Greenhorn

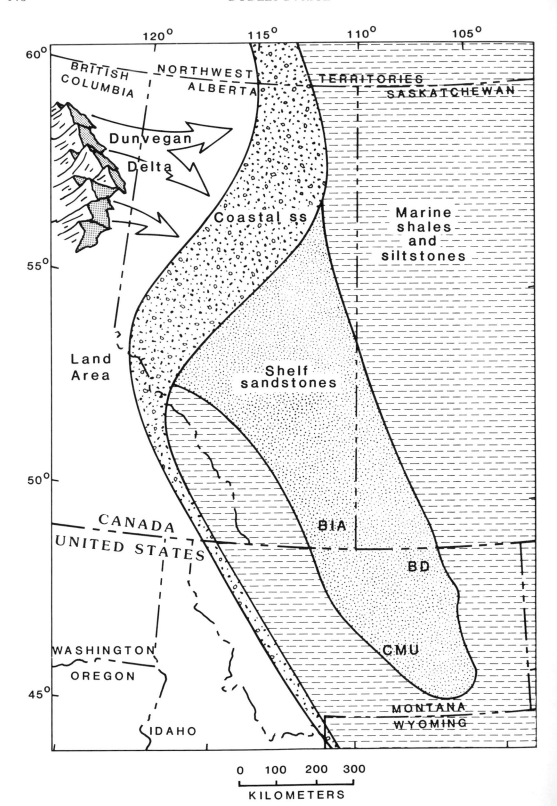

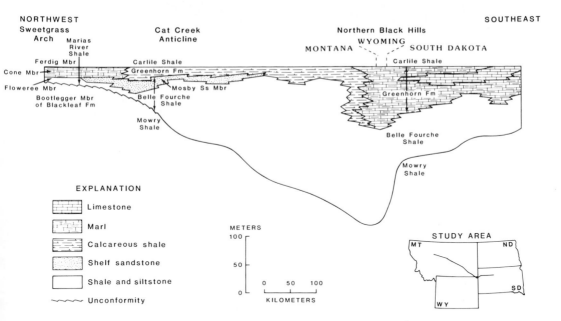

Fig. 4.—Diagrammatic cross section of lower Upper Cretaceous rocks from Sweetgrass Arch, northwestern Montana to central South Dakota.

Formation (Merewether *et al.*, 1976). The lateral change in facies from noncalcareous to calcareous shales is probably controlled by the absence of clastic input from the east.

Shelf siltstones and sandstones correlative with the upper part of the Belle Fourche Shale appear to be laterally continuous with coastal sandstones assigned to the Dunvegan Formation in northern Alberta (Fig. 3). A sandstone tongue of the main body of the Dunvegan Formation mapped with the Kaskapau Formation, namely the Pouce Coupe Sandstone, is the time equivalent of the Mosby Sandstone Member in central Montana (Warren and Stelck, 1940; Stelck and Wall, 1954) (Fig. 1). The Dunvegan attains a maximum thickness of more than 300 m and occupies a large part of the Cenomanian section (Belle Fourche equivalent) in northeastern British Columbia and northwestern Alberta (Fig. 1). The Dunvegan consists of a series of informal sandstone members that are interbedded with siltstone, shale, and coal. The number and thickness of the individual sandstone members decrease in a radial pattern to the west, south, and east in northern Alberta (Williams and Burk, 1964). The sandstones grade laterally into siltstones, shales, and minor sandstones assigned to the Blackstone Formation.

The Mosby Sandstone Member and equivalent rocks are overlain by calcareous rocks assigned to the Green- horn Formation in central and eastern Montana, the Cone Member of the Marias River Shale in western Montana, and equivalent units in other areas (Figs. 1 and 4). Calcareous rocks were deposited over a large part of the Western Interior seaway during late Cenomanian and early Turonian time and mark a sudden change in deposition from coarser grained, noncalcareous rocks to fine-grained calcareous rocks.

Although siltstones and sandstones were deposited over the entire area of the southward-projecting lobe, the cleanest and coarsest grained sediments appear to be concentrated on paleotectonic highs such as the Bow Island Arch in southeastern Alberta, Bowdoin Dome in north-central Montana, and the Central Montana Uplift. On the Bowdoin Dome, gamma-ray, density, and resistivity logs indicate that three coarsening-upward cycles make up the Mosby Sandstone Member in this area (Nydegger *et al.*, 1980) (Fig. 5). The total Mosby is as much as 60 m thick, whereas the individual cycles can be as much as 25 m thick. Although cores are not available, interpretation of logs and cuttings indicate that the individual cycles grade upward from shale to very fine grained sandstone. The coarsening-upward cycles appear to become cleaner and to have better reservoir properties in an upward direction.

On the Bow Island Arch, the Second white specks sandstone is interpreted to have been deposited as a

Fig. 3.—Paleogeographic map of northwestern United States and southwestern Canada during deposition of Mosby Sandstone Member. Dunnegan delta area is modified from Williams and Burk (1964). BIA, Bow Island Arch; BD, Bowdoin Dome; CMU, Central Montana Uplift.

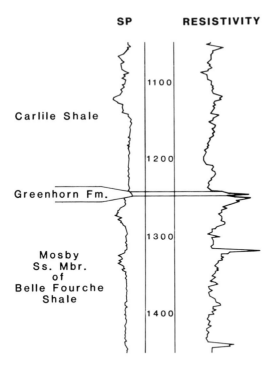

SP   RESISTIVITY

Carlile Shale

Greenhorn Fm.

Mosby
Ss. Mbr.
of
Belle Fourche
Shale

FIG. 5.—Electric log of lower Upper Cretaceous rocks on Bowdoin Dome. Location of well shown on Figure 2 (Sec. 22, T. 32 N., R. 31 E). Depth shown in feet below kelly bushing (1 ft = 0.3 m).

series of elongate offshore bars (Suffield Block Study Committee, 1972). The porous sandstone in the central part of the bars is as much as 6 m thick and forms northwest-southeast trending features that are about 55 km wide and 80 km long (Last, Kloepfer Ltd., 1974).

DESCRIPTION

The Mosby Sandstone Member consists of one or more very fine grained sandstone units that are interbedded with noncalcareous shale. Sandstone units are hereby defined as intervals composed predominantly (generally composing more than 75 percent of the total unit) of sandstone beds that form conspicuous ledges in an otherwise finer grained sequence.

Interbedded Rocks

The rocks immediately below, between, and above sandstone units of the Mosby are mainly noncalcareous, silty to sandy shale. Calcareous septarian sandstone and septarian limestone concretions as much as 0.6 m thick and 1.5 m wide are locally abundant. The medium gray to light brownish gray shale contains lenses of light yellowish gray to light olive gray siltstone and very fine-grained sandstone. Lenses are thin, discrete layers that are readily recognizable in the field within a dominantly shale section. The lenses generally increase in number and thickness upward to

the sharp base of the overlying sandstone unit and form a transitional interval that can be as much as 6 m thick (Fig. 6). The lenses are as much as 10 cm thick, thicken and thin laterally, and pinch out over a distance of several tens of meters. The lenses have sharp bases and exhibit parallel to wavy lamination as their primary sedimentary structure. Horizontal, smooth-walled burrows, as much as 1.3 cm in diameter, occur on the bottom surfaces of the sandstone in contact with the underlying shale. The oyster *Exogyra columbella* locally is scattered along bedding planes in the transitional interval in the upper part of the Mosby at Brush Creek Dome, and in the transitional interval below the upper sandstone unit that extends from McDonald Creek to Devil's Basin, below the sandstone unit that is laterally continuous between Mosby Dome and East Dome, and below the upper sandstone unit at East Dome (Figs. 7 and 8).

Sandstone Units

Correlation and lateral continuity.—The sandstone units form discontinuous bodies of rather limited extent that lie more or less at the same stratigraphic position. The bodies grade laterally in every direction into shale. Because of the nature of the outcrops, any particular body can be seen only in a two-dimensional section.

The dimensions of individual bodies composing the upper sandstone unit can be observed in a north-south direction from McDonald Creek to Devil's Basin (Fig. 8). Although outcrops are not continuous for the entire distance, individual bodies are considered to be approximately time-equivalent because they are at about the same distance below the Greenhorn Formation. The bodies along this north-south trend extend for distances of 3 to 5 km before grading laterally at each end into shale. The spacing between individual bodies is 0.5 to 1 km.

Bodies developed in stratigraphically older sandstone units between McDonald Creek and Devil's

FIG. 6.—Transitional interval beneath sandstone unit of Mosby showing upward increasing lenses of siltstone and sandstone, and overlain sharply by sandstone unit at top of hammer. Hammer is 33 cm long.

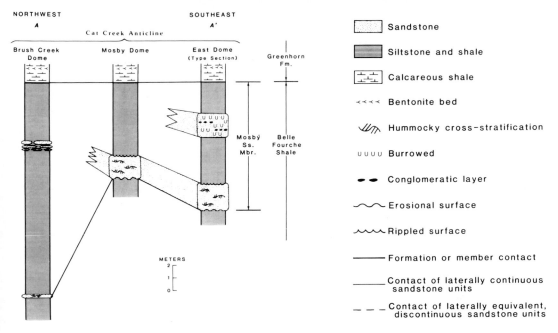

FIG. 7.—Cross section along line A-A' showing sections of Mosby Sandstone Member. Line of section shown on Figure 2.

Basin (Fig. 8) are probably more extensive laterally because they were not observed to grade into shale. Although these older units are at approximately the same stratigraphic position below the Greenhorn Formation, they cannot be correlated between different localities.

The lower sandstone unit exposed at East Dome is laterally continuous with the only sandstone unit developed on Mosby Dome (Fig. 7). The body extends for the entire length of the Mosby outcrop, a distance of 11 km in an east-west direction, and does not exhibit lateral facies changes. Eleven kilometers probably cannot, however, be used as a minimum east-west dimension of the bodies in general. This body, developed on Cat Creek anticline, does not display the vertical or lateral variation in internal sedimentary structures as do bodies that are at the same stratigraphic position to the south.

The upper sandstone unit at East Dome (Fig. 7) is distinctly different from the other units. The unit persists from the eastern limit of the outcrops westward for 5 km before thinning and grading into shale. An equivalent unit was not deposited elsewhere on Cat Creek anticline. Discontinuous lenses of sandstones, as much as 15 cm thick, that have features similar to those developed at East Dome, locally are at the same stratigraphic interval below the Greenhorn Formation at McDonald Creek, Yellow Water Creek, and Devil's Basin (Fig. 8).

Composition.—The sandstone units are made up of individual sandstone beds intercalated with silty to sandy shale (Fig. 9); or are amalgamated with little or no finer grained sediment between them (Fig. 10). The units often contain sandstone and limestone concretions similar to those described for interbedded rocks. The sandstone is very fine grained, yellowish gray, and locally cemented by calcium carbonate. Thin layers of pebble and granule conglomerate, as much as 5 cm thick, are locally present in the upper unit at East Dome and in discontinuous lenses in upper part of Mosby at McDonald Creek, Yellow Water Creek, and Devil's Basin (Figs. 7 and 8). The units range in thickness from 0.3 to 2.7 m (average 1.5 m) and maintain a fairly uniform thickness laterally where composed dominantly of sandstone. Individual beds are typically 0.1 to 1 m thick. Intercalations of finer grained sediment are generally less than a few centimeters thick.

Sedimentary Structures.—The sandstone units are either structureless or display hummocky cross-stratification, parallel lamination, ripple cross-lamination including climbing ripple cross-lamination, ripple marks, and shale intercalations (Figs. 9-16).

Hummocky cross-stratification is the most prevalent type of bedding in the sandstone units, particularly in the beds that are amalgamated (Figs. 10 and 11). An exception is the upper sandstone unit at East Dome in which parallel lamination and shale intercalations are the only observable interval structures. Hummocky cross-stratification is characterized by the following features: (1) the lower surface of individual sets is erosional and undulates to form broad troughs and hummocks that slope to as much as 15°, (2) the laminae within the sets are nearly parallel to the lower erosional surface, and (3) the dip directions of the erosional sur-

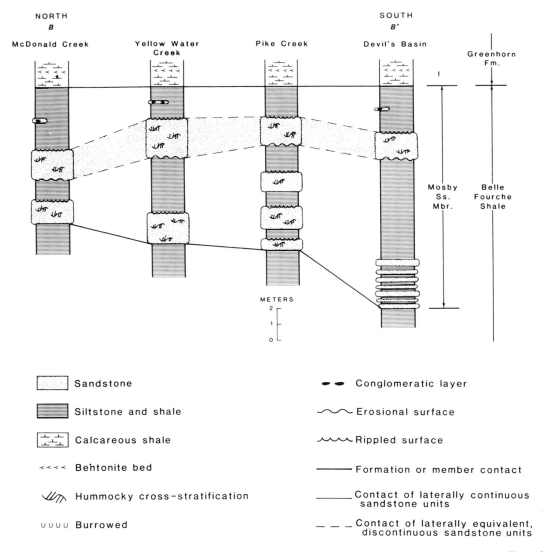

FIG. 8.—Cross section along line B-B' showing sections of Mosby Sandstone Member. Line of section shown on Figure 2.

faces and overlying laminae are not systematic. Wavelengths between troughs and hummocks range from 1 to 5 m and amplitudes are about 13 to 25 cm. The term hummocky cross-stratification was given by Harms and others (1975) to a type of bedding that has not been produced in flumes or in wave tanks, although similar bedding features were described prior to that time by Campbell (1966) and Goldring and Bridges (1973). The term hummocky has been used recently by deRaaf and others (1977), Hamblin and Walker (1979), Bourgeois (1980), and Cant (1980) to describe similar features in other deposits.

Ripple marks are commonly well preserved on the upper surfaces of beds exhibiting ripple cross-lamination. The ripples are usually symmetrical and rounded in profile and have long straight, sometimes bifurcating crests (Fig. 16). Wavelengths are 8 to 10 cm and heights are about 3 mm. There are also some interference ripple marks and asymmetrical ripples that have a mixture of cuspate and linguoid forms. Azimuths of the ripple crests are northeast-southwest for most localities (Fig. 17).

Individual sandstone units commonly display only one stratification type. Where the beds display more than one type of stratification or where several beds are amalgamated, the internal structures are usually arranged in a vertical sequence that suggests an upward decrease in energy level (Fig. 18). Hummocky cross-stratification usually passes upward into parallel lamination, and parallel lamination commonly passes up-

FIG. 9.—Sandstone unit of Mosby composed mainly of parallel-laminated sandstone beds intercalated with shale. Hammer is 33 cm long.

FIG. 11.—Hummocky cross-stratified sandstone unit of Mosby that has erosional undulating base. Hammer is 33 cm long.

ward into ripple cross-lamination. Intercalations of finer grained rock usually initiate or terminate a sequence, although they occasionally may be found within a sequence. In addition, stratification sequences may show rapid lateral facies changes. For example, amagamated beds displaying hummocky cross-stratification may pass laterally within a distance of a few meters into parallel laminated beds intercalated with shale.

Although the units show rapid lateral variation, predictable facies changes occur laterally within individual bodies (Fig. 19). The central part of the body is characterized by amalgamated beds that are mainly hummocky cross-stratified and that together are about 2 m thick. Shells concentrated in lenses and concretions occur in the lower part of the unit. Structures indicative of decreasing energy are present only in the upper part of the unit. Toward the margins of the body, the amalgamated beds pass laterally into several individual beds, often less than 30 cm thick, intercalated with shale. Parallel lamination is the primary sedimentary structure, although there also may be ripple cross-

lamination at the top of each bed. The parallel-laminated sandstone beds that have numerous shale intercalations grade into silty to sandy shale at the lateral termination of each body. The shale contains lenses of siltstone and sandstone similar to those in the transitional interval below and above the sandstone units where developed. In addition, fossiliferous concretions, with the same assemblage as is present at the base of the sandstone units, occur in the interval. These same lateral facies changes of sandstone beds grading into shale with fossiliferous concretions occur on a smaller scale within the individual bodies.

Boundary Surfaces.—All the sandstone beds that make up the units display very sharp bases that may be flat, gently undulating, or broadly scoured. The depth of scour may be as much as 30 cm, particularly on the basal bed of units that lack structure or that display hummocky cross-stratification. The bases of beds, especially those with scour, locally have sole marks including shell impression, bounce marks, grooves, and scratches. They appear to be unoriented.

FIG. 10.—Hummocky cross-stratified sandstone unit of Mosby that has sharp, planar base. Hammer is 33 cm long.

FIG. 12.—Sandstone unit of Mosby composed of amalgamated beds that are structurelss at base and that pass upward into parallel laminations. Hammer is 33 cm long.

FIG. 13.—Upper part of sandstone unit of Mosby that has climbing ripple cross-lamination. Marking pen is 14 cm long.

FIG. 14.—Sandstone unit of Mosby showing the following sedimentary structures: hummocky cross-stratification, parallel lamination, ripple cross-lamination, and interlaminations of shale. Hammer is 33 cm long.

The tops of the sandstone units, specifically those composed of amalgamated beds and occasionally those of individual beds, are gradational into the overlying shale (Fig. 20). The sandstone beds pass upward into lenses that decrease in number and thickness in a dominantly shale section. The end result is a transitional interval that is a mirror image of that which commonly underlies the units.

Body fossils.—Body fossils of marine invertebrates are locally abundant in the upper sandstone unit that extends from McDonald Creek to Devil's Basin and the sandstone unit that extends from Mosby Dome to East Dome (Figs. 7 and 8) and laterally equivalent rocks. The fauna includes a variety of ammonites and benthonic gastropods and bivalves.

The largest number of shells are concentrated in lenses or sandy limestone and calcareous sandstone concretions that occur in the lower part of the units or where the units grade laterally into shale (Fig. 21). The shells are randomly oriented, are disarticulated, and are not fragmented or abraded. The following fauna have been identified by W.A. Cobban, E.G. Kauffman, and N.F. Sohl:

Ammonites
  Metoicoceras mosbyense
  Metoicoceras muelleri
  Duveganoceras albertense
  Dunveganoceras parvum
Gastropods
  Pseudomelania hendricksoni
  Lunatia sp. cf. *L. dakotensis*
Bivalves
  Pycnodonte sp. cf. *P. kellumi*
  Crassostrea soleniscus
  Corbula sp.

Specimens of *Pseudomelania hendricksoni* form the bulk of the fossils and some concentrations are composed almost entirely of this one species (Fig. 22). Specimens of *Pycnodonte* sp. cf . *P. kellumi* are also locally abundant, but they are usually segregated in individual layers within the same concretion or lens from individuals of *Pseudomelania hendricksoni*.

FIG. 15.—Bedding surface on sandstone unit in Mosby showing straight-crested symmetrical ripples rounded in profile. Ripples that have dissimilar crest orientations are on different bedding planes. Pocket knife is 9 cm long.

FIG. 16.—Rippled surface on sandstone unit in Mosby showing with abundant horizontal burrows, including *Thalassinoides* and *Chondrites*. Pocket knife is 9 cm long.

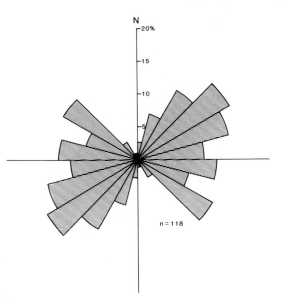

FIG. 17.—Rose diagram showing predominance of northeast-southwest trending ripple crests in Mosby at most localities on Figure 2. Exception is Yellow Water Creek where ripple crests generally trend northwest-southeast. n = numbers of measurements.

In addition to the assemblage that is concentrated in lenses and concretions, specimens of the following bivalves have also been collected along bedding surfaces and are identified by E.G. Kauffman as follows:

  Callistina sp. aff. *C. taffi*
  Callistina sp. aff. *C. lamarensis*
  cf. *Cymbophora* sp.
  cf. *Cyprimeria* sp.

Specimens of cf. *cymbophora* sp. and cf. *Cyprimera* sp. are size-sorted and abraded.

Trace Fossils.—Although body fossils are generally confined to the lower parts of the sandstone units, trace fossils are usually restricted to the rippled surfaces near the tops of individual beds that are intercalated with shale or near the tops of units that consist of amalgamated beds (Fig. 16). The assemblage usually includes the following: horizontal *Ophiomorpha* and *Thalassinoides* as much as 2 cm in diameter and *Chondrites*; and sparse unnamed smooth-walled, vertical burrows and *Corophioides*.

The upper sandstone unit at East Dome (Fig. 7) is extensively burrowed with *Corophioides, Planolites,* and horizontally oriented *Ophiomorpha*. In addition, the stratigraphically equivalent discontinuous lenses at McDonald Creek, Yellow Water Creek, and Devil's Basin are bioturbated (Fig. 8).

Petrology.—The grain size of sandstone in the Mosby, measured in thin section, generally ranges from 0.05 to 0.09 mm (silt to very fine grained sand). A slight decrease in grain size upward is sometimes observed in units composed of amalgamated beds. The grains are subangular and the sandstones are well to very well sorted. The grain size of conglomeratic layers in the upper unit at East Dome is 1.5 mm (granule).

The sandstones are composed of the following framework grains: 70-80 percent quartz, 10-15 percent rock fragments, 2-5 percent chert, about 5 percent potassium feldspar, and less than 1 percent of plagioclase, mica, glauconite, and dolomite. Quartz consists of mostly monocrystalline grains. Rock fragments include metasedimentary grains of quartz and muscovite, polycrystalline mica, silty chert, chert-cemented sandstone, and siltstone. Bone fragments, shark's teeth, and granules of chert, quartzitic rock fragments, and untwinned potassium feldspar make up the conglomeratic layers in the upper unit at East Dome (Fig. 7). Detrital matrix and clay-rich laminae are also abundant in the conglomeratic sandstone.

Most of the samples contain abundant authigenic quartz, which occurs as overgrowths on detrital quartz grains. In addition, some sandstones are completely cemented by sparry calcite that commonly formed subsequent to the quartz overgrowths.

INTERPRETATION

Fine-grained sediments of the Belle Fourche Shale are interpreted to have been deposited on a marine shelf along the western side of an epicontinental seaway. A west-to-east cross section of lower Upper Cretaceous rocks shows that there is a gradual thickening of the Belle Fourche Shale from the Sweetgrass Arch to the northern Black Hills (Fig. 4). A similar thickening takes place in the Niobrara through Eagle interval in this same geographic area (Shurr, this volume). Asquith (1970) concluded from studies of fine-grained Upper Cretaceous rocks in the Powder River Basin in northeastern Wyoming that thicker sequences within approximate time-stratigraphic intervals indicated areas of significant submarine topography, probably a shelf-slope break in a shallow marine environment. He determined from cross sections using bentonite markers on electric logs that flat lying units were deposited on the shelf and that units that have inclined time-stratigraphic markers were deposited on a slope environment.

Along the line of section shown in Figure 4, the thickness of the section between the bentonite bed at the base of the Greenhorn Formation and the group of bentonite beds at the base of the sandy shale interval comprising the upper part of Belle Fourche Shale (see Fig. 23) remains fairly constant (about 45 m) over the area in which the Mosby Sandstone Member is developed. In addition, the same interval maintains a uniform thickness in an approximate north-south line of section from the Bowdoin Dome, where coarsening-upward cycles are present, to the area of the east-west section. This uniformity of thickness suggests that the Mosby was deposited on a broad shelf that had little or no paleoslope.

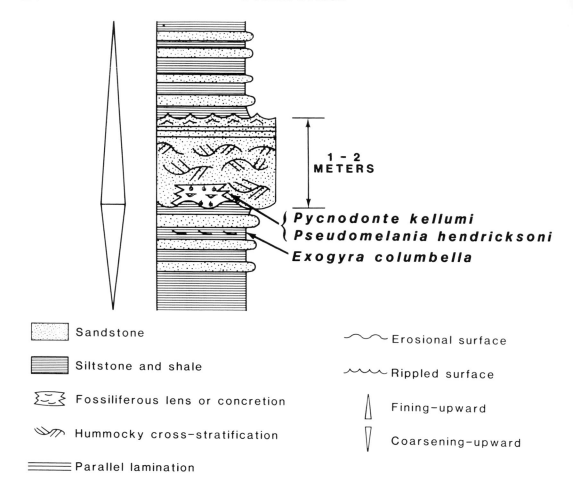

Legend:
- Sandstone
- Siltstone and shale
- Fossiliferous lens or concretion
- Hummocky cross-stratification
- Parallel lamination
- Erosional surface
- Rippled surface
- Fining-upward
- Coarsening-upward

1 – 2 METERS

Pycnodonte kellumi
Pseudomelania hendricksoni
Exogyra columbella

FIG. 18.—Typical vertical sequence of sandstone unit in Mosby composed of amalgamated beds and adjacent rocks.

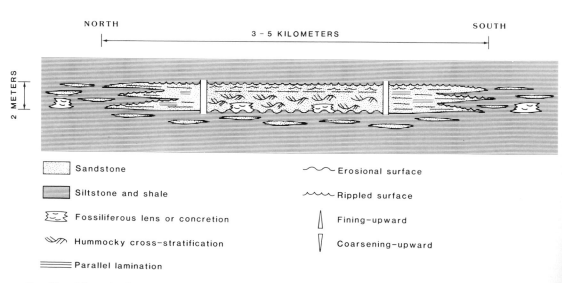

NORTH   3 – 5 KILOMETERS   SOUTH

2 METERS

Legend:
- Sandstone
- Siltstone and shale
- Fossiliferous lens or concretion
- Hummocky cross-stratification
- Parallel lamination
- Erosional surface
- Rippled surface
- Fining-upward
- Coarsening-upward

FIG. 19.—Schematic diagram of sandstone body in Mosby showing lateral facies.

STORM-GENERATED SANDSTONE UNITS 155

FIG. 20.—Sandstone unit in Mosby that has transitional top to shale. Hammer is 33 cm long.

FIG. 22.—Well-preserved shells of gastropod *Pseudomelania hendricksoni* from fossiliferous lens in Mosby. Coin is 2 cm in diameter.

From the eastward pinchout of the Mosby to the northern Black Hills (Fig. 4), the upper interval of the Belle Fourche Shale bounded by bentonite beds gradually increases in thickness. In addition, the interval from the top of the Mowry Shale to the group of bentonite beds in the lower part of the Belle Fourche thickens significantly from the Sweetgrass Arch to the northern Black Hills. The thickening can be explained in two ways: (1) tectonism exceeded the rate of sedimentation such that a relatively deep bathymetric basin was produced (Asquith, 1970), or (2) the rate of sedimentation kept pace with tectonism such that only a minor topographic feature was present and that the Black Hills area was in the central part of a relatively continuous shelf. Evidence is not available to substantiate either explanation for the thicker parts of the Belle Fourche. However, the geometry of time-stratigraphic markers bounding the upper part of the Belle Fourche indicates that the Mosby was deposited on a broad, essentially flat western shelf. The clastic input was minimal on the eastern shelf and a thinner sequence of calcareous rocks were deposited in that area.

Sediments comprising the Mosby Sandstone Member are interpreted to have been derived from the Dunvegan delta in northeastern British Columbia and northwestern Alberta (Fig. 3). Coarse clastics were

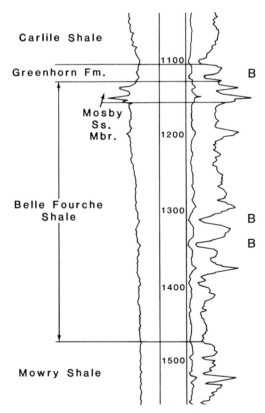

FIG. 23.—Typical electric log expression of lower Upper Cretaceous rocks on Cat Creek anticline. Location of well shown on Figure 2 (Sec. 2, T. 14 N., R. 29 E.). Depth shown in feet below kelly bushing (1 ft = 0.3 m). B, bentonite beds.

FIG. 21.—Fossiliferous lens composed mainly of gastropods in lower part of sandstone unit of Mosby. Pocket knife is 9 cm long.

eroded from highlands in north-central British Columbia, were carried to the coast in eastward-flowing rivers, and were deposited in a major deltaic complex during a large part of Cenomanian time. Continued uplift of the source area in late Cenomanian time resulted in the erosion of the upper parts of the delta (Stelck and Wall, 1955). The reworked deltaic sediments were partially the source for the isolated sandstone beds interbedded with the lower part of the Kaskapau Formation. One of these sandstones, the Pouce Coupe Sandstone, is the time-equivalent of the Mosby Sandstone Member in central Montana (Fig. 1). The shoreline prograded to the southeast at this time, and sediments were deposited a considerable distance offshore (Stelck and Wall, 1955). Evidence for southward transport of the sediments will be presented later.

An alternative source area for the Mosby sediments was western Montana and Idaho. Because the Mosby grades to the west into siltstones and shales (Fig. 3), it is considered unlikely that sand would have bypassed the coastal area along the western shoreline and have been deposited on the shelf to the east.

The following characteristics of sandstone beds and/or units in the Mosby are considered of importance in assessing the proccesses responsible for transportation and distribution of the sediments on the shelf:

(1) The sandstone units are composed of sharp-based beds that are separated by shale or are amalgamated.

(2) The vertical succession of bedforms commonly indicates that depositional conditions were of high energy initially and progressively waned, although they sometimes fluctuated. The units also display rapid lateral variation in stratification type.

(3) Trace fossils are generally horizontal forms restricted to rippled surfaces in the upper part of sandstone beds and/or units.

(4) Units are strongly lenticular instead of laterally continuous.

(5) Shells concentrated in lenses and concretions are not fragmented or abraded.

(6) Sole marks and shells are not preferentially oriented.

The first three characteristics indicate that the sandstone units were deposited by infrequent, high-energy events of short duration. The last two features suggest that the shells and associated sediments comprising the units were not transported far or by unidirectional currents, waves, or other agents during final stages of deposition. The conclusion is made that the sediments were initially transported across the shelf by geostrophic currents. Subsequently, the sediments were locally redistributed by storm waves.

Sediments were moved from the coastal environment through the "littoral energy fence" to the open shelf in one of two ways during deposition of the Mosby. Swift (1976) proposed that during progradation, sediments are transferred to the shelf by river-mouth bypass. Seaward-moving currents generated by either ebb-tidal jets or flood-stage jets of a river mouth are responsible for this bypass.

The other mechanism of sediment bypass is by density currents, ebb currents, or bottom return flows resulting from storms. Rises in sea level of short duration, referred to as storm tide or surge (Harris, 1958), result in considerable erosion of the coastal environment (Ball et al., 1967; Hayes, 1967; Perkins and Enos, 1968). Hayes (1967) inferred that sediment eroded along the Texas coast by Hurricane Carla was transported and deposited by a seaward-flowing density current driven by the returing storm surge. A graded bed of sand and silt as much as 9 cm thick was deposited in depths of water as much as 36 m deep. The bed extended at least 23 km offshore and covered a minimum area of 1,000 km². In another example, Murray (1970) measured offshore-flowing bottom currents on the Gulf Coast shelf during Hurricane Camille that were capable of transporting sediments. In the ancient record, Brenchley et al. (1979), Hamblin and Walker (1979), and Cant (1980) have postulated the deposition of sandstone beds that have features similar to those of the Mosby, by ebb currents, density currents, and bottom return flows resulting from storms. During the deposition of the Mosby, this mechanism may have moved sediment tens of kilometers from the coastal zone onto the shelf.

Turbidity currents are another process by which it has been proposed that sediments can be moved onto the shelf. In deep water, turbidity currents may be generated from large slumps in which rapidly deposited sediment moves down a slope and gradually turns into turbulent flow. An example is the Grand Banks slump of 1929 (Heezen and Drake, 1964). A turbidity-current origin is doubtful for the Mosby because pronounced submarine topographic relief was not present. In addition, the units are strongly lenticular instead of laterally continuous, and consistent vertical sequences of sedimentary structures generally present in turbidites (Bouma, 1962) are missing. These last two features could also indicate that turbidity currents were not the transporting agent; the features are interpreted, however, to be the result of later redistribution instead of their original mode of transport.

Once the sediments were moved onto the shelf, the interaction of waves and geostrophic currents was probably responsible for their transport on the shelf. The oscillatory motion of waves is capable of lifting the sediment into suspension and making it susceptible to transport by geostrophic currents resulting from storms (Komar, 1976; Madsen, 1976). In the Western Interior seaway of Late Cretaceous time, southward-flowing currents parallel to the shoreline have been proposed by many authors for long-distance transport of sediment on the open shelf (Berg, 1975; Porter, 1976; Spearing, 1976; Brenner, 1978). Simlar south- to southeast-flowing currents probably operated during deposition of the Mosby and are represented by:

(1) Southward-projecting lobe of shelf sandstones that grades laterally (west, east, and south) into finer grained siltstones and shales (Fig. 3).

(2) Northwest-southwest elongation of sand ridges on Bow Island Arch (Suffield Block Study Committee, 1972) and on Bowdoin Dome.

(3) Thinning of sand ridges southward across the Bowdoin Dome.

(4) General northeast-southwest trend of wave ripple and current-modified wave-ripple crests (Fig. 17). This orientation indicates that oscillation was northwest-southeast and that wave approach was probably from the northwest.

Deposition by waves and geostrophic currents resulted in the accumulation of fine-grained sediment on the shelf area radiating southward from the Dunvegan delta (Fig. 3). Intercalations of silt and sand as discontinous lenses probably represent thin storm layers (Reineck and Singh, 1972; deRaaf et al., 1977). The increase in number and thickness of these lenses in the upper part of the Belle Fourche section may be interpreted to indicate a general shoaling over the entire area.

The proportion of coarser grained sediment (very fine grained sand) appears to be concentrated on ancestral paleotectonic highs, such as the Bow Island Arch, Bowdoin Dome, and Central Montana Uplift. These features had the same relative positions and trends as do present-day surface features. Episodic tectonic movement of these features resulted in the relative rise and fall of sea level and probably greatly affected sedimentation patterns. The concentration of sand on these features can be explained in two ways:

(1) Coarser grained sediment was initially concentrated on topographic highs by geostrophic storm flows.

(2) Sand, silt, and mud were uniformly distributed over the entire southward-projecting lobe, and the sand was later concentrated by winnowing on paleotectonically controlled shoals.

As a result of winnowing and/or original concentration by depositional processes, the sands were deposited on ancestral positive features as coarsening-upward cycles. Where developed in the subsurface in the vicinity of Bowdoin Dome and Bow Island Arch, the coarsening-upward cycles form elongate sand ridges that are a few to several tens of kilometers in dimension (Suffield Block Study Committee, 1972). The bars are comparable in areal geometry to sand ridges developed in the Shannon Sandstone Member of the Gammon Shale in southeastern Montana (Shurr, this volume) and of the Cody and Steele Shales in the Powder River Basin (Spearing, 1976), and to sand-ridge fields on modern continental shelves (Swift, 1976).

The upward increase in the percentage of coarser sediment and the upward succession of stratification types indicate increasing energy levels, probably due to progressive shoaling and tectonic uplift. The sand is interpreted to have been concentrated on minor topographic highs by flow expansion and deceleration over these features. The deposits generally coarsen upward because current perturbation and wave agitation were intensified upward as the sand ridge grew.

Coarsening-upward cycles capped by beds of very fine grained sandstone are interpreted to have been deposited originally on the Central Montana Uplift because: (1) coarsening-upward transitional intervals commonly are immediately beneath the sandstone units, (2) the sandstone units sharply overlie, sometimes with scour, the transitional intervals, and (3) shells and associated sediments comprising the sandstone units probably were not moved far during the final stages of deposition.

The coarsening-upward cycles postulated for the study area formed sand ridges in shallow, normal marine waters. The sequence of body fossils usually present only in the upper part of the Mosby Sandstone Member indicates the progressive upward shoaling of the sand ridges.

Exogyra columbella generally occurs in the transitional interval beneath the uppermost sandstone unit in the Mosby. *Exogyra* is a normal marine oyster that is widely distributed in shelf habitats at water depths of less than 50 to 60 m (Kauffman, written commun., 1980; Sohl, personal. commun., 1980).

The rest of the fauna associated with the Mosby commonly is concentrated in lenses or concretions near the base of the upper sandstone units. The lack of fragmentation and abrasion of individual shells suggests that they have not been transported far, and, therefore, they can be used for paleoecological interpretation.

The fauna concentrated in lenses and concretions is dominated by the gastropod *Pseudomelonia hendricksoni* which is interpreted to have lived on shallow, vegetated flats, perhaps in depths of water as shallow as 5 to 15 m (Sohl, written commun., 1974; Kauffman, written commun., 1979). This gastropod also has been reported from the Floweree Member on the Sweetgrass Arch, upper part of Belle Fourche Shale in northern Black Hills, and the upper cycle of Mosby Sandstone Member on Bowdoin Dome (Cobban, 1951; Cobban, written commun., 1976). Other benthonic forms, such as *Psycnodonte* sp. aff. *P. kellumi* and *Crossostrea soleniscus*, also occur locally in the concentrations, and they commonly are segregated in layers. These genera also suggest shallow, normal marine settings (Kauffman, written comm., 1980), and they probably lived on the deeper margins of the flats where there was much less turbulence.

The sandstone units are interpreted as representing the upper part of coarsening-upward cycles that were locally redistributed by storms. Storm activity has been proposed by Ball (1967), Hobday and Reading (1972), Brenner and Davies (1973), Goldring and Bridges (1973), Dzulynski and Kubicz (1975), Kelling and Mullin (1975), Brenchley et al. (1979), Hamblin and Walker (1979), Bourgeois (1980), and Cant (1980), as the process responsible for deposition of other ancient sequences.

Due to the progressive shoaling of the sand bars, the shelf bottom became increasingly affected by storm waves. The very fine-grained sand and intercalated finer grained sediment forming the upper part of the shoals were entrained by the wave surges, and the bottom was scoured into an undulating surface of broad troughs and hummocks. As the bottom surge diminished, the suspended sediment quickly dropped into laminations essentially parallel to the undulating surface, producing hummocky cross-stratification. Harms *et al.* (1975) have speculated further on the formation of hummocky cross-stratification based on observations of the sea floor made during periods of strong storm activity. They stated that storm-dominated shallow seas have complex surface patterns and that the bottom surges are varied in direction, which explains the apparent lack of preferred orientation of the sole marks, cross stratification, and shells in the Mosby. The waning stages of the storm are represented by upward progression of stratification types deposited under lower energy conditions and by the local occurrence of slightly finer grained sand or silt. Post-storm activity and the return to normal-weather conditions are indicated by (1) reworking of sediments by oscillatory waves, (2) horizontal burrowing along rippled surfaces, and (3) deposition of interbeds of finer grained silt and mud.

The concentration of shells at the base of sandstone units, and laterally where the units grade into shale, was probably controlled by pressure gradients created by passing wave surges (Powers and Kinsman, 1953; Gernant, 1970; Brenner and Davies, 1973; Specht and Brenner, 1979). At depths where the bottom was affected by storm waves, a pressure gradient was established between water-saturated sediment and the overlying water column as the wave passed. The heavier material, such as shells, settled out of flow first and was concentrated. As Gernant (1970) suggested, shells do not have to be moved more than a few meters by this mechanism, which best explains the excellent preservation of unoriented, concentrated shells in the Mosby.

The geometry of the initially formed sand ridges that have coarsening-upward cycles was probably controlled by paleotectonic activity on ancestral structural features, and the sand ridges were tens of kilometers wide and long. The smaller dimensions of the sandstone bodies resulting from local redistribution by storm activity and their lateral facies were probably controlled by local bottom topography and by water depth, which were influenced by tectonic uplift.

In the central part of the sandstone body, which was probably in the shallowest water, the bottom surface was repeatedly reworked by each storm event. The reworking resulted in the deposition of several, amalgamated beds of very fine grained sandstone that have little or no finer grained material (Fig. 19). The dominant stratification type in the amalgamated beds is hummocky cross-stratification. Sedimentary structures and other features that indicate waning of storm activity and the return to normal-weather conditions are present only near the top of the amalgamated beds which form a unit. The unit, then, represents several storm events and has no record of intervening fair-weather deposition.

The amalgamated beds in the central part pass laterally on the flanks of the body into several individual beds that are intercalated with shale (Fig. 19). The beds are commonly parallel-laminated; the top of each bed commonly has ripple cross-lamination. The parallel laminations suggest deposition in an environment that was of somewhat lower average energy than that of hummocky cross-stratification. Normal deposition is represented by the intercalations of shale. Each bed is interpreted to represent the deposits of a single storm event. The presence of sedimentary structures near the base of each bed that represent lower levels of energy than does hummocky cross-stratification, and the preservation of sediments deposited during fair weather suggest deeper water because of local topography.

Laterally, the sandstone units grade into shale that has lenses of siltstone and sandstone and fossiliferous concretions (Fig. 19). This facies is interpreted to have been deposited in slightly deeper water where the bottom was relatively unaffected by storm waves. The shells were concentrated by pressure gradients created in the finer grained sediments by passing wave surges.

At each locality, there is commonly more than one sandstone unit (Figs. 7 and 8). The uppermost unit, except at East Dome, is interpreted to have been deposited in the shallowest water because (1) the beds are often amalgamated, indicating repeated reworking by reoccurring storm events, (2) the unit generally has more basal scour, suggesting the intersection of storm waves with the bottom surface, (3) stratification types predominate that indicate a high-energy regime and, therefore, shallow water, and (4) body fossils occur in an upward succession that suggests progressive shoaling. The lower sandstone units are thinner, are more extensive laterally, and are interpreted to have been deposited in deeper waters less influenced by storms.

The upper sandstone on East Dome, and discontinuous lenses in the upper part of the Mosby at McDonald Creek, Yellow Water Creek, and Devil's Basin (Figs. 7 and 8) are extensively burrowed, and contain abundant clay matrix, shale partings, and layers of pebble and granule conglomerate that have shark's teeth and bone fragments. This type of lithology is restricted to the upper part of the Mosby and occurs, where present, at about the same stratigraphic interval below the Greenhorn Formation.

The presence both of pebble- and granule-size material and of finer grained sediment indicates the influx of a broad variety of sizes from the source area and transport at times by high-velocity currents. The sorting of the material by size and burrowing suggests that the original deposit was reworked during a period of slow sedimentation. This unit is interpreted to have been reworked following a relative rise in sea level prior to the Greenhorn transgression. Because of the sea-level rise, sedimentation was starved, infauna colonized, and the burrowed sediments were reworked by fair-weather processes on shallower positive features, such as the Cat Creek anticline.

The Greenhorn Formation represents a sudden change in depositional pattern—from fine-grained clastics redistributed by storm events on a shallow marine shelf in upper part of Belle Fourche to calcareous rocks. The calcareous rocks probably indicate deposition during a major transgression and worldwide rise in sea level during late Cenomanian time (Hancock and Kauffman, 1979). The associated fauna of *Pseudoperna bentonensis* suggests cooler, deeper waters than the older clastic facies in water depths probably less than 100 m (Kauffman, written commun., 1980).

CONCLUSIONS

Thin, widespread sandstone units of the Mosby characterized by hummocky cross-stratification were deposited on a storm-dominated, shallow-marine shelf in an overall progradational sequence. The sediments were transported as much as 1,100 km in a southward direction, parallel to the shoreline, by geostrophic currents induced by strong winds associated with winter storms. The sequence of events leading to deposition of the Mosby on Central Montana Uplift is shown in Figure 24 and is briefly described below. The sediments were deposited initially on regional to subregional paleotectonic highs as coarsening-upward

sand ridges. Sediments in the upper part of the coarsening-upward cycles were locally redistributed by storm activity. The resulting sandstone units consist of individual beds that are separated by intercalations of finer grained material or that are amalgamated and have sedimentary structures indicative of decreasing energy upwards. The lateral facies and geometry of individual bodies are controlled by bottom topography and by water depth.

The thin, storm-generated sandstone units form "sweet spots" for shallow biogenic gas in a section typified by low-permeability reservoirs (Rice and Shurr, 1980). Wells in widespread accumulations of gas are being developed in coarsening-upward sand ridges in the Mosby, and in the laterally equivalent Second white specks sandstone on the Bowdoin Dome in north-central Montana, and on the Bow Island Arch in southeastern Alberta (Last, Kloepfer Ltd., 1974; Nydegger *et al.* 1980). Natural gas is being produced at depths of less than 600 m, and productive sandstones have an estimated average porosity of 15-17 percent and a permeability of as much as 2-3 md. Similar reservoir properties should exist in sandstones of the Mosby on the Central Montana Uplift and other shelf sandstones that were deposited during storm events.

FIG. 24.—Sketches showing sequence of events resulting in deposition of the Mosby Sandstone Member on Central Montana Uplift. *A*, Deposition of coarsening-upward sequences during fair-weather conditions by southward-flowing geostrophic currents. *B*, Intense storm conditions resulting in erosion of upper part of coarsening-upward sequences and suspension of the sediments. *C*, Deposition of sandstone units during waning stages of storm activity. *D*, Reworking of sediments during fair-weather conditions.

ACKNOWLEDGEMENTS

W. A. Cobban and N. F. Sohl of the U.S. Geological Survey and E. G. Kauffman of the University of Colorado identified the fossils and provided valuable information on their paleoecology. D. L. Gautier of the U.S. Geological Survey assisted with the petrographic study. J. C. Harms of Harms and Brady Geological Consultants, Inc. provided stimulating discussion on the origin of hummocky cross-stratification and the importance of storm-related deposition in the shallow-marine environment. H. E. Clifton of the U.S. Geological Survey and D. J. P. Swift of Plano Research Center, ARCO Oil and Gas Company critically reviewed the manuscript and provided many helpful suggestions. Laurie Baer drafted the illustrations. The U.S. Department of Energy funded the study.

REFERENCES

ASQUITH, D.O., 1970, Depositional topography and major marine environments, Late Cretaceous, Wyoming: Am. Assoc. Petroleum Geologists Bull., v. 54, p. 1184-1224.

BALL, M. M., 1967, Carbonate sand bodies of Florida and the Bahamas: Jour. Sed. Petrology, v. 37, p. 556-591.

_____, E. A. SHINN, AND K. W. STOCKMAN, 1967, The geologic effects of Hurricane Donna in south Florida: Jour. Geology, v. 75, p. 583-597.

BERG, R. R., 1975, Depositional environment of Upper Cretaceous Sussex Sandstone, House Creek Field, Wyoming: Am. Assoc. Petroleum Geologists Bull., v. 59, p. 2099-2110.

BOUMA, A. H., 1962, Sedimentology of some flysch deposits—A graphic approach to facies interpretation: Elsevier, Amsterdam, 168 p.

BOURGEOIS, J., 1980, A transgressive shelf sequence exhibiting hummocky stratification; The Cape Sebastion Sandstone (Upper Cretaceous), southwestern Oregon: Jour. Sed. Petrology, v. 50, p. 681-702.

BRENCHLEY, P. J., G. NEWALL, AND I. G. STANISTREET, 1979, A storm surge origin for sandstone beds in an epicontinental platform sequence, Ordovician, Norway: Sed. Geology, v. 22, p. 185-217.

BRENNER, R. L., 1978, Sussex sandstone of Wyoming—Example of Cretaceous offshore sedimentation: Am. Assoc. Petroleum Geologists Bull., v. 62, p. 181-200.

_____, 1980, Construction of process-response models for ancient epicontinental seaway depositional systems using partial analogs: Am. Assoc. Petroleum Geologists Bull., v. 64, p. 1223-1243.

BRENNER, R. L., AND D. K. DAVIES, 1973, Storm-generated coquinoid sandstone; genesis of high-energy marine sediments from the Upper Jurassic of Wyoming and Montana: Geol. Soc. America Bull., v. 84, p. 1685-1697.

CAMPBELL, C. V., 1966, Truncated wave-ripple laminae: Jour. Sed. Petrology, v. 36, p. 825-828.

CANT, D. J., 1980, Storm-dominated shallow marine sediments of the Arisaig Group (Silurian-Devonian) of Nova Scotia: Canadian Jour. Earth Sciences, v. 17, p. 120-131.

COBBAN, W. A., 1951, Colorado Shale of central and northwestern Montana and equivalent rocks of Black Hills: Am. Assoc. Petroleum Geologists Bull., v. 35, p. 2170-2198.

_____, 1952, Cenomanian ammonite fauna from the Mosby Sandstone of central Montana: U.S. Geol. Survey Prof. Paper 243-D, p. 45-54.

_____, C. E. ERDMANN, R. W. LEMKE AND E. K. MAUGHAN, 1976, Type sections and stratigraphy of the members of the Blackleaf and Marias River Formations (Cretaceous) of the Sweetgrass Arch, Montana: U.S. Geol. Survey Prof. Paper 974, 66 p.

DERAAF, J. F. M., J. R. BOERSMA, AND A. VAN GELDER, 1977, Wave-generated structures and sequences from a shallow marine succession, Lower Carboniferous, County Cork, Ireland: Sedimentology, v. 24, p. 451-483.

DZULYNSKI, A. AND A. KUBICZ, 1975, Storm accumulations of brachiopod shells and sedimentary environments of the Terebratula beds in the Muschelkalk of Upper Silesia (southern Poland): Ann. Soc. Geol. Pol., v. 45, p. 57-169.

GERNANT, R. E., 1970, Paleoecology of the Choptank Formation (Miocene) of Maryland and Virginia: Maryland Geol. Survey Rpt. Invest. 12, 90 p.

GOLDRING, R., AND P. BRIDGES, 1973, Sublittoral sheet sandstones: Jour. Sed. Petrology, v. 43, p. 736-747.

HAMBLIN, A. P., AND R. G. WALKER, 1979, Storm-dominated shallow marine deposits: the Fernie-Kootenay (Jurassic) transition, southern Rocky Mountains: Canadian Jour. Earth Sciences, v. 16, p. 1673-1690.

HANCOCK, J. M., AND E. G. KAUFFMAN, 1979, The great transgressions of the Late Cretaceous: Jour. Geol. Soc., v. 136, p. 175-186.

HARMS, J. C., J. B. SOUTHARD, D. R. SPEARING, AND R. G. WALKER, 1975, Depositional environments as interpreted from primary sedimentary structures and stratification sequences: Soc. Econ. Paleontologists and Mineralogists, Short Course No. 2, 161 p.

HARRIS, D. L., 1958, Meteorological aspects of storm surge generation: Jour. Hydraulics Div., Proceed. Amer. Soc. Civil Engineers, v. 84, Paper 1859, 25 p.

HATTIN, D. E., 1975, Petrology and origin of fecal pellets in Upper Cretaceous strata of Kansas and Saskatchewan: Jour. Sed. Petrology, v. 45, p. 686-696.

HAYES, M.O., 1967, Hurricanes as geological agents: Case studies of Hurricanes Carla, 1961, and Cindy, 1963: Texas Bur. Econ. Geol. Rept. Invest. 61, 54 p.

HEEZEN, B. C., AND C. L. DRAKE, 1964, Grand Banks slump: Am. Assoc. Petroleum Geologists Bull., v. 48, p. 221-223.

HOBDAY, D. K., AND H. G. READING, 1972, Fair weather versus storm processes in shallow marine sand bar sequences in the late Precambrian of Finmark, north Norway: Jour. Sed. Petrology, v. 42, p. 318-324.

JOHNSON, W. D., JR., AND H. R. SMITH, 1964, Geology of the Winnett-Mosby area, Petroleum, Garfield, Rosebud, and Fergus Counties, Montana: U.S. Geol. Survey Bull. 1149, 91 p.

KELLING, G. AND P. R. MULLIN, 1975, Graded limestones and limestone-quartzite couplets; possible storm-deposits from the Moroccan Carboniferous: Sed. Geology, p. 161-190.

KOMAR, P. D., 1976, The transport of cohesionless sediments on continental shelves, *in* D. J. Stanley and D. J. P. Swift, (eds.), Marine Sediment Transport and Environmental Management: John Wiley and Sons, New York, p. 107-125.

LAST, KLOEPFER LTD., 1974, Suffield evaluation drilling program—A report submitted to the Suffield Evaluation Committee: Province of Alberta, 75 p.

LUPTON, C. T. AND W. LEE, 1921, Geology of the Cat Creek oil fields, Fergus and Garfield Counties, Montana: Am. Assoc. Petroleum Geologists Bull., v. 5, p. 252-275.

MADSEN A. S., 1976, Wave climate of the continental margin elements of its mathematical description, *in* D. J. Stanley and D. J. P. Swift, (eds.), Marine Sediment Transport and Environmental Management: John Wiley and Sons, New York, p. 65-87.

MEREWETHER, E. A., W. A. COBBAN, AND C. W. SPENCER, 1976, The Upper Cretaceous Frontier Formation in the Kaycee-Tisale Mountain area, Johnson County, Wyoming: Wyoming Geol. Assoc. Guidebook 28th Ann. Field Conf., p. 33-44.

MUDGE, M. R., 1972, Pre-Quaternary rocks in the Sun River Canyon area, northwestern Montana: U.S. Geol. Survey Prof. Paper 663-A, 138 p.

MURRAY, S. P., 1970, Bottom currents near the coast during Hurricane Camille: Jour. Geophys. Research, v. 75, p. 4579-4582.

NORWOOD, E. E., 1965, Geological history of central and south-central Montana: Am. Assoc. Petroleum Geologists Bull., v. 49, p. 1824-1832.

NYDEGGER, G. L., D. D. RICE, AND C. A. BROWN, 1980, Analysis of shallow gas development from low-permeability reservoirs of Late Cretaceous age, Bowdoin Dome area: Jour. Petroleum Tech., v. 32, p. 2111-2120.

PERKINS, R. D., AND P. ENOS, 1968, Hurricane Betsy in Florida-Bahama area—Geologic effects and comparison with Hurricane Donna: Jour. Geology, v. 76, p. 710-717.

PETERSON, J. A., 1966, Sedimentary history of the Sweetgrass Arch: Billings Geol. Soc. 17th Ann. Guidebook, p. 112-134.

PORTER, K. W., 1976, Marine shelf model, Hygiene Member of Pierre Shale, Upper Cretaceous, Denver Basin, Colorado, *in* Studies in Colorado Field Geology: Colorado School Mines Prof. Contr. No. 8, p. 251-263.

POWERS, M. C., AND B. KINSMAN, 1953, Shell accumulations in underwater sediments and their relation to the thickness of the traction zone: Jour. Sed. Petrology, v. 23, p. 229-234.

REEVES, F., 1927, Geology of the Cat Creek and Devil's Basin oil fields and adjacent areas in Montana: U.S. Geol. Survey Bull. 786-B, p. 39-96.

REINECK, H. E., AND I. B. SINGH, 1972, Genesis of laminated sand and graded rhythmites in storm-sand layers of shelf mud: Sedimentology, v. 18, p. 123-128.

RICE, D. D., AND W. A. COBBAN, 1977, Cretaceous stratigraphy of the Glacier National Park area, northwestern Montana: Bull. Canadian Petroleum Geology, v. 25, p. 828-841.

_____, AND G. W. SHURR, 1980, Shallow, low-permeability reservoirs of northern Great Plains—Assessment of their natural gas resources: Am. Assoc. Petroleum Geologists Bull., v 64, p. 969-987.

ROBINSON, C. S., W. J. MAPEL, AND M. H. BERGENDAHL, 1964, Stratigraphy and structure of the northern and western flanks of the Black Hills Uplift, Wyoming, Montana, and South Dakota: U.S. Geol. Survey Prof. Paper 404, 134 p.

SPEARING, D. R., 1976, Upper Cretaceous Shannon sandstone; an offshore, shallow-marine sand body: Wyoming Geol. Assoc. Guidebook, 28th Ann. Field Conf., p. 65-72.

SPECHT, R. W., AND R. L. BRENNER, 1979, Storm-wave genesis of bioclastic carbonates in Upper Jurassic epicontinental mudstones, east-central Wyoming: Jour. Sed. Petrology, v. 49, p. 1307-1322.

STELCK, C. R., AND J. H. WALL, 1954, Kaskapau foraminifera from Peace River area of western Canada: Research Council Alberta Rept. No. 68, 38 p.

_____ AND _____, 1955, Foraminifera of the Cenamanian *Dunveganoceras* zone from Peace River area of western Canada: Research Council Alberta Rept. No. 70, 81 p.

STOTT, D. F., 1963, The Cretaceous Alberta Group and equivalent rocks, Rocky Mountain Foothills, Alberta: Geol. Survey Canada Mem. 317, 306 p.

SUFFIELD BLOCK STUDY COMMITTEE, 1972, A resource evaluation, Suffield Block: Prepared for Province of Alberta, 248 p.

SWIFT, D. J. P., 1976, Continental shelf sedimentation, *in* D. J. Stanley and D. J. P. Swift, (eds.), Marine Sediment Transport and Environmental Management: John Wiley and Sons, New York, p. 311-350.

WALKER, R. G., 1979, Shallow marine sands, *in* R. G. Walker, (ed.), Facies Models: Geoscience Canada, Reprint series 1, p. 75-89.

WARREN, P. S., AND C. R. STELCK, 1940, Cenomanian and Turonian faunas in the Pouce Coupe district, Alberta and British Columbia: Transactions Royal Society Canada, Sec. IV, p. 143-152.

WILLIAMS, G. D., AND C. F. BURK, JR., 1964, Upper Cretaceous, *in* R. G. McCrossan and R. P. Glaister, (eds.), Geological History of Western Canada: Alberta Soc. Petroleum Geologists, p. 169-189.

_____, AND C. R. STELCK, 1975, Speculations on the Cretaceous paleogeography of North America, *in* W. G. E. Caldwell, (ed.), The Cretaceous System in the Western Interior of North America: Geol. Assoc. Canada Spec. Paper No. 13, p. 1-20.

RETROGRADATIONAL SHELF SEDIMENTATION: LOWER CRETACEOUS VIKING FORMATION, CENTRAL ALBERTA

EDWARD A. BEAUMONT[1]
Basin Study Group,
Cities Service Company,
Tulsa, Oklahoma

ABSTRACT

The Viking Sandstone in much of central Alberta, including the Joffre-Joarcam area, consists of a series of overlapping sediment sheets that become progressively younger westward toward the Paleoshoreline. During a profound regression at the beginning of Viking deposition, streams flowed across the former shelf surface depositing sand in deltas, as now evidenced by the irregular-shaped reservoirs of eastern Alberta. An ensuing trangression, punctuated by minor regressions or stillstands, which reworked shoreline sediment (supplied to shorelines) into linear shelf sand bodies, formed the linear reservoirs (e.g., Joffre and Joarcam fields) west of a series of irregular-shaped sandstone reservoirs. During the transgression, the retrogradational nature of the sediments sheets, which contain the linear sand bodies, was formed.

Well log cross sections show, in addition to large-scale shoreward shingling, that the Viking thickens westward, pinches out westward, and that each sediment sheet produced during the overall transgression contains several northwest-trending, elongate "cleaner" sandstone bodies. Cores of these elongate sandstone bodies and their underlying beds commonly exhibit a coarsening-upward succession of: (1) silty marine shale, (2) intercalated silty shale and rippled sand, which in some places becomes a structureless bioturbated clayey sand, (3) glauconitic crossbedded sandstone, and (4) polymictic pebble conglomerate. Conglomerate also occurs randomly within the sequnce.

Final deposition of the elongate bodies on a shelf tens of miles from the paleoshoreline is documented by: (1) marine shale enclosing the Viking, (2) absence of consistent landward-seaward facies changes, (3) abundant glauconite, (4) an "offshore" trace fossil assemblage, and (5) distant eastward location with respect to contemporaneous strandline facies.

Sparse evidence, such as coal partings and plant fragments in the irregular-shaped sand bodies, supports the interpretation that they are drowned delta complexes. All evidence, including the shingling relationship of the sediment sheets containing the elongate sand bodies, can be explained by retrogradational shelf sedimentation. Modern sand ridges of the New Jersey Shelf may be analogous in many respects to the linear "cleaner" sandstone bodies of the Viking.

The Viking Formation, named by Slipper (1918), has been the object of much study because of the large amount of hydrocarbons it contains. Most Viking reservoirs are thought to be stratigraphic traps. As a result, many workers have realized that an understanding of the environments of deposition and diagenesis would aid in finding Viking reservoirs. Interpreted as everything from deep water turbidites (Beach, 1956; Roessingh, 1959) to barrier island sandstones (Thomas and Oliver, 1979; Shelton, 1973; and Tizzard and Lerbekmo, 1975), the environment of deposition of Viking sandstones always has been controversial. Most recent workers have interpreted the Viking to be a marine shelf sandstone (Evans, 1970; Simpson, 1975; Koldijk, 1976; and Boethling, 1977a, b); however, processes responsible for transport and deposition of coarse terrigenous clastics in a marine shelf setting are poorly understood. This problem can possibly be remedied by detailed comparative studies with modern shelf sands.

Sand body complexes are ubiquitous features on modern continental shelves. Some of the best documented are those found on the continental shelf of the East Coast of the United States. As a possible ancient analog for these sand complexes, the Viking offers some striking similarities.

This study begins with a detailed examination of facies associated with two hydrocarbon-bearing linear sand bodies at Joffre and Joarcam Fields (Fig. 1), then proceeds to deposition of the Viking Formation of the entire Alberta basin.

Twenty cores were examined from the Joffre and Joarcam Fields, as well as nearly 20 additional cores from various other Viking fields. More than 500 borehole logs were used to construct isopach maps and cross sections.

Stratigraphy.—The Viking Formation of Alberta of Upper Albian age is equivalent to the Newcastle or Muddy Formations of Wyoming and Montana (Fig. 2). It is underlain by the Joli Fou Shale and overlain by the Colorado Shale. These are equivalent to the Skull Creek and Mowry Shales of Wyoming and Montana, respectively. Lateral equivalents of importance to this

[1]Present Address: American Association of Petroleum Geologists, Box 979, Tulsa, OK 74101

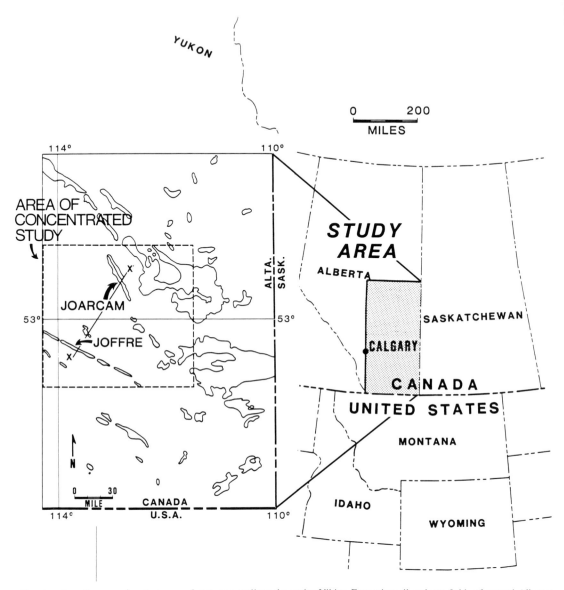

FIG. 1.—Location map for study area: Inset map outlines the major Viking Formation oil and gas fields of central Alberta; note contrast between irregular reservoirs to east and linear reservoirs to west.

study are the Bow Island of southern Alberta and southwestern Saskatchewan, the St. Walburg of western Saskatchewan, the Pelican of northern Alberta, and in part the Paddy of northwestern Alberta.

Thickness, internal stratigraphy, and contact relations.—Over most of Central Alberta, the Viking Formation ranges from 50 to 100 feet (18 to 36 meters) thick (Fig. 3). if thickens westward and southward from central Alberta. In southern Alberta the Viking is over 200 feet (75 meters) thick and correlates in part with the Bow Island Formation. The Viking becomes progressively thinner and finer grained from central

Alberta to the east and northeast where equivalents can be detected as silty shale on well logs (Rudkin, 1964).

Westward thinning of the underlying Joli Fou may be due, at least in part, to an unconformity at the base of the Viking and equivalents. Stelck (1958) indicates that in the northwest plains of Alberta a hiatus exists between the Paddy Sandstone and underlying Cadotte Shale. The Paddy is correlative in part with the Viking. In southwestern Saskatchewan, Evans (1970) notes that "a slight disconformity between the Joli Fou and Viking Formations is indicated by the abrupt disappearance in certain areas of silty-shale markers at the

CENTRAL ALBERTA	MONTANA AND WYOMING
COLORADO	MOWRY
VIKING	NEWCASTLE/ MUDDY
JOLI FOU	SKULL CREEK

(row labels at left: **UPPER ALBIAN**)

FIG. 2.—Upper Albian stratigraphic correlation chart for northern Rocky Mountains of the United States and Alberta Basin of Canada.

top of the Joli Fou Formation, combined with the thinning of this formation." Jones (1961) reports finding a sand-filled dessication crack in the upper part of the Joli Fou (Sun Eureka Well No. 2-7-31-22W3), about 10 feet (3 meters) below the Viking, suggesting shallow water to emergent conditions during late Joli Fou deposition. The upper contact of the Viking is conformable with other overlying formations. Stapp (1967) describes a regional disconformity at the base of the Newcastle Formation which is the Viking equivalent in the northern Rocky Mountains.

The existence of an unconformity, either an angular unconformity or disconformity, is significant. As will be discussed further on, the author believes that an

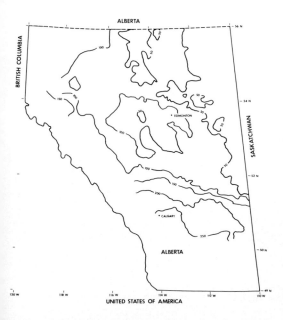

FIG. 3.—Isopach map of Viking Formation in the Alberta Basin. Contour interval is 50 feet.

unconformity was formed at the base of the Viking when a rapid regression of the sea occurred just prior to Viking deposition. This regression caused erosion into or nondeposition of underlying formations over much of the Alberta basin.

The Viking in the study area is made up of a series of shingled sediment sheets (Fig. 4). These are termed sediment sheets rather than sand sheets because each is made up of interbedded sand and shale. Each sheet also contains several distinct sand bodies (Fig. 5). Brenner(1978) also used the term sediment sheet in describing the Sussex Sandstone in the Power River Basin of Wyoming.

The "shingled" Viking sediment sheets have a retrogradational relationship as opposed to a progradational one; in other words, sheets become younger toward the source area rather than away from it (Figs. 4 and 6). In a study of the Viking in southwestern Saskatchewan, Evans (1970) also noted sediment sheets become younger in the same general direction.

SEDIMENTOLOGY

Offshore Facies.—The Viking in the study area consists of overlapping generally coarsening-upward sequences that average approximately 90 feet (30 meters) in thickness and are composed of five facies (Fig. 7).

Burrowed, silty gray shale containing wisps of very fine to fine-grained sandstone forms the basal facies of the sequence (Fig. 8). This facies is overlain transitionally by intercalated silty shale and ripple-bedded sandstone (ripple-bedded sandstone facies) having a flaser appearance (Fig. 9). The ripple-bedded facies makes up the bulk of the Viking in the study area. Identifiable burrows, dominated by horizontal forms, are common in the ripple-bedded sandstone facies; included are *Teichichnus, Rosellia, Arenicolites,* and *Planolites.* Common plant fragments and small, rare pieces of coal (less than ½ inch, 1 cm, in diameter) are found in this facies. In contrast, as will be discussed below, cores from the irregular-shaped sand bodies lying to the east contain more abundant plant fragments and coal partings in addition to pieces of coal.

Interchangeable in sequence position with the ripple-bedded standstone facies is bioturbated shaly, fine-grained sandstone (bioturbated facies) (Fig. 10). This unit probably is a more highly burrowed version of the ripple-bedded sandstone facies.

Overlying either of the two lithofacies described above is a trough cross-bedded sandstone (cross-bedded sandstone facies) (Fig. 11) which averages 20 feet in thickness. It makes up most of the reservoir of the Viking. In the study area, it is fine to very coarse-grained, well to moderately sorted, and glauconitic. Glauconite comprises up to 15% of the grains. Shale clasts are common and are often sideritic. Burrows are rare and, where present, they are usually vertical. The lower contact of this facies is generally sharp and the upper contact is generally gradational.

A polymictic granule to pebble-sized conglomerate

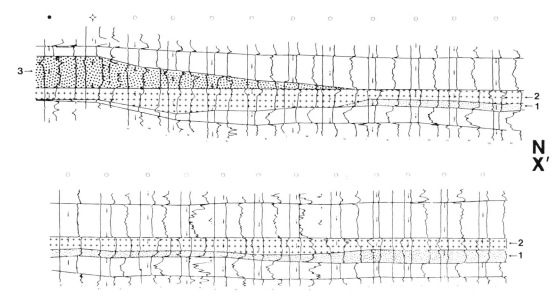

FIG. 4.—Electric log cross-section (two adjoining segments) extending between Joffre and Joarcam Fields, which illustrates shingled relationship of Viking sediment sheets. Viking Formation is contained within shaded section of cross-section. Different patterns represent different sediment sheets. Sheet labeled with number 1 was deposited first; sheet labeled 2, second; and sheet labeled 3, last. Location of cross-section is shown in Figure 1.

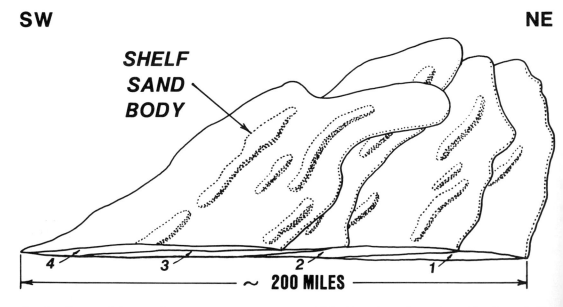

FIG. 5.—Conceptual diagram of "shingling" relationship between Viking Formation sediment sheets. Note that each sheet contains several shelf bodies.

LANDWARD   SEAWARD

PROGRADING SEDIMENT SHEETS

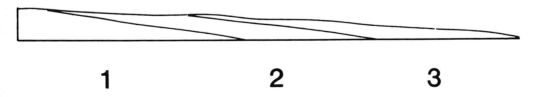

1   2   3

RETROGRADING SEDIMENT SHEETS

3   2   1

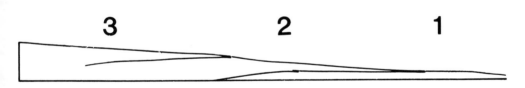

FIG. 6.—Diagram illustrating difference between prograding and retrograding sediment sheets.

(conglomerate facies) (Fig. 12) commonly forms the top of the sequence. It is poorly to moderately sorted and contains well-rounded pebbles composed predominantly of chert. Some pebbles of limestone and phosphate have been reported (Boethling, 1977b; Simpson, 1975). The matrix of the conglomerates is either sandstone or shale, and in places the matrix is alternating thin beds of sandstone and shale. The conglomerate facies is discontinuous and generally cannot be correlated between wells half-a-mile part. Bed thicknesses range from less than an inch to over eight feet. Conglomerates appear to have both random and ordered occurrences in the sequence with the random being more common. Conglomerates can be found enclosed by any of the other lithologies, in which case the lower contact may be abrupt, or gradationally overlying a sandstone.

Core cross sections perpendicular to the trends of Joffre and Joarcam field (Fig. 13) show that the Viking of the study area contains long linear bodies of the cross-bedded sandstone facies separated by ripple-sandstone facies (Fig. 14). Both the conglomerate and bioturbated facies have patchy distributions.

Environment of Deposition.—The Viking Formation of Central Alberta is interpreted to be an offshore sand body complex as opposed to a shoreline deposit based on the following evidence:

1) The Viking is underlain and overlain by marine shale.

2) Absence of landward-seaward facies changes as would be expected with a shoreline deposit.

3) The Viking contains abundant glauconite, concentrated in the cross-bedded sandstone facies. Soft clay grains such as glauconite are usually destroyed by the high energy of a foreshore shoreline setting. Most formations in the Rocky Mountains identified as shoreline deposits contain very little glauconite.

4) The Viking has an assemblage of trace fossils that fall into Seilacher's *Cruziana* shelf facies.

5) The Viking paleoshoreline (Paddy?) was thought to be far to the west of the study area at the close of Viking deposition (Fig. 16) as is discussed below.

Lithofacies of the Viking contain abundant evidence of deposition by episodic and sometimes fast flowing bottom currents. Episodic, storm-generated currents, which may have been augmented by tidal currents,

IDEALIZED VIKING LITHOLOGIC SEQUENCE

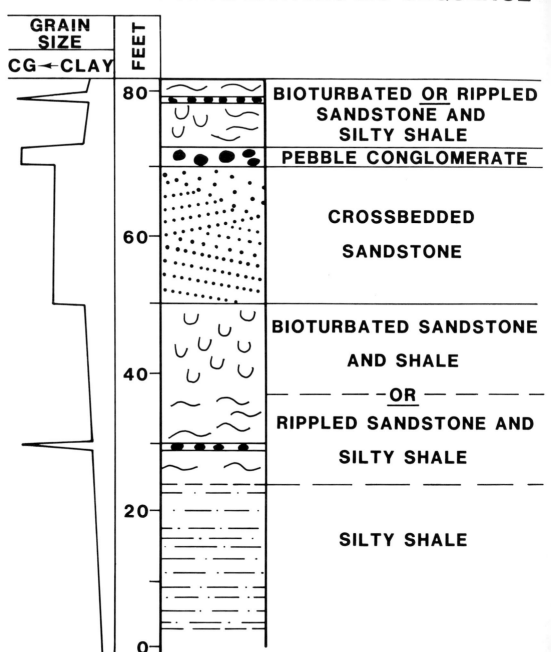

Fig. 7.—Generalized lithofacies sequence for the Joffre-Joarcam area.

were responsible for 1) interbedded sandstone and shale in the ripple-bedded sandstone facies, 2) formation of rip-up clasts which occur in the cross-bedded sandstone facies, and 3) random distribution of the conglomerates. The episodic nature of currents caused by storms allowed clay to settle out during times of quiescence and sand to be transported in ripple forms when currents swept the bottom. Rip-up clasts attest to

Fig. 8.—Silty gray shale containing wisps of sand typical of basal facies of sequence. Photograph is of exterior of four-inch core.

Fig. 9.—Intercalated silty shale and ripples typical of the ripple-bedded sandstone facies. Photograph is of exterior of a four-inch core.

the intensity of these currents as they scoured the bottom during some of the more powerful storms. Conglomerate, which is interpreted to have been initially brought to the area of Viking shelf sand deposition at low stands of sea level, was probably moved on the shelf during higher sea level stands by the more powerful storms. The random distribution of the conglomerate may be related, in part, to the distribution and intensity of the storms.

Paleogeography.—Figure 15 is an interpretation of the paleogeography at the close of Viking deposition. Modified by Boethling (1977), it is based on data from Stelck (1958), Rudkin (1964), Evans (1970), McGookey *et al.* (1972), Williams and Stelck (1975), and Simpson (1975). Basically, this reconstruction shows that most of the Alberta Basin was covered by a blanket of marine shelf sandstone complexes. The boundary between these sandstone complexes and shelf mudstones and siltstones was in eastern Alberta and southwestern Saskatchewan. The principal source of sediment was from the west, although a minor amount was contributed from the shield to the east (Fig. 15). The preserved paleoshoreline lay at least 200 miles (320 km) west of the study area.

FIG. 10.—Bioturbated shaley, fine-grained sandstone typical of the bioturbated facies. Photograph is of exterior of a four-inch core.

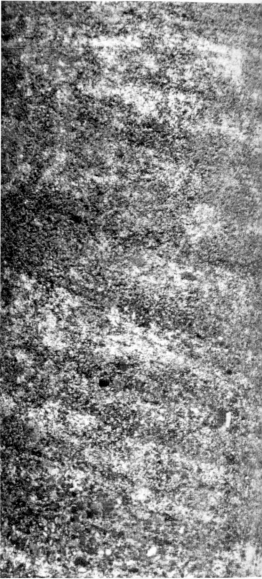

FIG. 11.—Glauconitic troughs typical of the cross-bedded sandstone facies. Photograph is of the exterior of a four-inch core.

REGIONAL DEPOSITION OF THE VIKING FORMATION

Two strongly contrasting models have been proposed to explain deposition of coarse sand-size sediment on modern marine shelves (Swift, 1976). One model occurs under conditions of regression or (possibly) very slow transgression where an upcoast delta "leaks" sediment to the shelf primarily by river mouth bypass. The other model calls for a relatively rapid transgression where river mouths become sediment traps, and coarse sediment is supplied to the shelf by shoreface erosion.

The process model suggested here for the Viking calls for a relatively rapid regression of the sea at the close of Joli Fou deposition followed by an ensuring transgression during which the Viking was deposited. This overall transgression was punctuated by a series of stillstands or minor regressions (Fig. 16). Sediment sheets with a retrogradational relationship formed during the stillstands or minor regressions. These retrogradational relationships of sediment sheets of the

FIG. 12.—Polymictic pebble-sized pebbles typical of the conglomeratic facies. Photograph is of the exterior of a four-inch core.

Viking study area offer an important clue to regional deposition of the Viking and its equivalents. Specifically, this model for Viking deposition can be broken down into two parts: a pre- to early-Viking regressive phase and a Viking transgressive phase (Fig. 17). During the regressive phase sea level fell rapidly, subaerially exposing most of the shelf. A sea level fall of two or three hundred feet would probably have exposed much of the shelf; it is believed that the water depth over most of the shelf may have been no more than several hundred feet. The consequent drop in base level allowed streams to erode the Joli Fou Shale over parts of this newly exposed area. Sediment was probably eroded from the more highly elevated western portions and deposited in delta complexes to the east. The apparent unconformity at the base of the Viking indicates that erosion took place before Viking deposition. Evidence of erosion includes truncated markers near the top of the Joli Fou and the progressive westward thinning of the Joli Fou. Much of this newly exposed area

probably experienced very little erosion because of the low gradient of the area's surface. Analogous modern trailing edge shelf surfaces also have extremely low gradients. Streams that eroded this surface during the regressive phase supplied sediment to the new shoreline forming deltas, beaches, and proximal shelf sand bodies.

Delta Complexes.—The existence of deltas in the Viking Formation is based primarily on the irregular shapes of some of the stratigraphic reservoirs along the eastern pinch-out of the Viking, the position of these fields with respect to the preserved later paleoshoreline (Fig. 15), and the abundant plant fragments and thin coal partings found in these irregular-shaped reservoirs. If the Viking of the Joffre-Joarcam area is mainly made up of linear shelf sandstone bodies and the preserved shoreline facies lies to the west, then one cannot help but wonder if the sandstone bodies comprising the irregular-shaped reservoirs east of the linear shelf sandstone bodies are not indeed deltas that were submerged by the ensuing transgression. Several cores from irregular-shaped fields, located in 10-12-27-SW4 at 2700 feet (930 m) and 2-15-34-10W4 at 2900 feet (970 m), contain thin coal partings and abundant plant debris.

Pebbles found in the conglomerate facies are interpreted to have been transported across the former shelf by streams and deposited initially as channel lags. Later this gravel was reworked during the ensuing transgressive episode.

The transgressive phase, which was punctuated by stillstands and minor regressions, began with a relatively rapid transgression. The initial sea level rise occurred so quickly that the original irregular shape of deltaic sand bodies deposited along the shoreline at the apex of the regressive phase was not altered appreciably by shoreface erosion. The large volume of sand they contained inhibited the shelf hydraulic regime from destroying their amoeboid shape after they were submerged.

Linear Reservoirs.—As the transgressive phase continued, shoreface erosion of sediment deposited during minor regressions and stillstands supplied sand to the shelf that was subsequently restructured into linear sand bodies, such as Joffre and Joarcam Fields, by the shelf hydraulic regime.

During minor regressions, the overlapping nature of the sediment sheets developed. Shelf current continually moved sediment seaward causing the seaward edge of a newly developed sheet to lap onto the landward edge of a previously developed sheet (Fig. 17).

McGookey *et al.* (1972) describe a similar model for deposition of the Newcastle or Muddy, the Viking Formation's equivalent in the northern part of the Rocky Mountain region of the U.S., as follows:

"The presence of the ubiquitous coaly beds above a generally sharp contact with the apparently deep-water Skull Creek Shale indicates

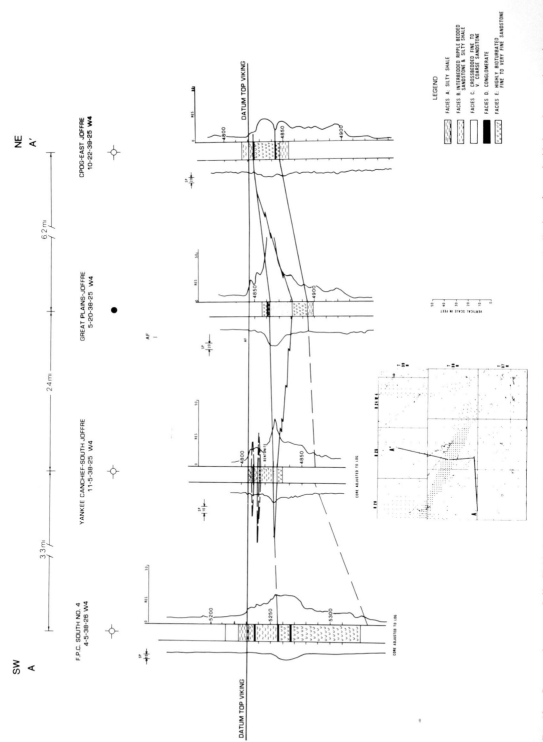

Fig. 13.—Cross-section across Joffre Field. Cored intervals for each of the wells are indicated by patterns. Note distribution of cross-bedded sandstone and conglomerate

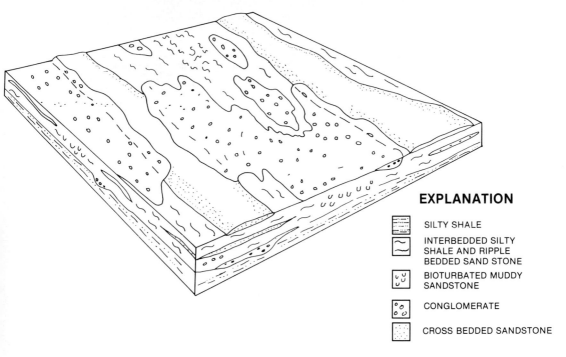

EXPLANATION

SILTY SHALE
INTERBEDDED SILTY SHALE AND RIPPLE BEDDED SAND STONE
BIOTURBATED MUDDY SANDSTONE
CONGLOMERATE
CROSS BEDDED SANDSTONE

FIG. 14.—Schematic diagram of Viking facies distribution in Joffre-Joarcam area.

withdrawal of marine waters from the entire region. The change from mixed Gulf of Mexico and boreal molluscan faunas in the Skull Creek Shale to a fauna that includes only boreal types of the Shell Creek-Mowry interval further emphasizes the abrupt and regional nature of the changes. Furthermore, the Shell Creek-Mowry [Colorado Shale equivalents] time-stratigraphic relationships indicate that the base of this unit is time-transgressive and thus must record a significant marine transgression. Accordingly, it would be expected that the underlying sandstone section predominatly would be the complex transgressive deposits associated with this transgression."

Modern Analog.—Although there are no documented examples of modern epicontinental shelf sediments, the shelf sediments of the East Coast of the United States seem to offer a suitable analog for sediments of the Viking Formation. One model suggests that after a rapid regression sand was deposited on the Atlantic shelf during an overall transgression. As also was probably the case with the Viking, this transgression may also have been punctuated by stillstands. On the Atlantic shelf, stillstands are evidenced by erosional scarps (Fig. 18). Submerged deltas deposited during the low stand at the edge of the shelf seem to be analogous to the irregular-shaped Viking reservoirs. Further landward, slightly younger linear shelf sand

bodies are found and appear to be analogous to the linear sand boides of the Viking.

CONCLUSIONS AND SUMMARY

In conclusion, a basically transgressive model with minor regressive phases best explains deposition of the Viking Formation of central Alberta. This conclusion is based on the following evidence:

1. An apparent unconformity at the base of the Viking supports the interpretation that sea level fell exposing the former shelf to erosion or non-deposition during the pre-Viking regressive phase.

2. The Viking consists of a series of retrograding sediment sheets.

3. Irregular-shaped reservoirs are located east of linear reservoirs which also lie east of the preserved paleoshoreline. If linear reservoirs such as those at Joffre and Joarcam are interpreted to be marine shelf sand bodies and the shoreline and source areas were to the west, then it seems reasonable to interpret the irregular-shaped reservoirs to the east (seaward) as submerged deltaic deposits as suggested by the presence of coal and abundant plant fragments.

An analogy is found on the shelf of the East Coast of the United States where irregular-shaped sands bodies interpreted by some to be submerged deltaic deposits lie seaward of linear marine shelf sand bodies.

4. Thick (8 feet, 2.5 meters) discontinuous conglomerate beds with pebbles one inch in diameter are

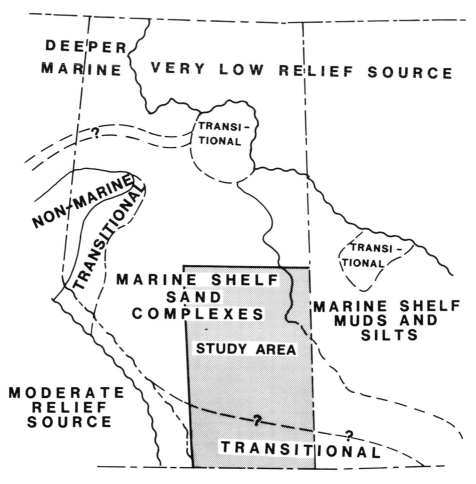

FIG. 15.—Paleogeographic map of Alberta Basin in Canada at the close of Viking deposition (modified from Boeth-ling, 1977b).

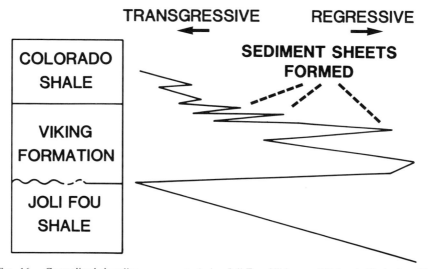

FIG. 16.—Generalized shoreline movements during Joli Fou, Viking, and Colorado Shale deposition.

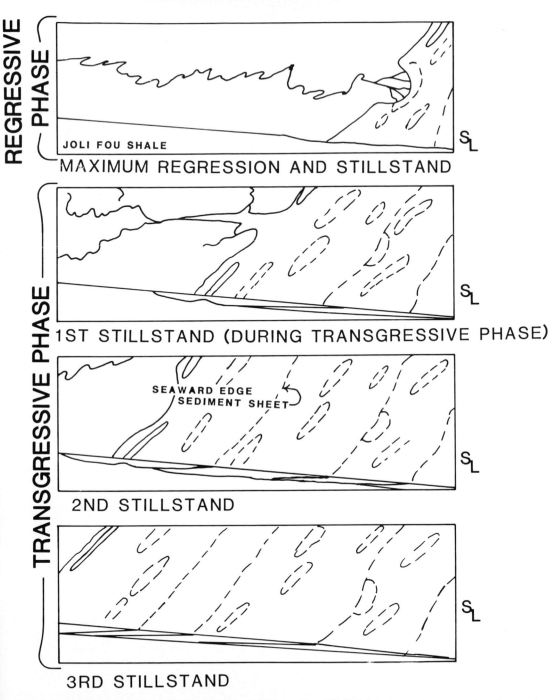

REGRESSIVE PHASE

JOLI FOU SHALE

MAXIMUM REGRESSION AND STILLSTAND

TRANSGRESSIVE PHASE

1ST STILLSTAND (DURING TRANSGRESSIVE PHASE)

SEAWARD EDGE
SEDIMENT SHEET

2ND STILLSTAND

3RD STILLSTAND

FIG. 17.—Diagram of model for Viking deposition which can be subdivided into two parts, a regressive phase and a transgressive phase. Retrogradation nature of sediment sheets developed during still-stands of transgressive phase. Solid and dash-dot lines mark shoreline and former shoreline locations. Dashed lines mark base of shoreface.

found preserved 200 miles (320 km) from the pre-served shoreline facies of the Viking which strongly suggest that streams must have carried the pebbles across the exposed shelf surface during a regression prior to Viking deposition. Pebbles brought onto the shelf were reworked and incorporated into the Viking during an ensuing transgressive phase. Without an initial regression before an overall transgressive deposi-

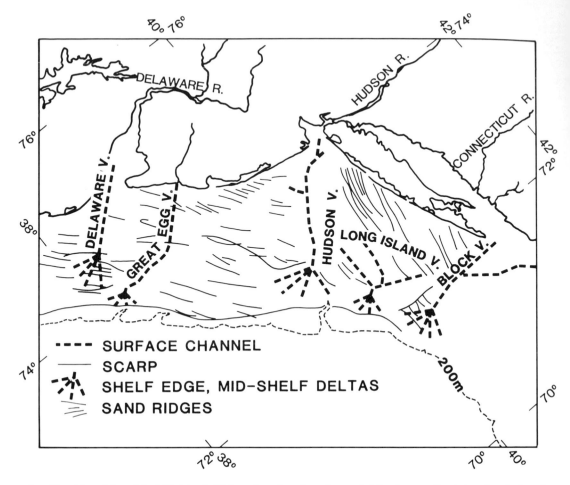

FIG. 18.—Map of part of the Mid-Atlantic Bight region of continental coastal plain of eastern North America. Sedimentary features shown on map are linear sand ridges, submerged deltas, and shoal retreat massifs (modified from Swift, 1976).

tional phase, it might be difficult to explain how so many pebbles were carried so far from shore and deposited in the Viking. Whether shelf currents were powerful enough to move the pebbles so far across such a broad flat surface is unknown.

ACKNOWLEDGEMENTS

I would like to thank Canada-Cities Service Ltd. for permission to publish this paper. This study was completed when I worked in the Basin Study Department of Cities Service Company, Tulsa, Oklahoma. Mike Fowler and John Hobson of that department participated in many discussions about the Viking Formation. Dario Sodero, Douglass Glass, Barry Weeks, Bob Skarako, and Mike Booth all helped me understand much about Viking geology. I would also like to thank Mark Thomas of Petro-Canada and Frank Simpson of Windsor University, Ontario, for discussing their work on the Viking with me.

REFERENCES

BEACH, F.K., 1956, Reply to DeWiel on turbidity current deposits: Alberta Soc. Petroleum Geologists Jour., v. 4, p. 175-177.
BOETHLING, F.C., 1977a, Increase in gas prices rekindles Viking Sandstone interest: The Oil and Gas Jour., Mar. 21, 1977, p. 196-200.
_____, 1977b, Typical Viking sequence: A marine sand enclosed with marine shales: The Oil and Gas Jour., Mar. 28, 1977, p. 172-176.
BRENNER, R.L., 1978, Sussex Sandstone of Wyoming—Example of Cretaceous offshore sedimentation: Am. Assoc. Petroleum Geologists Bull., v. 62, p. 181-200.

EVANS, W.E., 1970, Imbricate linear sandstone bodies of the Viking Formation in Dodsland-Hoosier area of southeastern Saskatchewan, Canada: Am. Assoc. Petroleum Geologists Bull., v. 54, p. 469-486.

JONES, H.L., 1961, The Viking Formation in southwestern Saskatchewan: Saskatchewan Dep. Mineral Res. Rep. 65, 80 p.

KOLDIJK, W.S., 1976, Gilby Viking 'B': A storm deposit, *in* M.M. Lerand (ed.), The Sedimentology of Selected Clastic Oil and Gas Reservoirs in Alberta: Canadian Soc. Petroleum Geologists, Calgary, p. 62-78.

McGOOKEY, D.P., J.D. HAUN, L.A. HALE, H.G. GOODELL, D.G. McCUBBIN, R.J. WEIMER, AND G.R. WULF, 1972, Cretaceous System, *in* Geologic Atlas of the Rocky Mountain Region: Rocky Mountain Assoc. Geologists, Denver, Colorado, p. 190-228.

ROESSINGH, H.K., 1959, Viking deposition in the southern Alberta plains: Alberta Soc. Petroleum Geologists, Guidebook 9th Ann. Field Conf., p. 130-137.

RUDKIN, R.A., 1964, Lower Cretaceous, *in* R.G. McCrossan and R.P. Glaister (eds.), Geological History of Western Canada: Alberta Soc. Petroleum Geologists, Calgary, p. 156-167.

SHELTON, J.S., 1973, Models of sandstone deposits: A methodology for determining sand genesis and trend: Oklahoma Geol. Surv. Bull., v. 118, p. 91-94.

SIMPSON, F., 1975, Marine lithofacies and biofacies of the Colorado Group (Middle Albian to Santonian) in Saskatchewan, *in* W.G.E. Caldwell (ed.), The Cretaceous System in the Western Interior of North America: Geol. Assoc. Canada Spec. Paper 13, p. 553-587.

SLIPPER, S.E., 1918, Viking gas field, structure of the area; Geol. Surv. Canada Summary Rep. 1917, part C, p. 6-7.

STAPP, R.W., 1967, Relationship of Lower Cretaceous depositional environment to oil accumulation, northeastern Powder River Basin, Wyoming: Am. Assoc. Petroleum Geologists Bull., v. 51, p. 2044-2055.

STELCK, C.R., 1958, Stratigraphic position of the Viking sand: Alberta Soc. Petroleum Geologists Jour., v. 6, p. 2-7.

SWIFT, D.J.P., 1976, Continental Shelf sedimentation, *in* D.J. Stanely and D.J.P. Swift (eds.), Marine Sediment Transport and Environmental Management: John Wiley and Sons, New York, p. 311-350.

THOMAS, M.B., AND T.A. OLIVER, 1979, Depth-porosity relationships in the Viking and Cardium Formations of central Alberta: Bull. Canadian Petroleum Geol., v. 27, p. 209-228.

TIZZARD, P.G., AND LERBEKMO, J.F., 1975, Depositional history of the Viking Formation, Suffield area, Alberta, Canada: Bull. Canadian Petroleum Geol., v. 23, p. 715-752.

WILLIAMS, G.D., AND STELCK, C.R., 1975, Speculations on the Cretaceous paleogeography of North America, *in* W.G.E. Caldwell (ed.), The Cretaceous System of the Western Interior of North America: Geol. Assoc. Canada Spec Paper 13, p. 1-20.

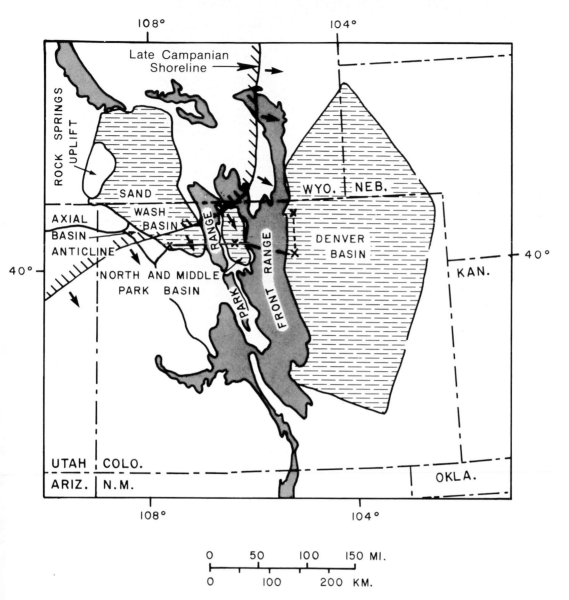

FIG. 2.—Index map showing present basins and positive areas and the approximate position of the late Campanian shoreline. "X" indicates study areas and location of cross-section of Figure 21. After McGookey et al., 1972.

STRATIGRAPHIC SETTING

Sand Wash Basin, northwestern Colorado.—The sequence studied in northwestern Colorado is along the southwest flank of the Sand Wash Basin (Fig. 2). Included in the interval is the uppermost 180 m of the Mancos Shale and the lower 390 m of the Mesaverde Group (Fig. 1). The upper Mancos Shale consists dominantly of shale with locally thick (30 + m) interbeds of sandstone. The lower of these sandstone interbeds was designated the "First Mancos sandstone" by Konishi (1959), and the upper one was formally named

the Loyd Sandstone Member of the Mancos Shale, also by Konishi (1959). *Baculites perplexus* (a middle form) (W. A. Cobban, oral commun., 1979) is found in both of these units, indicating that deposition occurred about 74-76 MYBP (Fig. 1).

The upper Mancos is transitional into the overlying Iles Formation (Hancock, 1925) of the lower Mesaverde Group. Marginal marine sandstones mark the lower boundary of the Iles. The lower of these was designated the "contact sandstone" by Hancock (1925); the upper one is identified as the "Rim Rock"

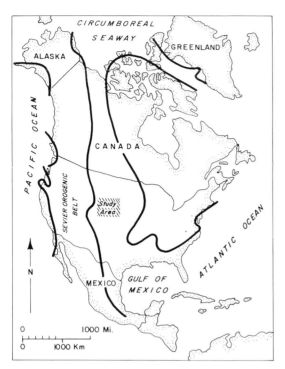

FIG. 3.—Location of Western Interior seaway during late Campanian time (after Gill and Cobban, 1973).

by Konishi (1959) and Tow Creek Member by Masters (1965). In our discussion we use the local name Tow Creek because of the confusion that has arisen from several different sandstones being referred to as "Rim Rock". The "contact sandstone" has been interpreted as a tongue of the Mesaverde, but studies by Kiteley (1980) show that an overall thinning of Mesaverde and lateral pinchout of the "contact sandstone" occurs to the south and west near Axial Basin anticlinal trend where lower Mesaverde brackish-water shales crop out. This pinch-out is related to early Laramide tectonism and a decrease in sediment supply away from distributaries which were located to the north and west (Figs. 4, 6). The Tow Creek is a tongue of the Mesaverde that interfingers westward with nonmarine beds in the main body of the Iles.

The dominantly nonmarine beds which comprise the main body of the Iles Formation are capped by a marine shale and sandstone called the Trout Creek Member (Hancock, 1925). The zonal fossil, *Exiteloceras jenneyi* (72 MYBP, Gill and Cobban, 1973) has been found in shales underlying the Trout Creek to the east near Oak Creek (Izett *et al.*, 1971). This horizon forms the upper limit of the stratigraphic interval studied for this report.

North Park Basin, central Colorado.—Beds exposed in North Park Basin that are correlative with those in the upper Mancos and Iles Formations in

northwestern Colorado (Fig. 1) consist of about 1,500 m of shale. Thin sandstones are interbedded within the Pierre Shale. Sandstones enclosed in the Pierre Shale in North Park Basin range from 2 to 28 m in thickness; each is underlain and overlain by from 60 to 300 m of marine shale. Sandstones included in this study were named, from older to younger, the Kremmling, Muddy Buttes, Hygiene, and Carter Sandstone Members by Izett *et al.* (1971). The Kremmling and Muddy Buttes are in the *B. perplexus* zone, the Hygiene is in the *B. Scotti* zone (about 73 MYBP), and the Carter is in the *Didymoceras stevensoni* zone (72.25 MYBP) (Izett *et al.*, 1971). Correlative units to the west, in ascending stratigraphic order, are the First Mancos Loyd, Tow Creek, and the uppermost part of the nonmarine facies of the Iles.

Western Denver Basin, Colorado.—The Pierre Shale in the western Denver Basin (Colorado) attains a maximum thickness of about 2,100 m. Of this thickness, the middle 1,200 to 1,300 m (Kiteley, 1978) is correlative with the upper Mancos and Iles Formations in northwestern Colorado (Fig. 1).

The interval studied in detail in the western Denver Basin includes about 650 m of interbedded sandstones and shales in the *Baculites scotti* to *Exiteloceras jenneyi* zones. Sandstone in the lower part of the interval was designated the Hygiene Sandstone Member by Fenneman (1905) for an exposure about 9.5 km west of the town of Hygiene in north-central Colorado. About 350 m higher in the section is a poorly exposed sandstone, the Terry Member, named by Schwennesen, Krampert, and Henly (in Ball, 1924). The Hygiene is equivalent to the Hygiene in North Park; the Terry is only slightly younger (about 0.5 my) than the Carter Member to the west. The interval below the Hygiene in the western Denver Basin is composed entirely of shale, and the Kremmling and Muddy Buttes have no counterparts east of North Park Basin.

CHARACTER AND DEPOSITIONAL ENVIRONMENTS OF UPPER CRETACEOUS SANDSTONES OF NORTHERN COLORADO

Sandstones in the Sand Wash Basin, Northwestern Colorado

"First Mancos Sandstone".—The "First Mancos sandstone" (Fig. 1) is a lenticular sand body lying about 100 m below the base of the Iles Formation (Figs. 1 and 7); to the west it onlaps a regressive nearshore sandstone designated the "unnamed sandstone: of *B. asperiformis* age (Kiteley, 1980). It is completely enclosed in silty shale of the upper Mancos Shale. Where exposed between localities 3 and 4 (Figs. 4 and 8), it attains a maximum thickness of approximately 6 m. The mean grain size of the "First Mancos sandstone" is very fine-grained sandstone, coarsening upward to fine, moderately well-sorted sandstone. Bedding consists almost entirely of tabular planar crossbeds in sets approximately 14 cm thick; trough crossbeds are only locally present at the top of the unit.

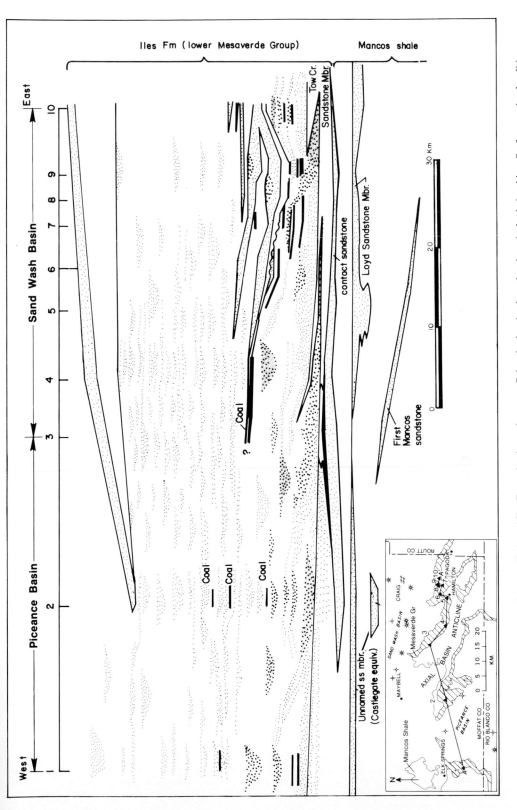

FIG. 4.—Facies diagram of the upper Mancos Shale and Iles Formation in northwestern Colorado showing lateral and vertical relationships. Surface-section localities indicated at top of cross-section are: (1) Coal Creek, (2) Deception Creek, (3) Morgan Gulch, (4) Duffy Mountain, (5) Konishi, (6) Castor Gulch, (7) Bellering Cow, (8) Stock Pass Gulch, (9) Horse Gulch, and (10) Poverty Gulch. Locations of sections are indicated on insert map. Facies indicated are discussed in detail in Kiteley (1980).

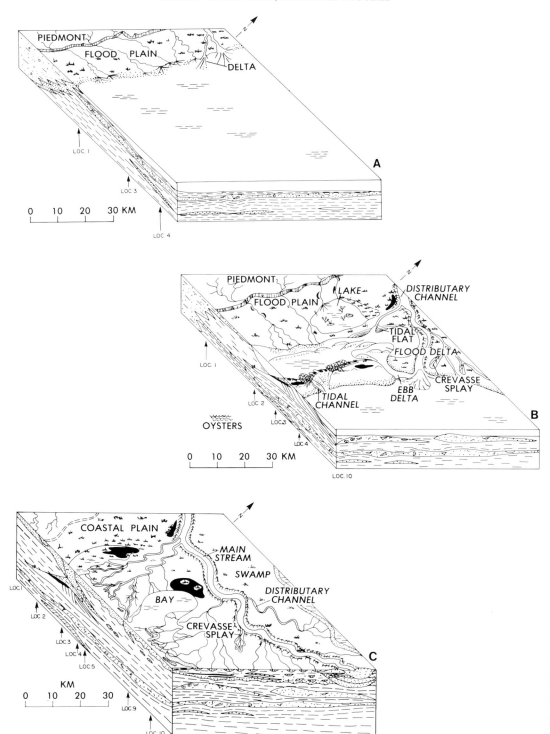

Fig. 5.—Block diagrams of the upper Mancos Shale and Iles Formation in northwestern Colorado showing the general distribution of depositional facies. *A*, Inundation of the region by a marine transgression and deposition of a shoreface sheet sand and mid- to outer-shelf sand ridges offshore. *B*, Deposition of and reworking of distributary mouth bars at the beginning of a regression. *C*, Progradation of small lobate (shoaling) deltas forming a coalesced delta-front sheet sand and distributary mouth bar complex at the peak of regression.

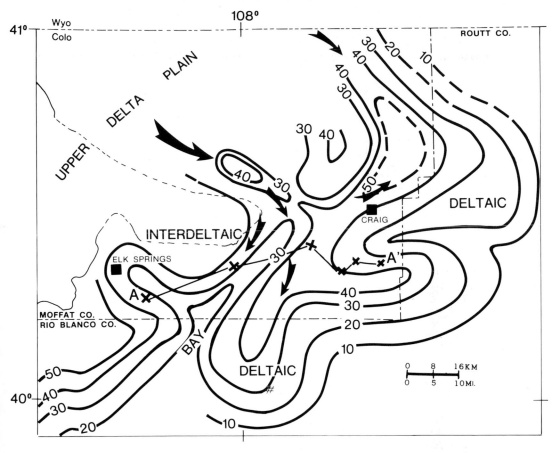

Fig. 6.—Map showing sand trends in the Iles Formation of northwestern Colorado, based on percentage of sandstone versus shale present in 29 wells and 15 outcrop sections. General paleogeography is indicated; dominant direction of sand transport (northwest to southeast) is shown by arrows. A-A' is approximate line of section of facies diagram in Figure 4.

Upper surfaces of bed sets are commonly rippled and contain rounded, iron-stained clay clasts. The dominant direction of transport, as measured on tabular planar cross-beds, was southwestward; a minor, opposed (northeastward) direction of transport was measured in a few places. Both the basal and the upper contacts with the underlying and overlying olive-gray Mancos shale are sharp. Fauna contained in the "First Mancos sandstone" include a diverse assemblage of trace fossils (i.e., *Asterosoma* form *Cylindrichnus* (Howard, 1972), *Ophiomorpha*, *Chondrites*, *Gyrochorte*, etc.), fragments of oysters, a few sharks teeth, and *Inoceramus* and *Baculites*.

The "First Mancos sandstone" is interpreted to have been deposited as a shallow offshore sand-ridge that accreted at an oblique angle to the shoreline as sand waves (represented by tabular-planar crossbedding) migrated nearly parallel to the elongation of the ridge. Trough crossbedding at the top of the unit suggests the occasional formation of dunes, probably during storms; ripples at the tops of sets indicate periods of lower flow velocities probably associated with fair

weather conditions. Rounded claystone pebbles at the tops of rippled surfaces indicate that reworking and redeposition of partially consolidated shelf mud occurred during a period of higher velocity flows. Trace fossils and other fauna in the "First Mancos" indicate open marine conditions and probable lower shoreface environments, based on comparisons with fauna found in sandstones in the Book Cliffs in Utah (Howard, 1972).

Processes that deposited the "First Mancos sandstone" may have been similar to those forming modern linear sand ridges on the shelf off the U.S. East Coast (Duane *et al.*, 1972), or in the North Sea off Holland (Houbolt, 1968; McCave, 1971). These modern ridges extend diagonally across the shelf and are separated at intervals of about 3 km; they have been interpreted to have formed either by strong tidal currents or storms (Walker, 1979). Those in the tide-dominated North Sea are elongated parallel to the tidal current direction. Bedforms consist of sand waves up to 7 m high with megaripples up to 1.5 m high on their crests (McCave, 1971).

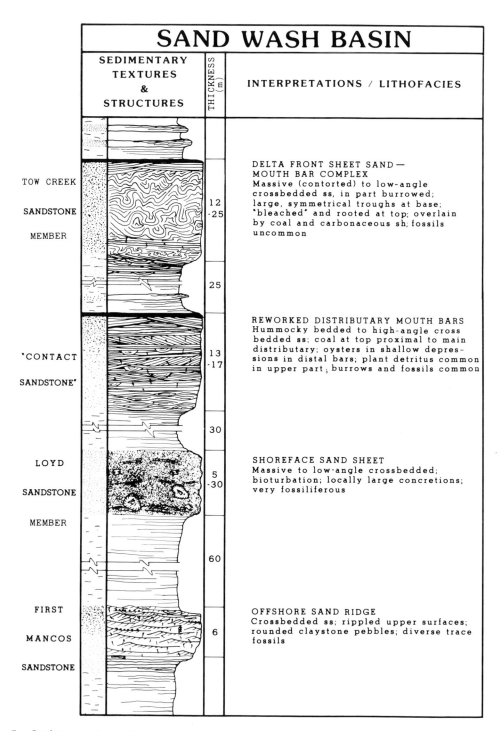

FIG. 7.—Sandstones and depositional facies of the upper Mancos and lower Iles Formations, Sand Wash Basin, Colorado.

FIG. 8.—"First Mancos sandstone" exposed on north flank of Axial Basin anticline in Sec. 25, T5N, R93W.

The largest sets of crossbeds (14 cm) in the "First Mancos" appear to be considerably smaller than those in modern sand ridges, but the general morphology closely resembles modern analogues. The dominance of unidirectionally oriented, planar-tabular cross-sets may indicate either reversing, rectilinear tidal flows, or storm-generated shelf flows. The presence of dunes and rounded claystone pebbles suggests periods of intense sediment transport, associated with storms, but an influence by tidal or oceanic currents cannot be ruled out at this time.

Loyd Sandstone Member.—The Loyd Sandstone Member, lying about 60 m above the "First Mancos" (Fig. 7), is very fossiliferous and grayish-green. It is underlain and overlain by marine shale and apparently forms a broad, laterally persistent sheet that is continuous with the main sandstone body of the Mesaverde Group to the west. Its green color results from the presence of chlorite which, according to Porter and Weimer (1979), is frequently associated with early diagenesis in Cretaceous sediments.

The thickness of the Loyd varies from about 5 m to 30 m. West of Hamilton (Fig. 4) the Loyd is particularly thick, but near Morgan Gulch (location 3, Fig. 4) it is represented by only a few calcareous rounded pebbles. In addition to its distinctive color, the Loyd is commonly characterized by a near absence of bedding (Fig. 9). Low-angle planar crossbeds are present locally, but the bulk of the sandstone appears homogenous. Numerous macroinvertebrates, indicative of high-energy conditions, are present throughout the unit (E. G. Kauffman, written communication, 1979), and both *Ophiomorpha* and *Thalassinoides* are common. Many fossils form the nuclei of calcareous concretions, which range from a few centimeters to about 4 m in diameter.

The inferred depositional environment of the Loyd was the wave-dominated high-energy zone of the shoreface (E. G. Kauffman, written communication, 1979). This interpretation is supported by the great abundance of suspension-feeding bivalves relative to deposit feeders and by the presence of *Ophiomorpha* and *Thalassinoides*. Homogenization of the sand probably occurred from movement of the macroinverte-

FIG. 9.—Loyd Sandstone Member. The unit exhibits little evidence of bedding due to intense burrowing and has a marked green color resulting from authigenic chlorite.

brates on the seafloor and from burrowing by infauna. The diversity of bivalves in the Loyd indicates normal marine conditions (E. G. Kauffman, written communication, 1979). The variable thickness of the sandstone and the lack of any evidence of subaerial exposure suggests that the Loyd was deposited as a broad sand sheet during a period of high influx of coarse clastics. Occasional periods of quiescence are indicated by mud trapped in some *Thalassinoides* tubes.

"Contact sandstone".—The "contact sandstone", 30 m above the Loyd (Fig. 7), consists of a series of lenticular sandstone bodies that overlie olive-gray shale (Fig. 10). The sandstones are about 13-17 m thick. Near Morgan Gulch (location 3, Fig. 4), where the sandstones are thickest, coals and carbonaceous shales overlie the unit, but to the south and west the sandstones thin, are burrowed throughout, and are overlain by bioturbated shales and sandstones. The westernmost units of the "contact sandstone" are found at Deception Creek (location 2, Fig. 4) where they pinch out into burrowed, olive-gray shale containing oysters. Many oysters also occur in the top of the sandstone at this locality in scoured depressions (Fig. 11A). Woody plant fragments and coalified logs also occur in pockets in the sandstone under these depressions.

The mean grain size in the "contact sandstone" coarsens upward from very fine to fine-grained sandstone. Grains are subangular and dolomite-cemented at the base, but near the top they are well rounded, very well sorted, and loosely packed. Bedding consists of long, low-angle, westward- to eastward- dipping laminations (Fig. 11B) in the upper part and mainly of hummocky bedding and ripples near the base. Tabular- and wedge-planar crossbeds commonly occur near the middle of the unit. The dominant paleocurrent flow direction, measured near the base of the sandstone near Morgan Gulch, was from northeast to southwest.

Ophiomorpha, in large numbers, occur primarily in the lower half of the sandstone; shell hash is commonly concentrated along bedding planes in the upper part. Specimens of *Inoceramus* and *Crassostrea sp. aff. C. subtrigonalis* (Evans and Shumard) and *Clione* have been identified from the sandstone at the Deception Creek locality (E. G. Kauffman, written communication, 1979).

Units comprising the "contact sandstone" are interpreted on the basis of their geometry, sedimentary structures, and facies relationships to be reworked distributary mouth bars. Brackish-water fauna, (*Crassostrea* and *Clione*, in the "contact sandstone" at Deception Creek; Fig. 11A; E. G. Kauffman, written communication, 1979) indicate the existence of a brackish bay landward of the bars which prograded from the north and west. Highly burrowed shales above are typical of marginal bay or lagoonal environments (Dickinson *et al.*, 1972).

Internal characteristics of the reworked distributary mouth bars indicate that marine processes were dominant. Hummocky-cross-stratification and ripples near the base suggest that deposition was initiated by eposodic movement of sand grains on the sea bottom.

A

B

FIG. 10.—Photographic panorama (A), and sketch (B) of the lenticular sandstone units of the "contact sandstone" (or I-1 sandstone) and the sheet-like Tow Creek Sandstone (I-2) of the Sand Wash Basin. View is to the northeast (perpendicular to depositional strike); location of exposure is between locations 4 and 5 (see Fig. 4). Several similar sandstone bodies comprise the "contact sandstone" between Deception Creek and Castor Gulch (Fig. 4).

FIG. 11.—"Contact sandstone" of the basal Iles Formation in the Sand Wash Basin. A—Oyster coquina (o, near top of pole) in shallow depressions at top of "contact sandstone" at Deception Creek (loc. 2, Fig. 4). B—Long, low angle laminations in the middle to upper part of the "contact sandstone."

At most localities the lower contact of the river mouth bars with the underlying shale is sharp, indicating that the bars accumulated in scoured depressions. A few localities, however, show gradational lower contacts which indicate more distal, prograding bars.

As a modern analogue, the reworked distributary mouth bars that produced the "contact sandstone" resemble generally those described by Coleman and Prior (1980) in areas like South Pass in the Mississippi Delta region. Coarse, fragmental plant and woody debris noted in the upper part of the "contact sandstone" are analogous to the "river-transported organic debris", or "coffee grounds" found at the tops of bars in the Mississippi River (Coleman and Prior, 1980). Other sedimentary structures in the "contact sandstone" are distinctly marine, however, indicating that the interdistributary bay was located far enough away from the main channel of the delta that river processes did not play an important role.

Tow Creek Sandstone Member.—The Tow Creek Sandstone Member (Fig. 7) is a coarsening upward sheet sandstone that is present both above and over a broad area that extended far to the east of the reworked distributary mouth bars. The Tow Creek lies about 25 m above the "contact sandstone" and varies in thickness from about 10 to 30 m. Thicknesses commonly vary considerably over very short lateral distances (10-20 m) (see Fig. 12). Maximum thicknesses of the Tow Creek are obtained in the vicinity of Hamilton; to the east, the sandstone grades into marine shale (Kiteley, 1980). To the west the unit interfingers with, and is overlain by, a thick interval of carbonaceous shales and lenticular sandstones of fluvial origin.

The basal contact of the Tow Creek is commonly marked by broad trough shaped scours that are generally symmetrical and filled with burrowed, very fine-

Ophiomorpha burrows associated with hummocks and ripples are commonly truncated, indicating rapid influx and deposition of sand, either during flood periods or storms, or a combination of the two. Overlying tabular- and wedge-planar bedding resulted from megaripple (sand wave) migration, probably at subtidal depths as indicated by numerous *Ophiomorpha*. Long, low-angle seaward-to landward-dipping laminations near the top of the sandstone indicate intense reworking of the tops of the bars by wave-swash (Thompson, 1937; Hunter *et al.*, 1972). The presence of coal or carbonaceous shale capping the bars indicates that freshwater marshes migrated seaward over the later abandoned distributary mouth bars. Small tidal channels filled with oyster debris at the tops of some of the more distal bars indicate that currents flowed across and into a bay behind the bars. This bay was quite large, open to tidal exchange, and filled predominantly with olive-gray shale and thinly interbedded and locally cross-bedded sand.

FIG. 12.—Tow Creek Sandstone Member; view looking east up Castor Gulch in Sec. 21, T5N, R91W. Thick, massive (m) parts were deposited by rivers as distributary mouth bars; adjacent thin parts (t) represent laterally spreading delta front sheet sands. Prodelta sands, silts, and shales and bay-fill sediments (d) in slope below Tow Creek.

grained sandstones, siltstone, and carbonaceous shales. Rounded claystone pebbles occur along bedding planes within these troughs. Troughs are eroded into either olive-gray marine shale containing carbonaceous debris, or siltstones interbedded with olive-gray shales.

Massively bedded, thick (up to 34 + m) sandstones comprise the upper part of the Tow Creek, which has an irregular, convex, upper surface (Fig. 12). The upper, massive parts of the Tow Creek are generally deformed internally and are coarser grained than underlying beds. Undeformed beds show very low angle, generally seaward-dipping, planar laminae and vertical burrows. These thick, massive sandstones of the Tow Creek grade laterally to about 10-meter thick, crossbedded sandstones (Fig. 12; and I-2 sandstone, Fig. 10) that closely resemble those found in the "contact sandstone". Paleocurrent measurements indicate that currents flowed east to west in the erosional troughs at the base of thick, massive sandstones and northeasterly to southwesterly (longshore) in the adjacent cross-bedded sandstones. *Ophiomorpha* and shell hash accumulations are common locally in upper parts of the Tow Creek.

The top of the Tow Creek is capped by "bleached" and rooted zones; oysters are present in carbonaceous (nonmarine) shales above the unit. *Teredolithus* borings in wood are present at the base of many of the lenticular sandstones within the nonmarine shales.

The Tow Creek Sandstone Member is interpreted as a delta front sheet sand-mouth bar complex that built seaward over olive-gray shales and interbedded siltstones deposited in interdistributary or prodelta environments. Thick accumulations of massive-appearing sandstone adjoining thinner sheet sandstones are interpreted as the deposits of distributary mouth bars, whereas the thinner sandstones probably represent the laterally spreading sheet sands which developed at the mouths of the distributaries. Recognition of the sheet sandstones and mouth bars is based mainly on their geometry and position relative to one another, as many of the sedimentary structures in the sheet sandstones resemble closely those in the "contact sandstone," deposited in shoreface and other high-energy open-marine environments.

Distributary mouth bars in the Tow Creek sandstone resulted from a decrease in velocity and competence of the stream as it discharged at the coast in the manner described by Coleman and Gagliano (1965) for the Mississippi River. The massive and deformed beds in the upper Tow Creek reflect this mode of deposition. The build-up of closely-spaced, sandy, mouth bars to a level exceeding that of the adjacent sheet sand (Fig. 12) suggests a shallow-water, laterally spreading, river-dominated delta with many distributaries, similar to those described by Fisk (1955). In contrast to the "contact sandstone", the main channel of the delta in the Tow Creek is interpreted to have been very close to the laterally spreading centers of deposition. Overlying

beds in the main body of the Iles support this interpretation, showing: (1) overall coarsening upward; (2) a predominance of fluvial facies; (3) pronounced coastal progradation; (4) thick delta plain sequences that are vertically distinct from thin destructive facies; and (5) low sand to mud ratio.

Sandstones in North Park Basin, Colorado

Kremmling Sandstone Member.—The Kremmling Sandstone Member (Fig. 13), where exposed in a prominent bluff at Kremmling, is a 21 m thick, broad sheet sandstone that dips gently to the north (Fig. 14A). It consists of thinly interbedded, lenticular sandstones, siltstones, and sandy shales (Fig. 15B) that coarsen upward into a more uniform, vertically burrowed, fine-grained sandstone at the top. The base grades into marine shale, whereas the upper surface is in sharp contact with the overlying marine shale (Fig. 14A).

Lenticular sandstones in the lower part of the Kremmling (Fig. 14B) are up to 6 m in length and have flat internal laminations and asymmetric (current) ripples at the tops of the beds. Bed sets are up to about 8 cm thick; ripples generally have spacings of about 10 cm. Each of the sandy interbeds is essentially unburrowed, but underlying and overlying siltstones and shales are intensely burrowed (Fig. 14B), mainly by profuse *Phycosiphon* and *Helminthoida* (K. Chamberlain, oral commun., 1981). *Diplocraterion*, *Schaubcylindrichnus*, and *Aulichnites*(?) were also observed. The top of the Kremmling is covered by abundant tracks, trails, and smooth-walled, sand-filled meandering tubes (Fig. 14C) which were left by deposit-feeding organisms that inhabited the muddy substrate above. *Gyrochorte* and *Corophioides* occur near the top of the Kremmling and inoceramids, in living position, also occur at this horizon.

The Kremmling Sandstone Member is interpreted to be a series of prodeltaic storm-deposited sandstones, siltstones, and shales, which were deposited as sheets in the deeper portions of the shelf. Characteristics of the Kremmling closely resemble those found in the offshore to lower shoreface transition of the Book Cliffs in Utah (Howard 1972). The relatively clean, parallel-laminated sandstones with rippled surfaces suggest storm or tidal influences and high rates of deposition. *Inoceramus* shells near the top indicate that the upper sand sheets were deposited in open marine environments.

Muddy Buttes Sandstone Member.—The Muddy Buttes Sandstone Member (Fig 13) forms a thin (2.0 m) sheet-like outcrop about 60 m above the Kremmling. The base of the sandstone is erosional and overlies olive-gray marine shale. The basal boundary is marked by a distinct bed of phosphate pebbles mixed with iron-stained pebbles of claystone (illite and kaolinite). In contrast to the Kremmling, the Muddy Buttes is composed entirely of silty, very fine-grained, well-bedded, rippled sandstone (Fig. 15). Bed sets are thin (1-4 cm) with a few vertical burrows. *Phycosiphon*, a deposit feeder (Chamberlain, oral commun., 1981) is very common throughout the sandstone. The upper surface of the sandstone is marked by tracks and traces, including *Rhyzocorallium*, and sand-filled meandering tubes. There are only minor differences between the trace fossils here and those at the top of the Kremmling. Ripples in the Muddy Buttes Sandstone Member consist of about 10-cm-amplitude ripples that alternate with smaller scale climbing ripples. Linguoid ripples occur at the tops of beds.

The Muddy Buttes Sandstone Member, like the Kremmling Sandstone Member, is interpreted as a prodelta deposit. However, there are differences between the two sand bodies that suggest minor differences in environmental setting. For example, the dominance of very fine-grained, sand-sized particles over silt or clay and the presence of vertical tubes in the Muddy Buttes may indicate periods of time of shoaler water conditions when suspension feeders are generally more abundant. The presence of *Phycosiphon*, however, indicates that sand deposition was sporadic, i.e., interrupted by many episodes of quiet-water (shale) deposition. *Rhyzocorallium* at the top of the sandstone suggests sublittoral depths, below average wave base in quiet offshore areas where well-sorted silts and sands are present (Frey, 1976).

The Kremmling and Muddy Buttes Sandstones correlate with the "First Mancos" sandstone and the unnamed sandstone in northwestern Colorado, respectively. These sandstones change upward to high-energy bedforms, recording a time equivalent water-shoaling event in two widely separated areas of the depositional basin. The shoaling event was probably controlled by a sea level drop that was uniform throughout the basin, although bottom conditions varied from place to place within the basin.

Hygiene Sandstone Member.—The Hygiene Sandstone Member (Fig. 13) lies about 300 m above the Muddy Buttes Member at the top of a thick interval of marine shale (Izett *et al.*, 1971). This 300 m of shale is correlative with only 30 m in the Sand Wash Basin to the west, suggesting that rates of subsidence were much greater in the east.

The Hygiene in North Park Basin is sheet-like (Fig. 18) and varies in thickness along strike from about 15 m to about 30 m. Several northwestward- to southeastward-trending thick sandstones with convex-upward tops are exposed between thinner and more uniform sandstones. The thick sandstone accumulations are separated by thin sandstones at intervals of 90 m, and their breadth is about 15-30 m.

The internal structures of the thick accumulations of sandstone are either massive to contorted and bioturbated or in some places composed of very thick sets (up to 2.4 m) of large-scale indistinctly foreset-bedded sandstone. The bedding in the thinner parts appears to be trough cross-beds in sets up to 0.3 m thick (Fig.

17A). Planar flat beds or hummocky cross-bedding cap the sequence (Fig. 17B). The unit coarsens upward from a sandstone, siltstone, and sandy shale about 12 m thick to a dominantly sandy interval about 17 m thick. *Schaubcylindrichnus* (white-walled burrows; Chamberlain, 1980), meandering worm tubes, *Helminthoida* (Chamberlain, oral communication, 1981) and *Thalassinoides* occur in the sandstone. The *Helminthoida* appear to be restricted to the lower part of the unit. Disarticulated oysters (Fig. 17C), bivalves, woody fragments, and organic debris occur in the middle to upper part of the sandstone along with *Ophiomorpha* and *Schaubcylindrichnus*; topmost beds contain *Inoceramus*.

Internal structures and geometry of the Hygiene Sandstone Member at Kremmling suggest deposition as a shallow-water (shoreface) probably delta-front sandstone. Thickening and thinning of the unit may indicate deposition in distributary channels and distributary mouth bars. Trough cross bedding in the sandstone and the presence of many brackish to shallow marine trace and body fossils indicate the close proximity of a strandline. This strandline probably lay to the north as indicated by the presence of coal in time-equivalent beds in North Park (Hail, 1965).

The transitional, burrowed, lower part of the Hygiene near Kremmling represents a prodelta environment where thin sand beds were deposited by storms on a muddy sea bottom. Extensive bioturbation by *Helminthoida* indicates a high content of organic nutrients in the sediment. The overlying thin to thick sand sheet is a part of the prograding delta front environment. Thick, massive and contorted, bioturbated sandstones are the result of intense burrowing by infauna in shoreface environments where organic detritus was high. The presence of disarticulated oysters, bivalves, and organic debris in the sand probably indicates periods of flooding when material was transported seaward out of brackish bays and marshes.

The upward change to a hummocky cross-bedded, *Inoceramus*-bearing sandstone indicates that conditions changed to open marine as water deepened over the delta front sandstone.

Carter Sandstone Member.—About 105 m of shaly siltstone and sandy shale separate the Hygiene Member from the overlying Carter Sandstone Member in North Park Basin (Fig. 13; Izett *et al.*, 1971). The Carter is an elongate south-southeast-trending sandstone body (Fig. 18A) that outcrops in a gently folded synclinal-like depression in the E ½ Sec. 5 T. 2 N. R. 80 W. The member consists of upper and lower parts which show distinctly different bedding characteristics. Bedding in the 15 meter-thick lower part consists mainly of trough and wedge-planar crossbeds (Fig. 18B) with a generally unidirectional southeastward transport. Hummocky cross-stratification and contorted units are also quite common in the lower part to the southeast (J. Harms, oral commun., 1981). Rounded claystone pebbles and *Teredolithus*-bored wood are present along

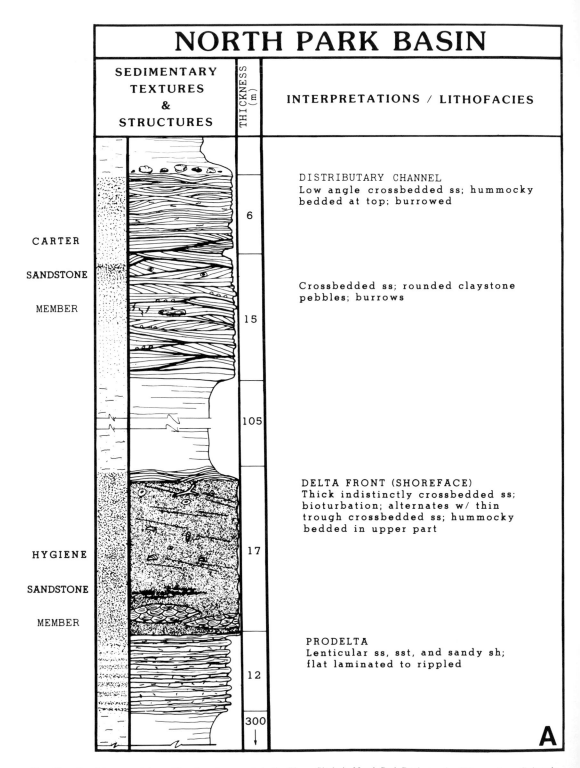

Fig. 13.—Sandstones and depositional environments in the Pierre Shale in North Park Basin north of Kremmling, Colorado.

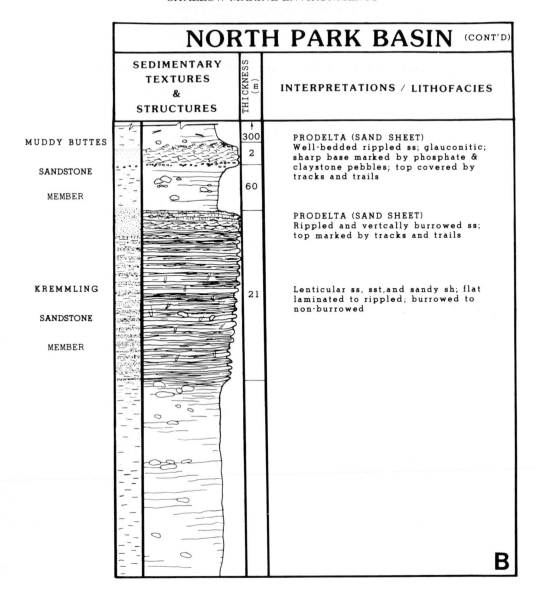

NORTH PARK BASIN (CONT'D)

SEDIMENTARY TEXTURES & STRUCTURES	THICKNESS (m)	INTERPRETATIONS / LITHOFACIES

MUDDY BUTTES

SANDSTONE

MEMBER

300
2
60

PRODELTA (SAND SHEET)
Well-bedded rippled ss; glauconitic; sharp base marked by phosphate & claystone pebbles; top covered by tracks and trails

PRODELTA (SAND SHEET)
Rippled and vertcally burrowed ss; top marked by tracks and trails

KREMMLING

SANDSTONE

MEMBER

21

Lenticular ss, sst,and sandy sh; flat laminated to rippled; burrowed to non-burrowed

B

some of the lower bounding surfaces of trough scours, and many burrows resembling *Macaronichnus segregatis* (Clifton and Thompson, 1978) occur randomly. No marine macroinvertebrates were found within the lower part at this locality, but *Schaubcylindrichnus*, *Trichichnus*, and *Asterosoma* are relatively common along with rare *Thalassinoides* and *Ophiomorpha irregularis* (K. Chamberlain, oral communication, 1981).

The six-meter thick upper part contains sandstones in three to four centimeter thick bed sets. These sets are either very low angle cross-bedded or hummocky cross-stratified (Fig. 18C); burrow types consist entirely of meandering tubes that resemble the *Macaronichnus segregatis* (Clifton and Thompson,

1978) found in the lower part. The upper contact of the unit is sharp, and marine shale containing ammonites within limey concretions and phosphate nodules overlies the Carter Sandstone Member (Izett *et al.*, 1971).

The geometry and sedimentary structures in the Carter Sandstone Member suggest deposition of a high-energy, possibly estuarine or distributary channel, in a shallow shelf setting during a low stand of the sea. The presence of many storm deposits (hummocky bedding) and contorted layers suggests periods of intense reworking as sand was brought into the area. *Schaubcylindrichnus* traces indicate water depths of about 9-18 m (K. Chamberlain, oral commun., 1981), and current readings show a dominant northwest to southeast flow. The abrupt change from troughs and wedge-

Fig. 15.—Thin, rippled beds of silty, very fine-grained sandstone comprising the Muddy Buttes Sandstone Member of the Pierre Shale.

Fig. 16.—Hygiene Sandstone Member in North Park showing variable thickness. Thick sandstones are massive to contorted and bioturbated (m), trough cross-bedded in thin parts (t).

Fig. 14.—Kremmling Sandstone Member in North Park Basin. *A*, Kremmling Sandstone Member; town of Kremmling, Colorado, in distance. View looks east, approximately parallel with depositional strike. *B*, Lenticular sandstone bed (current ripples at top) and interbedded, bioturbated siltstones and sandy shales. Length of ruler is 15 cm. *C*, Tracks, trails, and smooth-walled meandering tubes at top of Kremmling Sandstone Member.

planar cross-bedding in the lower part to flat bedded in the upper part probably resulted from lateral migration and eventual abandonment of the channel and tidal reworking of earlier formed channel deposits.

Marine burrows are present in both the lower and upper parts, but *Teredolithus*-bored wood, which is found only in the lower part, commonly indicates fresh water to saline conditions (Siemers, 1976). Burrowing infauna resembling *Macaronichnus segregatis* (Clifton and Thompson, 1978), are known only from modern

intertidal to shallow subtidal environments in zones of very active sedimentation. *Macaronichnus* tubes are thought to be a feeding structure of shallow marine polychaetes (Clifton and Thompson, 1978). Marine shales above, which contain ammonites and phosphate nodules (Izett *et al.*, 1971), indicate inundation of the area by a subsequent transgression and resumption of slow rates of deposition. The widespread distribution of shales overlying the unit, as represented by the Trout Creek Shale in northwestern Colorado and Lewis Shale above the time-equivalent Teapot Sandstone in Wyoming, suggests that this sea level change was basin-wide (eustatic).

Sandstones in the Western Denver Basin, Colorado

Hygiene Sandstone Member.—The Hygiene Sandstone Member in the western Denver Basin is a coarsening upward, southward-thickening sediment wedge that grades eastward and southward into marine shale (Kiteley, 1978). The unit is about 90 m thick along the Colorado-Wyoming border and thickens to about 180 m east of the Front Range (Fig. 2) in the vicinity of

Fig. 17.—Hygiene Sandstone Member in North Park Basin. *A*, Shallow trough cross-beds. *B*, Hummocky cross-bedding near top. *C*, Disarticulated oysters attached to Bryozoa.

Boulder (Kiteley, 1978). The overall grain size is finer to the south; the thickening near Boulder results from an increase of sandy shale and siltstone in the lower part accompanied by a decrease of sandstone near the top.

At Round Butte, near the Wyoming border, the unit consists of repetitive upward-coarsening cycles of sandy shale, siltstone, and sandstone that are from about 15 to 30 meters thick (Fig. 19). Two textural cycles are present at this locality; a possible third, younger cycle is exposed up-dip and south of the first

two. Each of the two older cycles consists of a thin (6.3 to 8 m) cross-bedded unit at the top (Fig. 20A) and a thick, basal flat-laminated to rippled unit, about 9 to 23 m thick.

The thick, basal flat-laminated to rippled part of each cycle within the Hygiene consists of thin (2 to 3 cm), lenticular sandstone beds that are intercalated with sandy shale and siltstone. Lenticular sandstone beds are very fine grained and commonly well bedded; they have sharp upper and lower contacts. Sandy shales and siltstones between sandstone interbeds are bioturbated. Abundant finely disseminated plant detritus is present, especially farther south in the vicinity of Boulder.

The thin, cross-bedded sandstones at the top are thick bedded, fine to medium grained, glauconitic, and have sharp to gradtional lower contacts. The contact at the top between marine shale and sandstone is very sharp. In addition to glauconite, sand grains consist of subrounded and moderately well-sorted quartz, feldspar, chert, and mica. Cross-bedding in the upper sandstones is medium- to high-angle planar-tabular (Fig. 20A) with a consistent dip direction to the south-southwest. Cross-bed sets are 0.4 to 0.5 m thick and are commonly truncated at the tops. Rounded clay clasts are a common feature along bedding planes. In plan view, cross-bed sets appear as long, sinuous crested, asymmetric megaripples or sand waves (Fig. 20B) with spacings of 3 to 5 m. Linguoid ripples cover the lee sides of the sand waves. Direction of migration of the ripples was also southward.

Ophiomorpha and *Asterosoma* are common within the thin upper cross-bedded sandstones, and horizontal smooth tubes, U-in-U burrows, and various tracks and trails occur on bedding surfaces and burrowing to a depth of about 3-4 cm is observed locally.

Thin cross-bedded sandstones in the upper Hygiene in the western Denver Basin are interpreted as the deposits of offshore sand-ridge complexes (termed "bars" by Porter, 1976, and "sand sheets" and "bars" by Kiteley, 1977). The basal flat-laminated to rippled units resemble those of the lower Hygiene and Kremmling Sandstones in the more westerly North Park Basin and probably represent deposition of storm-deposited sand, silt, and mud in a prodelta environment. The thin, cross-bedded sandstones at the top of the the thick basal units appear to have formed in an open-marine, high-energy shelf environment as elongate, discrete sand ridges. Unidirectional southward migration of the ridges over older prodelta deposits is indicated. A northerly source is suggested by the southward decrease in mean grain size.

The Hygiene probably accreted as successive layers of local blanket-type deposits built one above the other. The reason for remobilization and remolding of sands at the top of each cycle is not clear, but Nio (1976) has described ancient sand-wave complexes in Spain, southern England, and Switzerland, and relates their origin to the onset of a marine transgression. The sand

FIG. 18.—Carter Sandstone Member in E ½, Sec. 5, T2N, R8OW. *A*, View of Carter Sandstone Member, looking northeast. *B*, Trough and wedge-planar cross-bedding in lower 15 meters of the Carter. *C*, Hummocky cross-bedding in upper six meters of the Carter Sandstone Member.

waves described by Nio (1976) were 5-20 m high and resulted in large, cross-bedded sand bodies. Field and others (1981) and Nelson *et al.* (1981) describe similar sand wave complexes from the modern setting of the northern Bering Sea, which is also undergoing transgressive reworking.

Terry Sandstone Member.—The Terry Sandstone Member lies about 75 m above the Hygiene Member at Round Butte and is a friable sandstone and shale interval that has limited exposures. At Round Butte it consists of very fine-grained, rippled, and flat-laminated sandstone. Abundant limy concretions and phosphate nodules are present within the underlying and overlying shales. The Terry Member, slightly younger than the Carter Member to the west, was deposited in elongate northwest to southeast trending lobes that "climb stratigraphically in a westerly direction" (Moredock and Williams, 1976). The depositional environment of sandstone bodies in the Terry Member is difficult to interpret owing to the limited exposures, but Moredock and Williams (1976) have interpreted tidal channels, island sand ridges, and nearshore bars from their analyses of cores.

DISCUSSION

Relationship of Shelf Sand Bodies, Northern Colorado to Sea Level Changes

In this paper, an analysis has been made of time-contemporaneous units in widely separated areas of the Upper Cretaceous of northern Colorado in order to examine the relationship between these units and changes in sea level that have been attributed to eustatic controls. Since the early 1960's, the evidence in support of eustatic sea level changes in the Cretaceous has increased (i.e., Sloss, 1963, 1972; Hallam, 1963, 1971; Hays and Pitman, 1973; Flemming and Roberts, 1973; Rona, 1973a, b; Sleep, 1976; Vail and Wilbur, 1966; Vail *et al.*, 1977; Kauffman, 1973a, b, 1977; and Hancock and Kauffman, 1979). Most of these workers agree that eustatic sea level changes in the Cretaceous were caused by variations in the volume of mid-oceanic ridge systems which resulted from differential rates of sea floor spreading. Such changes are believed to have altered the capacity of ocean basins and resulted in transgressions and regressions on a global scale.

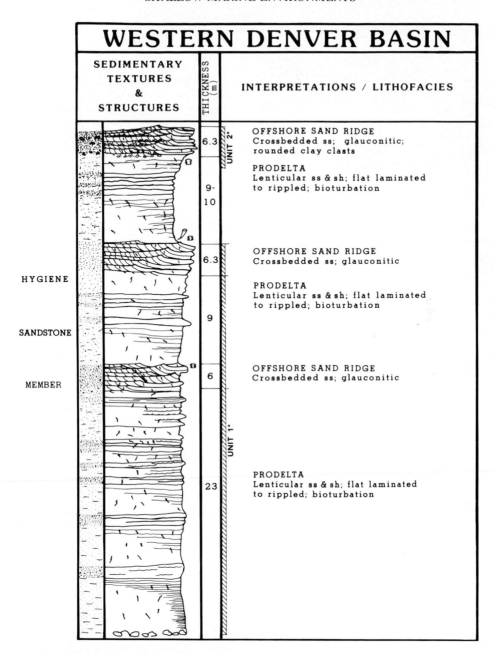

FIG. 19.—Repetitive upward-coarsening cycles of prodelta to offshore sand-ridges in the Hygiene Sandstone Member of the middle Pierre Shale at Round Butte near Ft. Collins in the western Denver Basin. Interval samples for mean grain size and sorting (see Fig. 22) are indicated by asterisks.

While some studies propose one or two major cycles of sea level rise and fall for the Cretaceous period, Kauffman (1977, 1979) and Hancock and Kauffman (1979) propose ten transgressive-regressive cycles for the Western Interior and parts of England. Our comparison of the trangressive-regressive history of the in- terval studied in northern Colorado with the model of Kauffman (1977, 1979) and Hancock and Kauffman (1979) shows a close agreement in timing of sea level changes (Fig. 21). In detail, however, facies across the northern Colorado Late Cretaceous shelf area show a variability that is not easily related directly to sea level

FIG. 20.—Hygiene Sandstone Member, western Denver Basin. *A*, Tabular-planar cross-bedding in upper six to eight meter cycles comprising the Hygiene. Current flow was from left to right. *B*, Top view of long, sinuous-crested megaripples or sand waves in the Hygiene Sandstone Member at the type locality west of Hygiene, Colorado.

changes. We attribute part of this variability to periods of non-deposition during times of rapid rise or fall of sea level or to "turbulence events" that accompanied sea level changes.

In the Sand Wash basin to the west, the vertical change upward is from a regressive sandstone unit, the "unnamed sandstone" (the lateral equivalent of the extensive Castlegate Sandstone Member of Utah), to a transgressive offshore sand ridge deposit, the First Mancos sandstone, which exhibits seaward to landward transport, to a shoaling shoreface sheet sandstone (the Loyd Member), and finally to delta front sandstones of the overlying Lower Mesaverde (the "contact sandstone" and the Tow Creek Member). This complete cycle is designated by Kauffman (1979) as the R_7 regression, the T_8 transgression, and the R_8 regression. The sequence that overlies the Tow Creek is represented by strike-elongated sandstone barrier bars or beaches that reflect transgressive reworking during a general rise in sea level (the early T_9 transgression). The subsequent regression (early R_9) is represented

entirely by continental beds of the upper Iles Formation. This regression, which has only been documented in northern Colorado, apparently resulted from local uplift in northwestern Colorado (Kiteley, 1983).

In North Park basin to the east, the unnamed sandstone (R_7 regression) is represented by a thick, upward-coarsening prodelta facies (the Kremmling Sandstone Member) which is overlain by marine shale. The Muddy Buttes Member above, in the *Baculites perplexus* zone, is scour based and finer in grain size than the Kremmling. The upward-fining of grain size suggests that the Muddy Buttes was deposited during a sea level rise at or below storm wave base. The Muddy Buttes, which is equivalent to the First Mancos sandstone in the Sand Wash Basin, approximately coincides with the T_8 transgression of Kauffman (1977, 1979). The Hygiene Sandstone Member at the top of a thick shale sequence above the Muddy Buttes was deposited as an upward-coarsening prodelta capped by delta front (channel mouth bars and lateral sheet sands) in a shoreface environment. Abrupt deepening of water over the delta (early T_9 transgression) front is indicated by hummocky-bedded storm deposits at the top. The Hygiene represents the R_8 regression of Kauffman (1977, 1979). The Carter Sandstone Member, which overlies marine shale above the Hygiene, is a shallow-water distributary channel with a marine-reworked top. The Carter is the lateral equivalent of continental beds in the Sand Wash Basin which were deposited during the early R_9 regression.

In the western Denver basin, deposits of the R_7 regression are absent due to non-deposition and only offshore marine shale representing the T_8 transgression is found in this interval. The R_8 regression is represented by the Hygiene Sandstone Member which consists of multiple stacked prodelta facies capped by southward accreting sand ridge deposits. Because of the close analogy between upper Hygiene sand ridge deposits and sand ridges in the first Mancos, which were deposited during the T_8 transgression, the implication is that upper Hygiene Sandstones in the western Denver basin may represent transgressive reworking of material supplied from prograding deltas located nearby. The early T_9 cycle is represented above the Hygiene by marine shale capped by onlapped, vertically stacked sandstones of the Terry that may have originated in settings that were similar to those of the Carter Sandstone Member in North Park Basin. The vertical stacking of these sandstones and their onlapped relationship, however, indicate that they were deposited during the subsequent transgression when oceanic currents impinged on the shelf.

During the major regression that occurred in the Tow Creek-Hygiene interval, an abrupt seaward shift of nearshore wave regime is evidenced by small symmetrical scour channels that occur at the base of delta front sandstones in the Sand Wash Basin. These channels are filled with sandstone and siltstone capped by carbonaceous shales and are directly overlain by thick

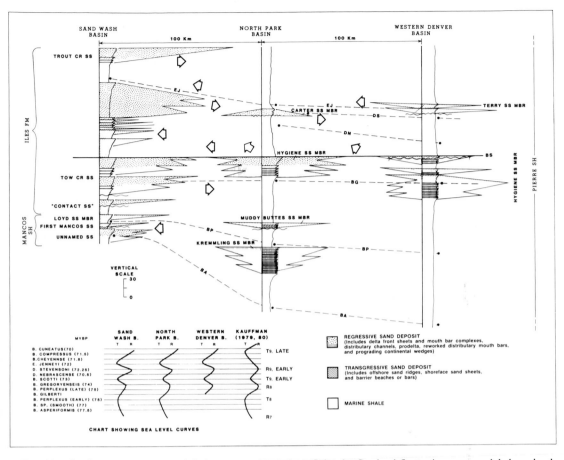

FIG. 21.—Sandstone occurrences and facies patterns in northern Colorado. Sea level fluctuations versus global sea level curve of Kauffman (1979, 1980) are shown in inset.

massive to contorted river mouth bars. Lower bounding surfaces of the small channels cut into underlying marine shales or prodelta, and unidirectional flow from west to east is indicated by current indicators within the channels. These relationships indicate that the fall of sea level briefly exceeded the rate of subsidence and that vegetation became established as the small channels were abandoned. Subsequently, river mouth bars prograded over the tops of these channels as the strandline retreated farther eastward. Finally, younger distributary channels began to cut through tops of the delta front sheet sand as sea level continued to drop. Offlap of successively younger deltas continued in this manner as base level fell to the east, and shoreface-deposited delta front environments (as represented by the Hygiene in North Park Basin) developed in shallow areas of the shelf.

During transgressions, deposition of much of the sandstone in the shelf area resulted from storms or oceanic currents impinging on the shelf. Storm depositon resulted from turbulence which was produced at or below storm wave base as water deepened during periods of rising sea level. Seilacher (1982) has termed these disturbances "turbulent events" and states that they are similar in several respects:

"1. That they reflect the onset, culmination, and waning of water turbulence during the event by distinctive erosional and depositional structures.

2. That they redistribute the organic and inorganic sediment material along a vertical (bottom to top) and horizontal (shallow to deep) gradient.

3. That they change the ecological situation for benthic organisms by altering the consistency and/or the food content of the bottom for a biologically relevant period after the event." (Seilacher, 1982, p. 162).

Based on the premise that "turbulence events" can be correlated within time-equivalent sequences in the ancient record, we can then assume that they reflect roughly synchronous deposition that occurred during rising sea level. Therefore, they can be directly related to sea level changes that occurred basinwide, or were eustatic in origin. Examples of deposits that represent roughly synchronous deposition during a turbulent event are the Muddy Buttes Sandstone Member and the time-equivalent First Mancos sandstone. Each of these units exhibit quite different internal structures, yet each was deposited at approximately the same time. Both are scour based, however, which indicates that deposition was initiated by turbulence on the shelf surface, and, therefore, both are correlatable and can be related to an "event" such as a general rise in sea level which is known to have occurred at this time.

In contrast, subsequent fall in sea level (the early R_8 regression) is indicated by the Loyd Sandstone Member, a thoroughly bioturbated (structureless) and extremely fossiliferous sandstone, that may represent the deposit of a "condensed horizon". Many shells in the Loyd are found in living position, indicating that mass extinction may have taken place, and some *Thalassinoides* tubes are filled only with mud, indicating that periods of sediment by-passing occurred. Such horizons as those found in the Loyd commonly occur during rapid sea level fall, and for that reason are termed "condensed horizons". Condensed horizons result from condensation of "coarser and more durable sediments" during sea level fall (Seilacher, 1982, p. 171). Very rapid rises, on the other hand, produce an inactive shelf surface, and no coarse sediment is left behind in the ancient record. This situation was what apparently occurred following deposition of the Tow Creek-Hygiene interval (Fig. 22).

In summary, we believe that relative periods of falling and rising sea level occurred in the late Cretaceous

shelf environment of northern Colorado which apparently coincide with eustatic sea level changes, and which profoundly affected shelf sedimentation in this region. Some units show regular cyclic change in vertical sequence, whereas others appear to be non-cyclic, but are in fact related to events that affected all parts of the shelf.

The relative rates of change of sea level in northern Colorado can be roughly determined from the sediments preserved in the record. The oldest regression (R_7) occurred during a long slow period of falling sea level when deltas built far seaward of strandlines located in Utah. The succeeding transgression (T_8) was also slow and characterized by storm activity and movement of sediment longshore by strong oceanic currents. The R_8 regression, which was characterized by offlap of successively younger deltas, resulted from a rapid fall in sea level and extensive seaward progradation; a "condensed horizon" characterized the initial phase of this regression. The subsequent transgression towards the west was rapid as indicated by the absence of preserved sandstone beds enclosed in marine shale. This transgression may have been initiated during late Hygiene time in the Denver Basin. The final regression (early R_9) was characterized by offlap of continental beds onto shelf surfaces; subsequent transgression (early T_9) began late in the R_9 cycle to the east and resulted in onlapping of successively younger transgressive deposits (Terry Member) onto older nearshore sandstones. This early R_9 regression apparently resulted from local uplift in northwestern Colorado and was not the result of a eustatic change as were earlier cycles (R_7, T_8, R_8, and early T_9).

Comparison of Ancient Shelf Sandstones in Northern Colorado to Modern Shelf Environments

This and previous studies document the presence of shallow marine (nearshore to outer shelf) sandstone bodies formed during the Late Cretaceous along the western margin of the Western Interior seaway. The identification of these sandstone deposits is based largely on facies relationships, marine fauna, and paleogeographic reconstructions of shoreline positions. Sediment source and mechanisms of formation of these sandstone bodies are, however, not well understood. Despite the abundance of recent studies on modern shelf sand bodies (e.g., Houbolt, 1968; Swift et al., 1972; Duane et al., 1972; Flemming, 1981; Field, 1980) it is difficult to identify modern analogues.

One difficulty in resolving the depositional history of Upper Cretaceous sand bodies is that apparently they passed through several stages of reworking. Original morphology and texture of a particular deposit are commonly erased by subsequent winnowing, bioturbation, remobilization, or a combination of these processes. Reworking of low-energy deposits by storm currents, for example, results in bedforms and internal structures which reflect only the conditions during the

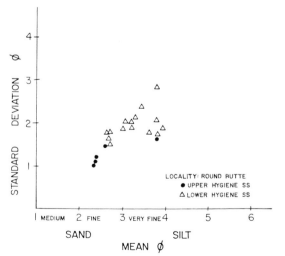

FIG. 22.—Mean grain-size and sorting of sands from the Hygiene Sandstone at Round Butte. Actual mean particle size is judged to be coarser than values shown because of abundance of very fine-grained alteration products. Lower Hygiene equals Unit 1 on Figure 19; upper Hygiene equals Unit 2 on Figure 19.

latest phase of reworking. This is especially true in sandstones such as the Hygiene of the western Denver Basin, for example, in which the upper part has been completely reworked and remolded, whereas the lower part is interpreted to be preserved in its original state. Alternatively, some of the nearshore deposits such as the Loyd Sandstone Member have been so thoroughly bioturbated by marine organisms that original bedding structures are rarely preserved.

Of the sandstones examined in this study, the thin cross-bedded sandstone units within the Hygiene Sandstone in the western Denver Basin and the First Mancos sandstone in the Sand Wash Basin are perhaps the best examples of continental shelf deposition. These sandstones are overlain by marine shale and underlain by either prodelta or offshore marine shale, and the basal contacts, where visible, are sometimes abrupt and probably erosional. The sandstones are well-sorted, dominantly fine to very fine-grained quartzose sandstones (Fig. 22) with locally abundant glauconite (Hygiene Member). Marine fossils are present and burrows indicative of high energy are very common.

One of the more striking features of the sandstones is the well-developed cross-stratification that characterizes the exposures of the Hygiene at Round Butte and Hygiene (type locality, Fig. 20) in the western Denver Basin. Sand waves with spacings on the order of 0.3-5 m are preserved on bedding planes. These bedforms are the response to a complicated set of factors of which water depth, grain size, and current velocity are the most significant (Southard, 1971), and it would be particularly useful if they could be interpreted to provide additional information on environmental conditions at the time of deposition. Rubin and McCulloch (1980) have assessed the relationship of bedform type with depth, grain size, and current velocity, based on studies of their own and by others, and their analysis provides, in part, a basis for estimating current conditions in ancestral marine sands such as the Hygiene. For example, sand in the Hygiene Member is roughly in the 0.19 to 0.22 mm range, and the types of bedforms are ripples and simple sand waves. Water depth is unknown, but if an outer shelf setting of 100 to 200 m water depth is assumed, then depth-averaged flow velocities probably were on the order of 70 to 100 cm/sec (Fig. 23). If the assumed water depth is correct or nearly so, this provides important information about the setting, as velocities of this magnitude do not occur during fair weather conditions on open shelves and imply an unusual tidal setting, the existence of a major coastal current, or the influence of periodic storm flows. For example, high velocities are well-documented from tidally-controlled passageways and reentrants such as Cook Inlet, Alaska (Bouma *et al.*, 1980), the North Sea (McCave, 1971), and San Francisco Bay (Rubin and McCulloch, 1980). Similar high velocities are commonly obtained beneath oceanic currents in restricted passages (e.g., the Bering Strait, Alaska; Coachman *et al.*, 1975; Field *et al*, 1981) or

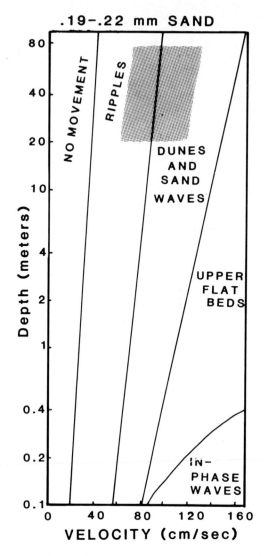

FIG. 23.—Depth versus velocity curve for sand in the 0.19–0.22 mm range showing types of bedforms found at different values (from Rubin and McCulloch, 1980). The stippled area represents the depth and bedform field for Hygiene sands in the Denver Basin and can be used to estimate required flow-velocities.

during storm events on the open shelf (Cacchione and Drake, 1979).

Sandstone bodies in the Hygiene Member are similar in thickness, composition, grain size, and sedimentary structures to other shelf sandstones described from the western portion of the Late Cretaceous Western Interior seaway (Spearing, 1976; Martinsen and Tillman, 1978). Although many of these sandstone bodies are recognized as shelf deposits, the processes of deposition and water depths are not well understood. For example, sandstones in the Powder River Basin in

Wyoming are thought to have been deposited at great distances from the shoreline (Spearing, 1976) based on interpretation of shoreline position and on the abundance of shale lying west (landward) of the sandstones. Deposition of sand bodies over 100 km from the shoreline may occur in some areas, but it requires the existence of a complicated shelf-bypassing system.

Our analyses in the western Denver Basin lead us to conclude that the shoreline itself was migrating back and forth across the wide shelf in response to major eustatic events and to local changes in sedimentation. Small deltaic systems are evident in many places across northern Colorado. Pulses of sedimentation associated with shifting depocenters resulted in abundant delta front and shoreface sands, some of which were subsequently modified in the shelf environment.

Studies of shelf sand bodies on the U.S. Atlantic shelf (Duane and others, 1972; Swift and others, 1972; Swift and Field, 1981; Field, 1980; Field and Duane, 1976; Stubblefield et al., this volume; Stubblefield and Swift, 1976) and from western Europe (Houboult, 1968; Caston and Stride, 1970; McCave, 1971; Kenyon, 1970) and from Alaska (Nelson et al., 1981; Field et al., 1981) provide a basis for comparison with Upper Cretaceous deposits examined in this study. In those areas it has been clearly recognized that the processes that initially deposit and those that subsequently modify shelf sand bodies are commonly quite distinct.

In each of the modern environments where shelf sand bodies have been examined in detail—the U. S. Atlantic Shelf, the northern Bering Sea, and the North Sea—they are recognized to be in part deposits that originated in a setting different than their present one.

Regardless of their present position on the shelf, all or part of each are interpreted as having formed in the nearshore environment. Their geometry and stratigraphic associations indicate formation as part of retreating estuarine, barrier island, or shoreface deposits. On the Atlantic shelf, a complete sequence of sand bodies has been recognized, ranging from those forming at present at the base of the shoreface to ones that are stranded on the shelf tens of kilometers from the shoreline (Field, 1980). Although sand bodies are separated from the shore and stranded on the shelf as a result of the Holocene rise in sea level and concomitant retreat of the coast, they continue to be shaped and modified on the shelf. Strong near-bottom currents generated by winter storms form sand waves and ripples. On the northern Bering Sea shelf, sand ridges lie in the path of a strong, unidirectional oceanic current, but here also it is the coupling of storm events with the oceanic current that results in bed load transport by sand wave migration across the ridges (Field et al., 1981). Sand bodies are well-known from tidally dominated environments (e.g., North Sea; Cook Inlet), but these environments require a unique setting, one that would probably be identifiable through facies and paleocurrent studies in Upper Cretaceous rocks of northern Colorado. Consequently, it is considered unlikely that the Hygiene is the result of strong tidal flows. More likely and supportable is a reconstruction that allows for one or two possibilities: (1) a migrating shoreline resulting in discrete sand packages on the shelf at shelf depths, and (2) reshaping of nearshore sand bodies by storm-driven or oceanic currents.

REFERENCES

BALL, M. W., 1924, Gas near Fort Collins, Colorado: Am. Assoc. Petroleum Geologists Bull., v. 8, p. 79–87.

BERG, R. R., 1975, Depositional environments of Upper Cretaceous Sussex Sandstone, House Creek Field, Wyoming: Am. Assoc. Petroleum Geologists Bull., v. 59, p. 2099–2110.

BOYLES, J. M., AND A. J. SCOTT, 1982, A model for migrating shelf bar sandstones in upper Mancos Shale (Campanian), northwest Colorado: Am. Assoc. Petroleum Geologists Bull., v. 66, p. 491–508.

BOUMA, A. H., M. L. RAPPEPORT, R. C. ORLANDO, AND M. A. HAMPTON, 1980, Identification of bedforms in lower Cook Inlet, Alaska: Sed. Geol., v. 26, p. 157–177.

CACCHIONE, D. A., AND D. E. DRAKE, 1979, Sediment transport in Norton Sound, Alaska: Regional patterns and GEOPROBE system measurements: U. S. Geol. Survey Open-File Report 79–1555, 88 p.

CAMPBELL, C. V., 1971, Depositional model—Upper Cretaceous Gallup beach shoreline, Ship Rock area, northwestern New Mexico: Jour. Sed. Petrology, v. 41, p. 395–409.

CASTON, V. N. D., AND A. H. STRIDE, 1970, Tidal sand movement between some linear sand banks in the North Sea off northeast Norfolk: Marine Geol., v. 9, p. M38–M42.

CHAMBERLAIN, D. K., 1980, The trace fossils of the Cretaceous Dakota hogback along Alameda Avenue, west of Denver, Colorado, in P. B. Basan (ed.), Trace Fossils of Nearshore Depositional Environments of Cretaceous and Ordovician Rocks, Front Range, Colorado: Rocky Mountain Sect., Soc. Econ. Paleontologists and Mineralogists, Guidebook for SEPM Field Trip No. 1, p. 30–39.

CLIFTON, N. E., AND J. K. THOMPSON, 1978, Macaronichnus segregatis: A feeding structure of shallow marine polychaetes: Jour. Sed. Petrology, v. 48, p. 1293–1302.

COACHMAN, L. K., ET AL., 1975, Bering Strait, the Regional Physical Oceanography: Univ. Washington Press, Seattle, 186 p.

COLEMAN, J. M., AND S. M. GAGLIANO, 1965, Sedimentary structures: Mississippi River delta plain, in G. V. Middleton (ed.), Primary Sedimentary Structures and Their Hydrodynamic Interpretation: A Symposium: Soc. Econ. Paleontologists and Mineralogists, Spec. Pub. 12, p. 133–138.

————, AND D. B. PRIOR, 1980, Deltaic sand bodies: Am. Assoc. Petroleum Geologists, Cont. Educ. Course Note Ser. No. 15, 171 p.

COTTER, E., 1975, Late Cretaceous sedimentation in a low-energy coastal zone: The Ferron Sandstone of Utah: Jour. Sed. Petrology, v. 45, p. 669–685.

DICKINSON, K. A., H. L. BERRYHILL, JR., AND C. W. HOLMES, 1972, Criteria for recognizing ancient barrier coastlines, *in* J. K. Rigby and W. K. Hamblin (eds.), Recognition of Ancient Sedimentary Environments: Soc. Econ. Pateontologists and Mineralogists, Spec. Pub. 16, p. 192–214.

DUANE, R. B., M. E. FIELD, E. P. MEISBURGER, D. J. P. SWIFT, AND S. J. WILLIAMS, 1972, Linear shoals on the Atlantic inner continental shelf, Florida to Long Island, *in* D. J. P. Swift, D. B. Duane, and O. H. Pilkey (eds.), Shelf Sediment Transport: Process and Pattern: Dowden, Hutchinson, and Ross, Stroudsburg, Pennsylvania, p. 447–499.

FENNEMAN, N. M., 1905, Geology of the Boulder District: U. S. Geol. Survey Bull. 265, 101 p.

FIELD, M. E., 1980, Sand bodies on coastal plain shelves: Holocene record of the U. S. Atlantic inner shelf off Maryland: Jour. Sed. Petrology, v. 50, p. 505–528.

_____, AND D. B. DUANE, 1976, Post Pleistocene history of the inner continental shelf: Significance to barrier island origin: Geol. Soc. America Bull., v. 87, p. 691–702.

_____, H. C. NELSON, D. A. CACCHIONE, AND D. E. DRAKE, 1981, Sandwaves on an epicontinental shelf: Northern Bering Sea: Marine Geol., v. 42, p. 233–258.

FISK, H. N., 1955, Sand facies of recent Mississippi delta deposits: 4th World Petroleum Cong., Rome, Proc., Sec. 1-C, p. 677–698.

FLEMMING, B. W., 1981, Factors controlling shelf sediment dispersal along the southeast African continental margin, *in* N. T. Trouer (ed.), Sedimentary Dynamics of Continental Shelves: Elsevier, New York, p. 259–278.

_____, AND D. G. ROBERTS, 1973, Tectono-eustatic changes in sea level and sea-floor spreading: Nature, v. 243, p. 19–22.

FREY, R. W., 1976, Approaches to ichnology, *in* C. K. Chamberlain and R. W. Frey (eds.), Seminar on Trace Fossils: Short Course Notes, U.S. Geol. Survey, Golden, Colorado, p. 1–32.

GILL, J. R., AND W. A. COBBAN, 1976, The Red Bird section of the Upper Cretaceous Pierre Shale, in Wyoming, *with a section on* new echinoids from the Cretaceous Pierre Shale of eastern Wyoming by P. N. Kier: U.S. Geol. Survey Prof. Paper 393-A, 73 p.

_____ AND _____, 1973, Stratigraphy and geologic history of the Montana Group and equivalent rocks, Montana, Wyoming, and North and South Dakota: U.S. Geol. Survey Prof. Paper 776, 37 p.

HAIL, W. J., 1965, Geology of northwestern North Park, Colorado: U.S. Geol. Survey Bull. 1188, 133 p.

HALLAM, A., 1963, Major epeirogenic and eustatic changes since the Cretaceous and their possible relationship to crustal structure: Am. Jour. Sci., v. 261, p. 397–423.

_____, 1971, Mesozoic geology and the opening of the North Atlantic: Jour. Geol., v. 79, p. 129–157.

HANCOCK, E. T., 1925, Geology and coal resources of the Axial and Monument Butte Quadrangles, Moffat County, Colorado: U.S. Geol. Survey Bull. 757, 134 p.

HANCOCK, J. M., AND E. G. KAUFFMAN, 1979, The great transgressions of the Late Cretaceous: Jour. Geol. Soc. London, v. 136, p. 175–186.

HAYS, J. D., AND W. C. PITTMAN, 1973, Lithospheric plate motion, sea level changes and climatic and ecological consequences: Nature, v. 246, p. 18–22.

HOUBOLT, J. J. H. C., 1968, Recent sediments in the southern bight of the North Sea: Geol. en Mijnbouw, v. 47, p. 245–273.

HOWARD, J. D., 1972, Trace fossils as criteria for recognizing shorelines in stratigraphic record, *in* J. K. Rigby and W. K. Hamblin (eds.), Recognition of Ancient Sedimentary Environments: Soc. Econ. Paleontologists and Mineralogists, Spec. Pub. 16, p. 215–225.

HUNTER, R. E., R. L. WATSON, G. W. HILL, AND K. A. DICKINSON, 1972, Modern depositional environments and processes, northern and central Padre Island, Texas: Gulf Coast Assoc. Geol. Soc., Padre Island Nat. Seashore Field Guide, 27 p.

IZETT, G. A., W. A. COBBAN, AND J. R. GILL, 1971, The Pierre Shale near Kremmling, Colorado, and its correlation to the east and the west: U. S. Geol. Survey Prof. Paper 684-A, 19 p.

KAUFFMAN, E. G., 1973a, Stratigraphic evidence for Cretaceous eustatic changes [abs.]: Geol. Soc. America, Abstracts with Programs, v. 5, p. 687.

_____, 1973b, Cretaceous Bivalvia, *in* A. Hallam (ed.), Atlas of Paleobiogeography: Elsevier Pub. Co., Amsterdam, p. 353–383.

_____, 1977, Geological and biological overview: Western Interior Cretaceous basin, *in* Field Guide: Cretaceous Facies, Faunas, and Paleoenvieonments across the Western Interior Basin: Mountain Geologist, v. 14, p. 75–99.

_____, 1979, Cretaceous: Treatise on Invert. Paleontology, Part A, p. A418–A487.

KENYON, N. H., 1970, Sand ribbons of European tidal seas: Marine Geol., v. 9, p. 25–39.

KITELEY, L. W., 1977, Shallow marine deposits in the Upper Cretaceous Pierre Shale of the northern Denver Basin and their relation to hydrocarbon accumulation, *in* Exploration Frontiers of the Central and Southern Rockies: Rocky Mountain Assoc. Geologists, Symposium, p. 197–211.

_____, 1978, Stratigraphic sections of Cretaceous rocks of the northern Denver Basin, northeastern Colorado and southeastern Wyoming: U.S. Geol. Survery Oil and Gas Invest. Chart OC-78.

_____, 1980, Facies analysis of the lower cycles of the Mesaverde Group (Upper Cretaceous) in northwestern Colorado: Unpub. M.S. Thesis, Univ. Colorado, Boulder, Colorado, 153 p.

_____, 1983, Paleogeography and eustatic-tectonic model of late Campanian Cretaceous sedimentation, southwestern Wyoming and northwestern Colorado, *in* M. W. Reynolds and E. D. Dolly (eds.), Mesozoic Paleogeography of West Central United States: Rocky Mountain Sec., Soc. Econ. Paleontologists and Mineralogists, p. 273–303.

KONISHI, K., 1959, Upper Cretaceous surface stratigraphy, Axial Basin and Williams Fork area, Moffat and Routt Counties, Colorado, *in* Symposium on Cretaceous Rocks of Colorado and Adjacent Areas: Rocky Mountain Assoc. Geologists, p. 67–73.

MARTINSEN, R. S., AND R. W. TILLMAN, 1978, Hartzog Draw, new giant oil field [abs.]: Am. Assoc. Petroleum Geologists Bull., v. 62, p. 540.

MASTERS, C. D., 1965, Sedimentology of the Mesaverde Group and of the upper part of the Mancos Formation, northwestern Colorado: Unpub. PhD Dissertation, Yale Univ., New Haven, Connecticut, 88 p.

McCAVE, I. N., 1971, Sand waves in the North Sea off the coast of Holland: Marine Geol., v. 10, p. 199–225.

McGOOKEY, E. P., ET AL., 1972, Cretaceous System, *in* W. W. Mallory (ed.), Geological Atlas of the Rocky Mountain Region: Rocky Mountain Assoc. Geologists, p. 190–228.

MOREDOCK, D. E., AND S. J. WILLIAMS, 1976, Upper Cretaceous Terry and Hygiene Sandstones—Singletree, Spindle, and Surrey Fields—Weld County, Colorado, *in* Studies in Colorado Field Geology: Prof. Contrib. Colorado School of Mines, Golden, p. 264–274.

NELSON, H. C., W. R. DUPRÉ, M. E. FIELD, AND J. D. HOWARD, 1981, Linear sand bodies in the Bering Sea epicontinental shelf: U.S. Geol. Survery Open-File Rep. 80–979.

NIO, S. D., 1976, Marine transgressions as a factor in the formation of sand wave complexes: Geol. en Mijnbouw, v. 55, p. 18–40.

OBRADOVICH, J. D., AND W. A. COBBAN, 1975, A time-scale for the Late Cretaceous of the Western Interior of North America: Geol. Assoc. Canada Spec. Paper 13, p. 31–54.

PORTER, K. W., 1976, Marine shelf model, Hygiene Member of the Pierre Shale, Upper Cretaceous, Denver Basin, Colorado, *in* Studies in Colorado Field Geology: Prof. Contrib. Colorado School of Mines, Golden, p. 251–263.

_____, AND R. J. WEIMER, 1979, Diagenesis in Hygiene and Terry Sandstones (Upper Cretaceous), Spindle Field, Colorado [abs.]: Am. Assoc. Petroleum Geologists Bull., v. 63, p. 510.

RONA, P. A., 1973a, Relations between rates of sediment accumulation on continental shelves, sea-floor spreading, and eustacy inferred from central North America: Geol. Soc. American Bull., v. 84, p. 2851–2872.

_____, 1973b, Worldwide unconformities in marine sediments related to eustatic changes of sea level: Nature Phys. Sci., v. 244, p. 25–26.

RUBIN, D. M., AND D. S. McCULLOCH, 1980, Single and superimposed bedforms: A synthesis of San Francisco Bay and flume observations: Sedimentary Geol., v. 26, p. 207–231.

SEELING, A., 1979, The Shannon Sandstone, a further look at the environment of deposition at Heldt Draw Field, Wyoming: Mountain Geologist, v. 15, p. 133–144.

SEILACHER, A., 1982, General remarks about event deposits, *in* Cyclic and Event Stratification: Springer-Verlag, Berlin, p. 161–174.

SIEMERS, C. T., 1976, Sedimentology of the Rocktown Channel Sandstone, upper part of the Dakota Formation (Cretaceous), central Kansas: Jour. Sed. Petrology, v. 46, p. 97–123.

SLEEP, N. H., 1976, Platform subsidence mechanisms and "eustatic" sea level changes: Tectonophysics, v. 36, p. 45–56.

SLOSS, L. L., 1963, Sequences in the cratonic interior of North America: Geol. Soc. America Bull., v. 74, p. 93–113.

_____, 1972, Synchrony of Phanerozoic sedimentary tectonic events of the North American Craton and the Russian Platform: 24th Internat. Geol. Congr. Proc., Sect. 6, p. 24–32.

SOUTHARD, J. B., 1971, Representation of bed configurations in depth-velocity-size diagrams: Jour. Sed. Petrology, v. 41, p. 903–915.

SPEARING, D. R., 1976, Upper Cretaceous Shannon Sandstone: An offshore shallow-marine sand body: Wyoming Geol. Assoc., Guidebook 28th Ann. Field Conf., p. 65–72.

STUBBLEFIELD, W. L., AND D. J. P. SWIFT, 1976, Ridge development as revealed by subbottom profiles on the central New Jersey shelf: Marine Geol., v. 10, p. 315–334.

SWIFT, D. J. P., AND M. E. FIELD, 1981, Evolution of a classic sandridge field: Maryland sector North American inner shelf: Sedimentology, v. 28, p. 461–482.

_____, J. W. KOFOED, F. P. SAULSBURY, AND P. SEARS, 1972, Holocene evolution of the shelf surface, central and southern Atlantic shelf of North America, *in* D. J. P. Swift, D. B. Duane, and O. H. Pilkey (eds.), Shelf Sediment Transport: Process and Pattern: Dowden, Hutchinson, and Ross, Stroudsburg, Pennsylvania, p. 499–575.

THOMPSON, W. O., 1937, Original structures of beaches, bars and dunes: Geol. Soc. America Bull., v. 48, p. 723–752.

VAIL, P. R. AND R. O. WILBUR, 1966, Onlap, key to worldwide unconformities and depositional cycles [abs.]: Am. Assoc. Petroleum Geologists Bull., v. 50, p. 638.

_____, ET AL., 1977, Seismic stratigraphy and global changes of sea level, *in* C. E. Payton (ed.), Seismic Stratigraphy—Applications to Hydrocarbon Exploration: Am. Assoc. Petroleum Geologists Mem. 26, p. 52–212.

WALKER, R. G., 1979, Shallow marine sands, in R. G. Walker (ed.), Facies Models: Geoscience Canada, Reprint Series 1, p. 75–89.

LOWER CRETACEOUS SHELF STORM DEPOSITS, NORTHEAST TEXAS

DAVID K. HOBDAY AND ROBERT A. MORTON

Bureau of Economic Geology, The University of Texas at Austin, Austin, Texas 78712

ABSTRACT

Graded sandstones separated by finer-grained sediments of a shelf and shoreface environment have been widely attributed to sporadic storm activity, with sand transport related to storm-surge ebb. According to this classic storm-surge ebb model, barrier-lagoon coasts are prerequisite. However, re-examination of hurricane data, as well as coring on the Texas shelf, suggests that emplacement of these storm-graded sediments was related to high energy bottom-return flow produced by wind forcing during the height of storms. Seaward runoff from backbarrier areas is interpreted to have had negligible influence on shelf sedimentation in the units studied.

Graded sandstones of the Lower Cretaceous Grayson Formation in northeast Texas, which bear close resemblance to the storm beds in the cored units, are interpreted to have been deposited off a coast dominated by small lobate deltas that lacked fringing barrier islands. Nearshore sandstones are relatively thick, vertically amalgamated, and dominated by hummocky cross stratification produced by the interaction of unidirectional currents and storm waves. In contrast, thinner sandstones deposited below storm wave base commonly include a basal shell layer with transported and aligned gastropods probably oriented by strong bottom currents. These sandstones show parallel lamination or rare low-angle foresets that grade upward into siltstone and mudstone with local intervening ripple cross lamination. Burrowing is restricted to the upper parts of the graded units and to the overlying fair-weather mudstones; however, burrowing is surprisingly sparse, possibly because of high rates of sedimentation.

INTRODUCTION

The Lower Cretaceous Grayson Formation (Washita Group) of northeast Texas (Fig. 1) shows a classical combination of features, many of which may be attributed to shelf storm processes. A pronounced shore-parallel facies zonation is observed. Thick, graded sandstones of the shoreface-inner shelf transition are complexly amalgamated, with evidence of simultaneous molding by powerful waves and currents. In contrast, discrete sandstone units of the inner shelf are separated by silty mudstones. The proportion of silty mudstones to sandstones increases basinward to make up almost the entire succession of the outer shelf.

Seaward transport of the sand from the shore zone and its emplacement as a graded bed may have been brought about by a single high-energy event. Alternatively, the graded texture may be a product of *in situ* reworking of previously deposited shelf sands, for example, by settling of clouds of sediment suspended temporarily by waves (Reineck and Singh, 1972) or by sorting through wave-induced pressure fluctuations (Powers and Kinsman, 1953). Judging from field evidence, however, the Grayson sands were deposited on the shelf during brief storm events of rapidly waning intensity, with reworking restricted to burrowing organisms in the uppermost few centimeters. The vertical and lateral sequences observed suggest that storms were infrequent and interrupted a fair-weather regime of shelf mud and autochthonous carbonate accumulation.

Publication authorized by the Director, Bureau of Economic Geology, The University of Texas at Austin.

SEAWARD TRANSPORT OF SAND ON SHELVES

As pointed out by Brenchley *et al.* (1979) the role of shore-parallel currents has been emphasized in accounts of shelf sedimentation (Mooers, 1976). However, there are several modern examples of offshore-directed currents (e.g., Vos, 1976; Murray, 1970; Forristall *et al.*, 1977), and numerous ancient examples where seaward transport is implied (e.g., Kelling and Mullin, 1975; Scott *et al.*, 1975; Brenchley *et al.*, 1979). One presently accepted model for seaward sand transport and its deposition as graded units stems from Hayes' (1967) Hurricane Carla studies. According to Hayes, the hydraulic head created by storm washover into backbarrier bays, lagoons, and marshes causes strong ebb currents to exit through washover channels and inlets. Supposedly coarse sediment entrained by these currents is carried across the shoreface and inner shelf as a density current (Hayes, 1967; Walker, 1979).

Re-examination of the data from Hurricane Carla (Morton, 1981) indicates that storm runoff from Laguna Madre was negligible. This runoff process was previously regarded by Hayes (1967) as responsible for the graded sand deposit observed on the Texas inner shelf. However, the thickest and most extensive development of the graded sand unit was off a stretch of coast where washover channels were entirely absent. In contrast with the Hayes model, Morton (1981) suggested that this sand was carried seaward by strong unidirectional bottom currents that occurred ahead of the storm center shortly after the strongest winds and before storm landfall.

The best evidence of these time-dependent relationships comes from field monitors in the Gulf of Mexico during the 1973 Tropical Storm Delia (Fig. 2).

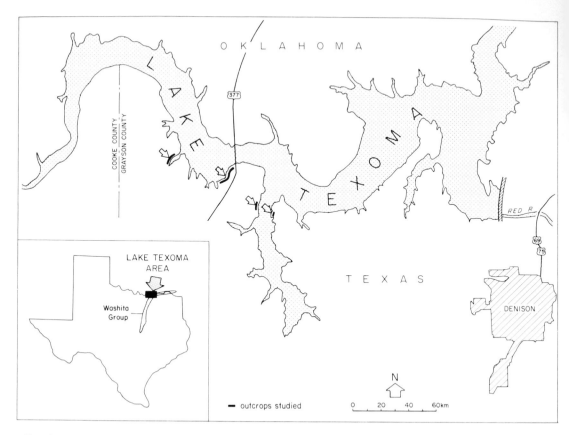

FIG. 1.—Locality map of detailed study area. Arrows indicate major cliff exposures on Lake Texoma, Grayson County, Texas.

Current velocities recorded throughout the water column (Forristall, *et al.*, 1977) showed a single-layer system with the entire water mass moving alongshore and slightly offshore with maximum velocities of nearly 2 m/sec. Peak bottom currents (Fig. 2) coincided with the onset of maximum surface currents which lasted for several hours following the maximum wind stress. These strong shelf currents occured before the storm crossed the coast and more than 30 hours before maximum surge heights were recorded by coastal tide gages. The timing of these events clearly precluded a storm-surge ebb mechanism and supports the interpretation of wind-forced bottom flows.

Graded sandstones of the Grayson Formation bear close resemblance to the Hurricane Carla deposits, and the non-barrier coast, elaborated upon below, rules out a conventional storm-surge ebb origin. In this paper we restrict our usage of the term "storm-surge ebb" to the floodwater runoff mechanisms proposed by Hayes (1967).

GRAYSON FORMATION

Geologic Setting

Stratigraphic and facies relationships in the Washita Group of the East Texas basin are provided by Hayward

and Brown (1967) and Scott *et al.* (1978). The East Texas basin was occupied by a shallow epicontinental sea that merged southward with the Central Texas carbonate platform and was bounded in the north by the Ouachita-Arbuckle highlands, the main source of terrigenous clastics. Changes in rate of terrigenous influx and basinal subsidence occasioned by hingeline flexuring caused complex migration and intertonguing of carbonate and siliciclastic facies tracts.

Grayson deposition was associated with an episode of rapid subsidence and accelerated sediment supply (Scott *et al.*, 1978). Foraminifer populations in the Main Street-Grayson succession indicate initial deepening followed by progressive shallowing (Albritton *et al.*, 1954). A cosmopolitan ammonoid fauna was regarded by Young (1972) as indicative of temporary breaching of the shelf-edge barrier reef to the south and the establishment of open circulation.

Shelf mudstones south of the Lake Texoma study area (Fig. 1) contain a diverse molluscan fauna, *Chondrites* burrows, and 1 to 2-cm thick graded siltstones with sole marks and climbing ripple cross-lamination (Scott *et al.*, 1978). These siltstone intercalations are regarded as the distal equivalents of graded sandstone beds to the north.

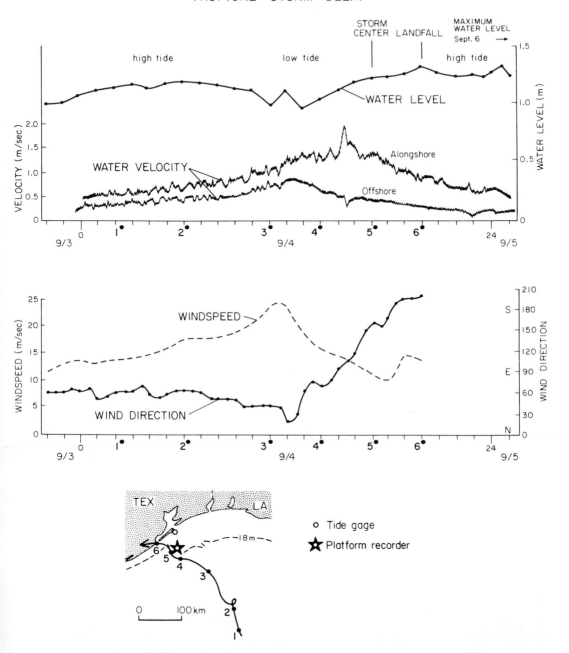

FIG. 2.—Temporal variations in wind and water measurements during Tropical Storm Delia, Gulf of Mexico, September 3–5, 1973. Water velocities, wind speeds, and wind directions from Forristall *et al.* (1977); water levels from unpublished records of the Galveston District, U.S. Army Corps of Engineers.

Lake Texoma Area

A 50 m succession through the Grayson Formation in cliff exposures at Lake Texoma (Fig. 3) consists of thick (80 to 300 cm), amalgamated sandstones at the base separated by subordinate siltstones, overlain by a siltstone dominated interval with thinner (2 to 40 cm) graded sandstones and bioclastic layers, capped by an upward-coarsening shoal-water delta sequence with coastal marsh lignite above. Most of the bioclastic deposits are thin, very localized, or form the base of

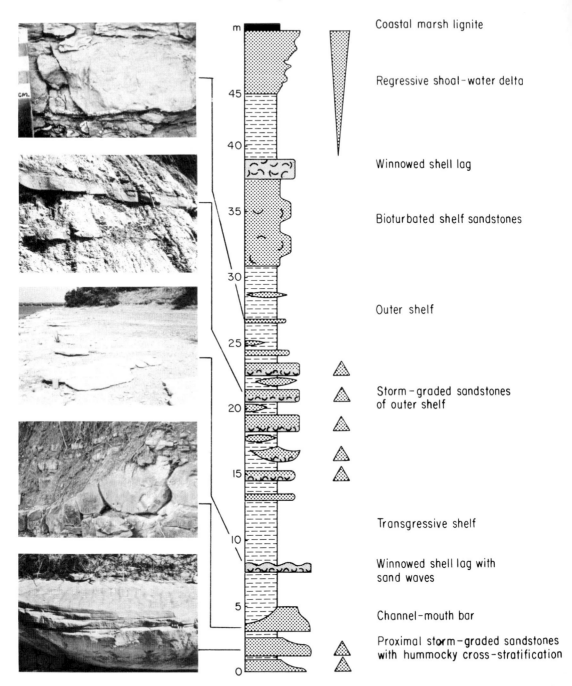

FIG. 3.—Vertical section through 50 m thick Lake Texoma succession (33° 51' N latitude, 96° 54' W longitude) illustrating the more significant sedimentary features, with generalized interpretations. Location shown in Figure 1.

sandstone beds; but one prominent and laterally continuous shell bed probably marks a significant transgressive hiatus between the thick, basal, nearshore sandstones deposited above storm wave base and the overlying aggradational shelf and regressive deltaic deposits (Fig. 3). A somewhat similar distinction between a shoreline association of thicker amalgamated beds and an open shelf association of interbedded sandstones and mudstones was recognized by Goldring and Bridges (1973).

FIG. 4.—Erosively-based nearshore sandstone unit with massive lower part overlain by hummocky cross-stratification, followed by parallel lamination and a rippled, burrowed upper surface.

Sandstones deposited above storm wave base.— Hummocky cross-stratification (Fig. 4) is the dominant feature of the thick, erosively-based sandstones at the base of the Grayson Formation. This structure, which commonly makes up the entire unit apart from a slightly coarser-grained, massive basal unit with mudstone intraclasts, has been shown by Walker (1979) to be characteristic of storm wave activity below fair-weather wave base (normally about 10 m, but probably less in the shallow, semi-restricted East Texas basin). The contact between superimposed or interfering sets of hummocky cross-stratification is in places accentuated by a few burrows, mainly *Rosselia*, indicating a break in deposition (Howard, 1972; Cotter, 1975). Intervening mudstones contain rippled and parallel-laminated silty lenses and are moderately burrowed in part.

Superimposed, abruptly based sandstones containing hummocky cross-stratification suggest multiple episodes of high-energy deposition and wave reworking in a storm-dominated shoreface environment (Fig. 5). The hummocky cross-stratification merges upward into parallel lamination overlain by ripple cross-lamination. Ripples (Fig. 3) include linguoid and caternary current forms, but linear symmetrical to slightly asymmetrical forms of presumed wave origin are most common. These stratification types probably represent deposition by lower energy, short-period waves as compared with those responsible for the hummocky cross-stratification, and may have formed during the waning phase of the storm. The upper few centimeters of sandstone are invariably burrowed. *Thalassinoides* is the most common trace fossil. In some cases the ripple forms and cross-lamination were almost entirely obliterated by burrowing. Wave ripples show a preferred trend slightly oblique to an inferred general east-west shoreline.

Compositional and textural immaturity of the fine-grained sandstone, together with abundant plant remains, suggests that sand was rapidly deposited off river mouths and reworked by waves. One such channel mouth bar and abandoned distributary channel (Fig. 3) that survived reworking conformably overlies siltstones above a thick sequence of hummocky cross-stratified sandstones. It is 3 m thick and markedly lenticular with a scoured upper surface. An axial distributary channel 120 cm deep and 8 m wide trends due south and is filled with poorly bedded, organic-rich, burrowed mudstone, siltstone, and sandstone. The overlying plant-rich mudstones with rare lagoonal mollusks record an episode of shallow inundation signalling marine transgression. The laterally persistent

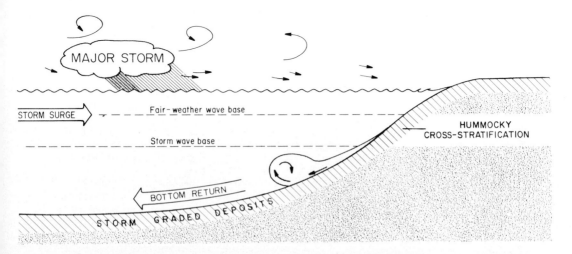

FIG. 5.—Development of hummocky cross-stratification above storm wave base and seaward transport of sand by bottom-return flow. Modified after Walker, 1979.

bioclastic layers above (Fig. 3) may represent a wave lag. Oysters and other mollusks are disarticulated, abraded and shingled, but are disrupted in places by large crustacean burrows. Large, internally symmetrical wave ripples or sand waves (Fig. 3) with a spacing of up to 120 cm and heights of 20 to 30 cm indicate waves of large orbital diameter. Similar features have been reported from the northeast Pacific shelf at depths of between 80 and 105 m where they trend parallel to the coast and are believed to have been generated by storm swell (Yorath et al., 1979). Water depths during deposition of the Grayson Formation were undoubtedly very much less. Considering the close relationship to the underlying nearshore and shorezone deposits, together with the limited thickness of strata (about 35 m) between this bioclastic unit and the lignite indicating coastal emergence at the top of the succession (Fig. 3), water depth is unlikely to have exceeded 30 m.

Sandstones deposited below storm wave base.—The Grayson succession above the bioclastic sand wave unit comprises a number of sharp-based sandstone beds with sole marks and elongate fossils oriented offshore. Internal structures are largely restricted to parallel lamination, with some low-angle foresets (Fig. 6) and ripple cross-lamination, including ripple drift. The sandstone to siltstone to mudstone gradation at the tops

FIG. 6.—Graded bed with shelly, erosive base, parallel lamination, and very low angle foresets, with a sporadically burrowed upper surface overlain by more intensely burrowed siltstone.

of beds is abrupt or spread over a thickness of several centimeters. Vertical patterns are thus strongly reminiscent of turbidites (Scott et al., 1975; Walker, 1979). Sandstone thickness varies between 2 and 40 cm. Many units persist laterally over distances of hundreds of meters whereas others are distinctly lenticular, an extreme case being isolated gutter casts (Whitaker, 1973), which are long, narrow, steep-sided scour and fill structures containing abundant shell debris. Gutter casts (Whitaker, 1973) have been described in deposits ranging from deep water turbidites to shallow subtidal environments.

Shell debris filling scours and gutter casts and extending as a thin, but persistent sheet at the base of some of the more continuous beds comprises a mixed nearshore and lagoonal molluscan assemblage with numerous individuals of opportunistic species (T. Hansen, personal communication). Valves are disarticulated and a majority are convex upward. Most striking is the general north-south alignment of *Turritella* (Fig. 7) approximately perpendicular to the inferred shoreline trend (Fig. 8). These sandy nearshore suspension feeders typically show high dominance (T. Hansen, personal communication), but where they are concentrated at the base of thin (3 to 8 cm), graded beds in an overwhelmingly muddy shelf succession, this implies a considerable distance of seaward transport. Upward convexity of associated clam shells argues against a turbidity current origin (Whitaker, 1973). Tool marks and rare flutes (Scott et al., 1975) show a shore normal trend similar to that of the high-spired gastropods.

The unidirectional orientation of gastropod shells (Figs. 7 and 8) indicate transport by high velocity offshore flow possibly analogous to rip currents. The massive, shelly base and overlying parallel-laminated sand were deposited in the upper flow regime, the fine grain size favoring the development of plane beds at relatively low flow power (Allen, 1970; Brenchley et al., 1979). Low-angle foresets may represent "washed out" dunes of the transition from upper to lower flow regime. Alternatively, in those examples where they occur near the lateral termination of sandstone beds, they suggest progradation of a subdued sand platform. Upward gradation from parallel-laminated sandstone into siltstone or, less commonly, from parallel lamination through cross lamination into homogeneous mudstone, indicates rapidly waning flow. Some sandstones show an abrupt contact with the overlying silty mudstones, but very rarely is there evidence of intervening erosion.

Fewer than 10 percent of the graded sandstones are intensely bioturbated in their upper parts. The remainder contain sporadic horizontal burrows or no trace fossils at all. The "parallel to burrowed" pattern described by Howard (1972) and further documented by Goldring and Bridges (1973), Cotter (1975), Kumar and Sanders (1976), among others, is widely accepted as indicative of intermittent sedimentation in the lower

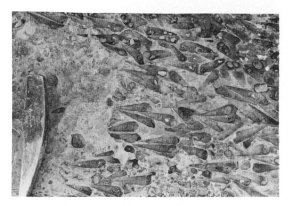

FIG. 7.—The underside of a graded bed showing gastropods oriented approximately normal to the paleoshoreline.

shoreface or offshore environment.

A core (Fig. 9) of a graded bed from the Texas inner shelf, which may be analogous to the Grayson sediments, shows more pronounced burrowing than is evident in most of the Grayson units. This modern shelf-storm deposit is characterized by an erosional lower contact which is overlain by size graded sediments. The basal shell-lag gravel grades upward into shelly sand and sand which in turn grades into mud with sand-filled burrows (Fig. 9). The physical setting of this core in relation to other shelf-storm deposits is presented by Morton (1981).

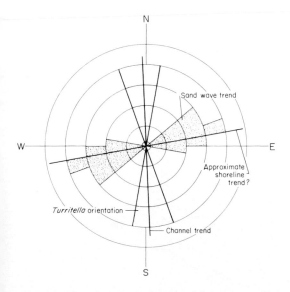

FIG. 8.—Diagram showing orientation of various components of the Grayson sequence of environments. Shown is the orientation of the distributary channel illustrated in Fig. 3, strike of sand waves, *Turritella* orientation, and inferred shoreline trend in the Lake Texoma area.

Shelf Mudstones

In the Grayson, shelf mudstones intervening between successive sandstone units are remarkably free from intense bioturbation, although isolated burrows are present throughout. Perhaps this is indicative of an

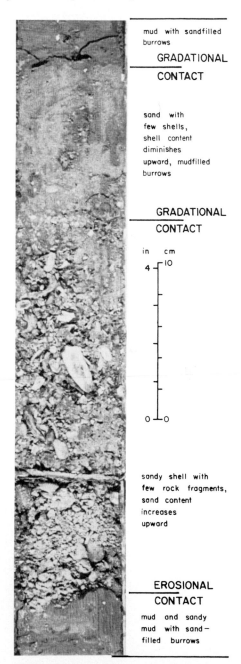

FIG. 9.—Graded bed preserved at a depth of 90 cm in a core from the Gulf of Mexico off Matagorda Island, Texas. Core taken in 11 m of water at the base of the lower shoreface. Modified from Morton (1981).

unusually rapid rate of fair-weather suspension sedimentation. Occasional ammonites, articulated clams, and oyster valves are present along with fine shell debris. Locally, there are layers of bored *Gryphaea* valves suggestive of episodes of very slow sedimentation. Glauconite, derived from fecal pellet precursors, tends to be associated with these *Gryphaea* bands.

Separating the offshore graded sandstones and mudstones from the regressive deltaic sequence above is a strongly bioturbated unit containing sandstones up to 1 m thick with scattered mollusks. This is followed by 6 m of homogeneous to slightly laminated prodelta mudstone which grades up into poorly sorted silty sandstone with abundant organic "coffee grounds." Above this are clinoform delta foresets with thin coal partings and layers of *Turritella* with no preferred alignment. The delta prograded in a direction slightly west of south. The upper surface of the foreset unit is extensively root disrupted and is overlain by impure lignite, probably representing delta-plain marsh.

PALEOENVIRONMENTAL RESUME

The Early Cretaceous East Texas basin was part of a storm-dominated epicontinental sea or proto-gulf which was comparable to the modern Gulf of Mexico in its vulnerability to hurricanes and tropical storms (Scott *et al.*, 1975). Tidal currents were ineffective except possibly where reinforced by local wind stress in bays. There were no well-developed barrier-lagoon systems to provide a runoff generated storm-surge ebb transport mechanism, and bathymetric relief was insufficient to provide significant gravity resedimentation (Scott *et al.*, 1975).

By analogy with the northern Gulf of Mexico we propose that large-scale, storm-generated bottom-return flows were responsible for the seaward transport of sediment which forms the graded sandy beds. Based on cross sections through the Washita Group by Scott et al. (1978), it is suggested that bottom gradients may have been comparable to the steeper portions of the modern Texas shelf, where wind-forced bottom currents are particularly effective.

Sand was provided mainly by small lobate deltas, but there is no evidence that riverine discharge persisted significantly beyond the shoreline. During storms, breaking waves and high orbital velocities were important in suspending sediment that was carried seaward by bottom currents. The thick, graded, hummocky cross-stratified sandstones were generated by storm waves (Walker, 1979) and waning unidirectional currents acting simultaneously. These nearshore sands were deposited rapidly but were also subject to reworking by later storms. The lower frequency of discrete sandstone beds deposited by bottom currents below

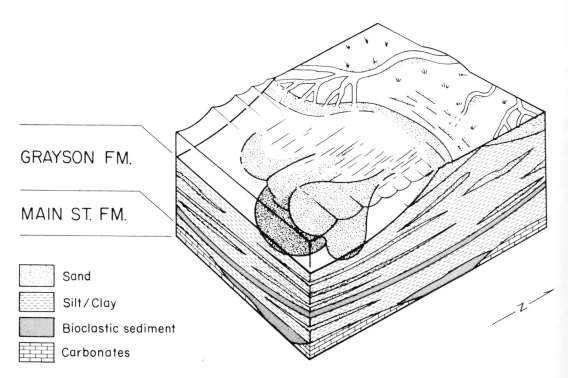

GRAYSON FM.

MAIN ST. FM.

Sand

Silt/Clay

Bioclastic sediment

Carbonates

FIG. 10.—Schematic reconstruction of the Grayson shore zone and shallow shelf depositional setting, with seaward transport of sand during storms on a delta-dominated coast.

storm wave base suggests that only the strongest storms were effective to this depth. The only reworking was by organisms and was generally insignificant.

In summary (Fig. 10), seaward transport of sand from the shore zone was probably restricted to relatively infrequent storm events and was a product of wind forced bottom-return flow quite independent of any flood-water storage and ebb runoff.

ACKNOWLEDGMENTS

Field work in northeast Texas was conducted as part of the National Uranium Resource Evaluation, Sherman Quadrangle, Texas and Oklahoma, under U.S. Department of Energy Contract No. DE-AC13-76GJ01664, Bendix Subcontract No. 78-144-E. Jim Howard, Charles Winker, Thor Hansen, and Dawn McKalips provided helpful suggestions in the field.

REFERENCES

ALBRITTON, C.C., JR., W.W. SCHELL, C.S. HILL, AND J.R. PURYEAR, 1954, Foraminiferal populations in the Grayson Marl: Geol. Soc. America Bull., v. 65, p. 327-336.

ALLEN, J.R.L., 1970, Physical Processes of Sedimentation: George Allen and Unwin, London, 248 p.

BRENCHLEY, P.J., G. NEWALL, AND I.G. STANISTREET, 1979, A storm surge origin for sandstone beds in an epicontinental platform sequence, Ordovician, Norway: Sedimentary Geol., v. 22, p. 185-217.

COTTER, E., 1975, Late Cretaceous sedimentation in a low-energy coastal zone: The Ferron Sandstone of Utah: Jour. Sed. Petrology, v. 45, p. 669-685.

FORRISTALL, G.S., R.C. HAMILTON, AND V.J. CARDONE, 1977, Continental shelf currents in Tropical Storm Delia: Observations and theory: Jour. Phys. Oceanography, v. 7, p. 532-546.

GOLDRING, R., AND P. BRIDGES, 1973, Sublittoral sheet sandstones: Jour. Sed. Petrology, v. 43, p. 736-747.

HAYES, M.O., 1967, Hurricanes as geological agents: Case studies of Hurricane Carla, 1961, and Cindy, 1963: Univ. Texas Bur. Econ. Geol. Rep. Invest. 61, 54 p.

HAYWARD, O.T., AND L.F. BROWN, JR., 1967, Comanchean (Cretaceous) rocks of central Texas: Permian Basin Sec., Soc. Econ. Paleontologists and Mineralogists Pub. 67-8, p. 51-64.

HOWARD, J.D., 1972, Trace fossils as criteria for recognizing shorelines in stratigraphic record, in J.K. Rigby and W.K. Hamblin (eds.), Recognition of Ancient Sedimentary Environments: Soc. Econ. Paleontologists and Mineralogists, Spec. Pub. 16, p. 215-225.

KELLING, G., AND P. MULLIN, 1975, Graded limestones and limestone-quartzite couplets: Possible storm deposits from the Moroccan Carboniferous: Sedimentary Geol., v. 13, p. 161-190.

KUMAR, N., AND J.E. SANDERS, 1976, Characteristics of shoreface storm deposits: Modern and ancient examples: Jour. Sed. Petrology, v. 46, p. 145-162.

MOOERS, C.N.K., 1976, Wind-driven currents on continental margins, in D.J. Stanley and D.J.P. Swift (eds.), Marine Sediment Transport and Environmental Management: John Wiley and Sons, New York, p. 29-52.

MORTON, R.A., 1981, Formation of storm deposits by wind-forced currents in the Gulf of Mexico and the North Sea: Internat. Assoc. Sedimentologists Spec. Pub. 5, p. 385-396.

MURRAY, S.P., 1970, Bottom currents near the coast during the Hurricane Camille: Jour. Geophys. Res., v. 75, p. 4579-4582.

POWERS, M.C., AND B. KINSMAN, 1953, Shell accumulations in underwater sediments and their relation to the thickness of the traction zone: Jour. Sed. Petrology, v. 23, p. 229-234.

REINECK, H.E., AND I.B. SINGH, 1972, Genesis of laminated sand and graded rhythmites in storm-sand layers of shelf mud: Sedimentology, v. 18, p. 123-128.

SCOTT, R.W., H. LAALI, AND D.W. FEE, 1975, Density-current strata in Lower Cretaceous Washita Group, north-central Texas: Jour. Sed. Petrology, v. 45, p. 562-575.

_____,D.W. FEE, R. MAGEE, AND H. LAALI, 1978, Epeiric depositional models for the Lower Cretaceous Washita Group, north-central Texas: Univ. Texas Bur. Econ. Geol. Rep. Invest. 94, 23 p.

VOS, R.G., 1976, Observations on the formation and location of transient rip currents: Sedimentary Geol., v. 16, p. 15-19.

WALKER, R.G., 1979, Shallow marine sands, in R.G. Walker (ed.), Facies Models: Geoscience Canada Reprint Ser. 1, p. 75-89.

WHITAKER, J.H. McD., 1973, "Gutter casts," a new name for scour-and-fill structures: With examples from the Llandoverian of Ringerike and Malmoya, southern Norway: Norsk. Geol. Tidsskrift, v. 53, p. 403-417.

YORATH, C.J., B.D. BORNHOLD, AND R.E. THOMSON, 1979, Oscillation ripples on the northeast Pacific continental shelf: Marine Geol., v. 31, p. 45-58.

YOUNG, K., 1972, Cretaceous paleogeography: Implications of endemic ammonite faunas: Univ. Texas Bur. Econ. Geol. Circ. 72-2, 13 p.

ENVIRONMENT OF DEPOSITION AND RESERVOIR PROPERTIES OF THE WOODBINE SANDSTONE AT KURTEN FIELD, BRAZOS CO., TEXAS

JAMES R. TURNER AND SUSAN J. CONGER
Pruet Oil Co., 310 Beck Building, Shreveport, Louisiana 71101
and Gulf Exploration and Production Co., P. O. Box 1635, Houston, Texas 77001

ABSTRACT

A combination of stratigraphic and diagenetic events has trapped oil in thin-bedded, clayey sandstones of the Upper Cretaceous Woodbine-Eagleford Formations. Five sandstone units occur in Kurten Field and are designated from top to bottom as "A" through "E". Foraminifera and nannofossils indicate these units to be late Turonian. The "C" and "D" units are elongate north to south, 4.5 miles wide, over 10 miles long, and 40 feet thick. The "B" and "E" units are thinner and trend northeast to southwest. Grain size coarsens upward in the "B", "C", and "D" units, averaging 0.14 mm and ranging from 0.09 mm to 0.18mm. Grain size fines upward in the "E" unit. The sandstone's average composition is 66% quartz, 1% feldspar, 2% rock fragments, and 28% matrix. Sedimentary structures in the "B", "C", and "D" units grade upward from laminated and bioturbated siltstones to clean sandstones with flaser cross-beds. The "E" unit consists of repeated bedsets about 1 foot thick of massive to faintly laminated sand, overlain by wavy to undulatory laminated sand, which is overlain by marine shale. Bedsets commonly show sharp to irregular basal contacts. Sedimentary structures, bioturbation, and microfauna indicate that the units are offshore bars which have been generated by a combination of river-mouth by-passing, storm-generated sheet flows, and longshore currents.

The porosity is largely diagenetic and occurs in the clayey beds. It appears to have been formed by fresh water leaching along an erosional unconformity overlain by the Austin Chalk. Permeability becomes progressively poorer away from the unconformity, and a permeability barrier ultimately forms a poorly defined updip limit for the field, making Kurten a combination diagenetic and stratigraphic trap. Relatively widespread occurrences of offshore bars suggest that similar traps may be fairly common in ancient shelf sediments.

INTRODUCTION

Kurten Field has a producing area of almost 100 square miles with estimates of reserves up to 100 million barrels of oil. It produces from a depth interval of 7800 to 8800 feet. The field remained undiscovered in one of the most mature petroleum exploration provinces in the United States until 1976 (Fig. 1). The trapping mechanism is the result of a fortunate coincidence of stratigraphic and diagenetic processes operating in sediments deposited in an offshore, shallow marine environment. Structure shows only gentle monoclinal dip to the southeast. Interpretation of conventional seismic lines across the field does not indicate the hydrocarbon trap. The type of trap at Kurten truly qualifies as one of a subtle variety which may be fairly common, though still hidden in the rock record, and may contain a significant portion of the undiscovered reserves remaining in the United States.

REGIONAL STRATIGRAPHY

Kurten Field produces from the Upper Cretaceous Woodbine-Eagleford section which unconformably overlies the Lower Cretaceous Buda Limestone and unconformably underlies the Upper Cretaceous Austin Chalk (Table 1). The Woodbine-Buda contact is unconformable at its outcrop on the west side of the East Texas Basin and in the subsurface along the west flank of the Sabine Uplift on the east side of the basin (Granata, 1963). Although a number of cores were taken at Kurten Field, none included the Lower Cre-

taceous-Woodbine contact; however, electric logs detect a sharp interface at the contact. The Buda is eroded in wells south of the field in Washington and Waller Counties, implying that the Buda-Woodbine contact is probably unconformable at Kurten. This unconformity appears to coincide with the mid-Cenomanian global low stand of sea level recognized by Vail, et al. (1977).

The contact with the overlying Austin Chalk appears unconformable at Kurten, around the margins of the East Texas Basin, and on the Sabine uplift. The contact probably becomes conformable toward the center of the basin (Eaton, 1956) and downdip from the Lower Cretaceous shelf edge southeast of Kurten. The Woodbine-Eagleford is about 700 feet thick at Kurten and progressively thins south of the field until it is erosionally missing in wells in Washington and Waller Counties where the Austin Chalk directly overlies the Georgetown Formation. This unconformity is believed to coincide with a worldwide low stand of sea level during the upper Turonian and lower Coniacian (Cooper, 1977; Vail et al., 1977).

The age of the Woodbine is generally believed to be mid-Cenomanian to Turonian. The formation consists mainly of sandstones and shales varying in thickness from 0 to 1,200 feet. It is nonmarine in the northern and central parts of the basin and marine in the south and southwest (Eaton, 1956). In the East Texas Basin, the Woodbine is usually subdivided into the Dexter and Lewisville members: the Dexter is the lower member consisting of massive quartz sandstone with little vol-

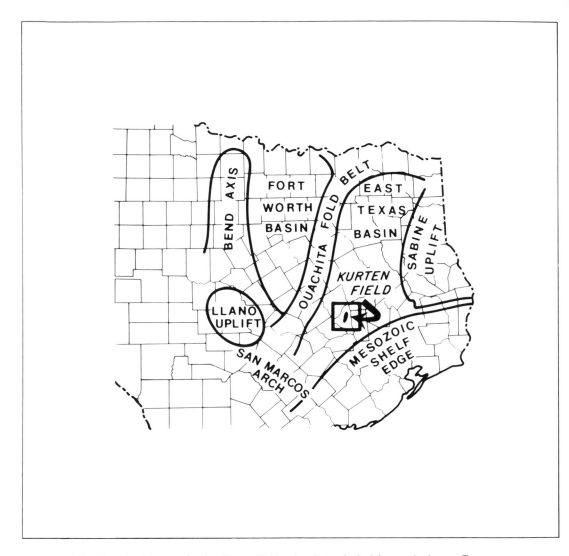

Fig. 1.—Location map showing Kurten Field and major geological features in the east Texas area.

canic material; the Lewisville is the upper member consisting of calcareous sandstones locally containing volcanic material (Nichols *et al.*, 1968).

The depositional systems of the interval were recognized by Oliver (1971) who delineated the Dexter fluvial system, the Freestone Delta, and the Lewisville strand plain system (Fig. 2A). The Dexter fluvial system is described as a series of meandering and braided flood plain and channel deposits that prograded from northeastern Texas. The Dexter fluvial system grades downdip into the Freestone Delta. The deltaic system extends across the central portion of the East Texas Basin where it occurs as a series of thick-bedded sandstone units totaling more than 400 feet thick. This system prograded rapidly southwestward over the Lower Cretaceous shelf and probably covered at least part of

the present-day Sabine Uplift. Progradation apparently was into fairly shallow water as evidenced by the extensive reworking of delta sands west and northwest into the Lewisville strand plain system. The Woodbine was the first clastic sedimentation since the Early Cretaceous Hosston (Neocomian) and marked the end of the great carbonate producing period of the Lower Cretaceous. The sediment source was the low grade metamorphic and sedimentary rocks of the Ouachita Mountains of southern Oklahoma and Arkansas. There also appears to be a significant amount of volcanic detritus in the Woodbine probably derived from active vents in southern Arkansas and from vents occurring around the East Texas Basin (Hunter and Davies, 1979).

The Eagleford overlies the Woodbine. The nature of

TABLE 1.—GEOLOGIC COLUMN SHOWING A PORTION OF THE CRETACEOUS SYSTEM IN EAST TEXAS
(MODIFIED AFTER NICHOLS ET AL., 1968)

MESOZOIC ERA															
CRETACEOUS SYSTEM															
COMANCHE SERIES				GULF SERIES											
ALBIAN STAGE				CENOMANIAN STAGE			TURONIAN STAGE	CONIACIAN–SANTONIAN		CAMPANIAN				MAESTRICHT	
FREDER-ICKSBURG GROUP		WASHITA GROUP		WOODBINE GROUP		EAGLEFORD GROUP		AUSTIN GROUP		TAYLOR GROUP				NAVARRO GROUP	
							EAGLE FORD								
GOODLAND-EDWARDS	KIAMICHI	GEORGETOWN		MANESS BUDA GRAYSON	DEXTER	LEWISVILLE	HARRIS	COKER	SUBCLARKSVILLE	AUSTIN CHALK	LOWER	UPPER	LOWER	UPPER	

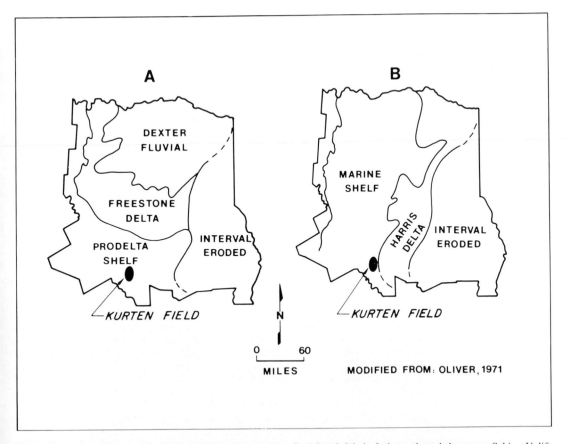

FIG. 2.—*A*, Distribution of Woodbine Depositional System fluvial and deltaic facies and eroded area on Sabine Uplift. Kurten Field is located in the area of prodelta shelf deposits. *B*, Progradation of Harris Delta from Sabine Uplift toward Kurten Field.

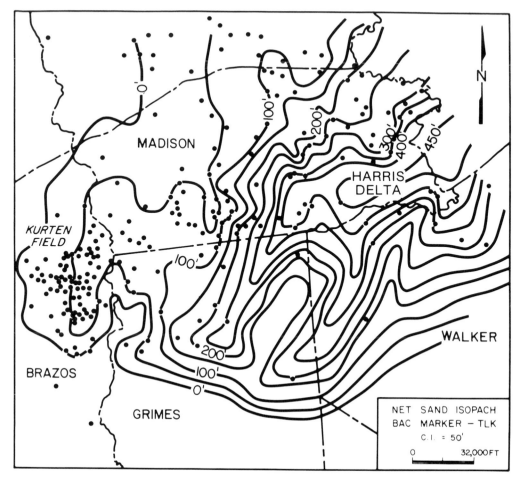

Fig. 3.—Net sandstone isopach of Woodbine-Eagleford interval is from BAC marker to the top of the Lower Cretaceous Buda Limestone (Fig. 4). Harris Delta and Kurten Field.

the contact is uncertain, but it is probably disconformable in parts of the basin. The Eagleford consists of micaceous gray to dark gray shale with fine to medium-grained porous to nonporous sandstone members (Nichols *et al.*, 1968). It is largely marine, varies in thickness from 0 to 800 feet thick, and locally contains fish teeth and glauconite. Eagleford deposition was contemporaneous with movement of the Sabine Uplift which became positive in East Texas and western Louisiana during the Turonian. The uplift gradually shifted its structural axis westward throughout the time of Eagleford deposition and became quiescent during deposition of the Austin Chalk (Coniacian) (Nichols, 1964).This shift resulted in onlap on the eastern flank and offlap on the western flank of the uplift, concurrent with erosion of the underlying Woodbine. Previously deposited Freestone Delta sediments were incorporated into a southwesterly prograding delta system that developed on the west flank of the uplift termed the

Harris Delta System (Fig. 2B) (Granata, 1963; Nichols, 1964).

The Harris Delta extends from the west flank of the Sabine uplift to northern Grimes County for a distance of more than 160 miles. It has a fairly abrupt southern boundary that follows the back side of the Lower Cretaceous shelf edge banks, parallel to the Angelina-Caldwell Flexure. The delta is composed of a sandstone, which varies in thickness from 0 to 450 feet, and is characterized by a relatively straight channel system. The delta advanced by cutting through its mouth bar systems in a southwesterly direction. A continuous sandstone provides a conduit for the migration of hydrocarbons northward to a point where it is truncated and overlain by the Austin Chalk at the East Texas Field.

The productive sandstones at Kurten Field are 12 miles west of the Harris Delta (Figs. 2B and 3) and occupy the upper third of the Woodbine-Eagleford sec-

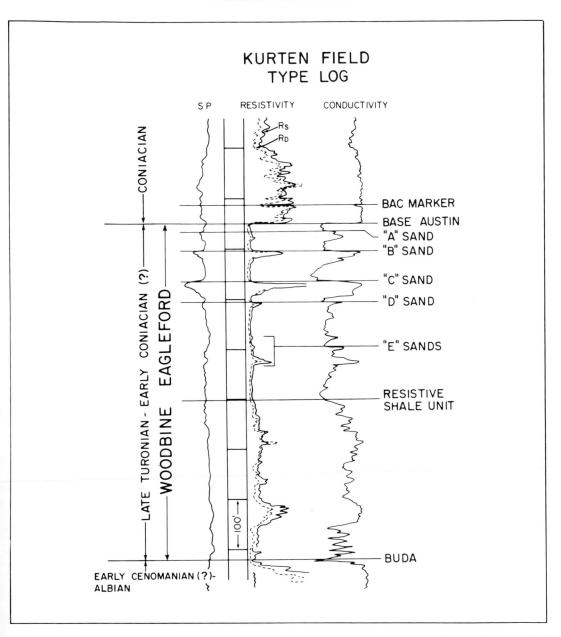

Fɪɢ. 4.—Type log of Kurten Field illustrating electric log markers and ages. Composited from Amalgamated Bonanza Lloyd and Cayuga Cobb No. 2. Locations shown in Figure 9.

tion immediately below the base of the Austin Chalk. They are designated from top to bottom as "A" through "E" (Fig. 4). Cross-section A-A' (Fig. 5) illustrates the relationship of the Kurten sandstones to the Harris Delta. Cross-section B-B' (Fig. 6) illustrates the truncation of the sandstone units to the east and the resistive shale to the west. The locations of the cross sections are shown in Figure 7. The datum is the BAC marker (Fig. 4), a reliable marker occurring throughout

Brazos, Grimes, Madison, and Leon Counties. The "C", "D", and "E" sandstone units at Kurten correlate with the main sandstone body of the Harris Delta. They thicken to the south and are truncated at the southern boundary of Kurten Field. They are overlain by the Austin Chalk. The "A" and "B" sandstone units are younger, possibly derived from reworked Harris Delta sediments during a marine transgression. Thinning of the interval between the base of the Austin

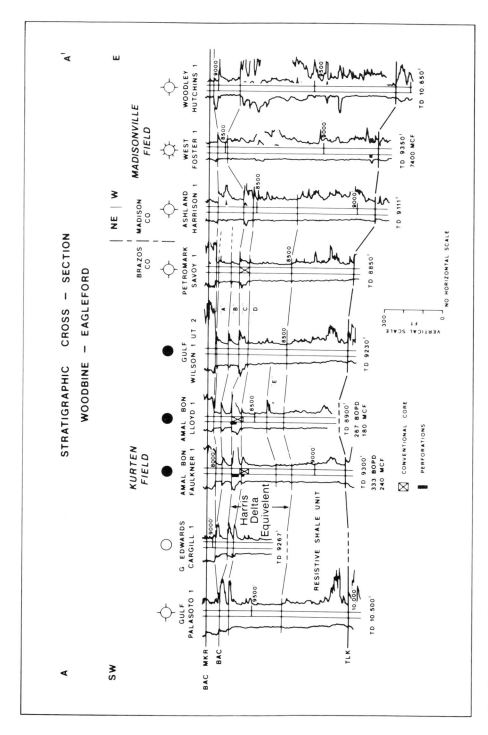

FIG. 5.—Cross-section A-A' showing relationship of Kurten sandstones to Harris Delta and erosional truncation of Kurten sandstones to the south. Datum is BAC marker. Location of cross-section is shown in Figure 7.

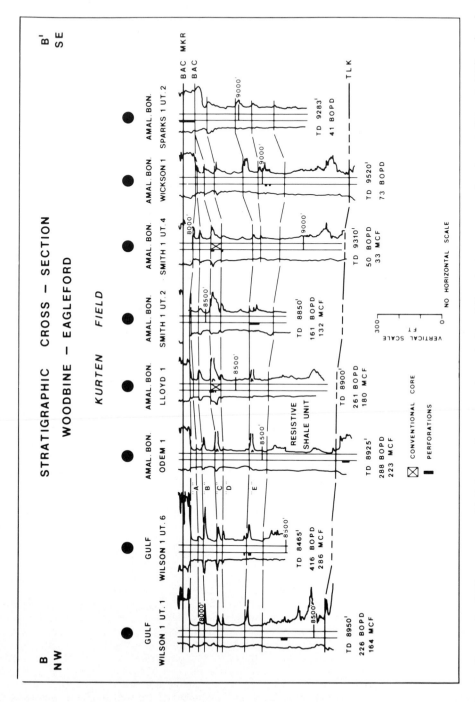

FIG. 6.—Cross-section B-B' across Kurten Field from northwest to southeast. A. B. and C sandstones are truncated to the east. Datum is BAC marker showing asymmetrical sandstone bodies with steep sides facing east.

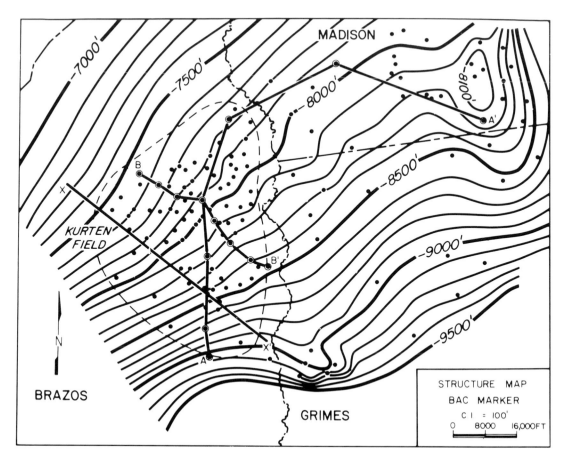

Fig. 7.—Structure map on BAC marker. The location of Kurten Field and the surrounding area is indicated. The positions of cross-sections A-A' and B-B' and of seismic line X-X' are shown.

Chalk and the top of the Harris Delta is believed to result from deposition onlapping topography during this transgression as shown in the West No. 1 Foster in Figure 5. The topography is significant because production occurs where sandstones thin or are truncated along the flanks of the Harris Delta in western Madison County. Sandstones deposited in a similar fashion on and around buried topography form hydrocarbon traps in western Canada and in the mid-continent area of the United States (Martin, 1966).

A resistive shale unit in the lower half of the Woodbine-Eagleford is shown in Figure 5. The unit varies from 300 to 400 feet thickness and is recognized by a mean increase in shale resistivity from 2 to 4 ohmmeters along a distinct boundary. Well log and biostratigraphic correlations do not give a clear indication whether the resistive shale is Woodbine or Eagleford, but it is probably Turonian and possibly represents prodelta shales of the Freestone Delta. Since the Kurten sandstones are contemporaneous with the Harris Delta, which post-dates the Freestone Delta, they equate to

the Eagleford Formation. Even so, operators in the area have generally termed the productive sandstones Woodbine. This paper makes no effort to resolve the difficulty in recognizing the Woodbine-Eagleford boundary and, following general industry usage, Woodbine will be used to indicate the interval between the base of the Austin Chalk and the top of the Lower Cretaceous, unless otherwise noted.

STRUCTURE

Structure at Kurten Field contoured on the BAC marker, shows monoclinal dip to the southeast of 140 feet per mile, illustrating the non-structural nature of the trap (Fig. 7). Regionally, the monocline is interrupted by a northeast-southwest trend of salt structures extending through Madison, Grimes, Brazos, and Burleson counties. In Madison County, a salt swell appears to underlie Madisonville Field, and a salt diapir is responsible for Day Field. To the southwest in Grimes County, a wider salt diapir with less relief creates the trap at Hill Field. Seismic line X-X' (Fig. 8)

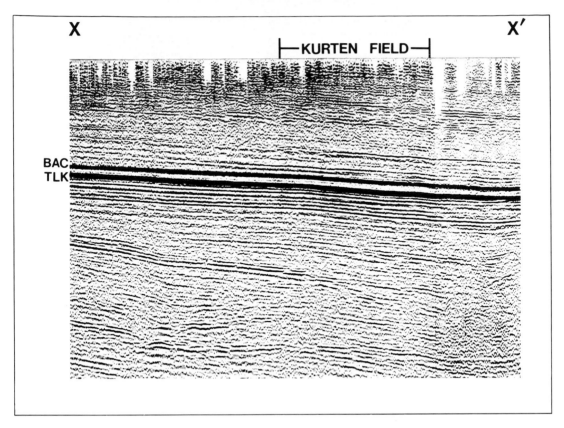

FIG. 8.—Seismic line X-X' showing structure and reflection character in the dip direction at Kurten Field. BAC is "base of Austin Chalk," TLK is "top Lower Cretaceous." Location of section is shown in Figure 7.

illustrates the structural and seismic character of the field and the lack of visibility of the field on conventional twelve fold lines. The location of X-X' is shown in Figure 7. Reflectors showing the base of the Austin Chalk and the top of the Lower Cretaceous are clearly visible on the line. The location of Kurten Field is shown on the line, and no geophysical indication of the trap is present between the two reflectors.

METHODS FOR LITHOLOGICAL DETERMINATIONS

Five slabbed conventional cores from Kurten Field were examined for composition, textures, and sedimentary structures. Wells and cored intervals are listed in Table 2. Compositional estimates were made by analysis of 20 thin sections using standard point counting procedures and 100 points per slide. Slides were impregnated with blue-dyed epoxy for porosity observation and stained with alizarine red for identification of carbonate cement. The long axes of 100 monocrystalline quartz grains per slide were measured for textural analysis and grain size statistics were calculated in phi units. The clay fraction was identified by x-ray diffraction and scanning electron microscopy. Porosity, permeability, and fluid saturation measurements

were made at commercial laboratories. Production and reservoir statistics were taken from Gulf Exploration and Production Company records. Subsurface control consists of standard electric logs, cores, and regional seismic lines.

PALEONTOLOGICAL DETERMINATION METHODS

To provide samples for paleontologic analysis, pieces of shale from one to three-foot intervals were composited from slabbed conventional cores. Core samples were examined from the wells listed in Table 2 as well as the Cayuga No. 1 Cobb and Amalgamated Bonanza No. 2 Guest. Ditch cuttings from 30-foot intervals were composited from the Gulf No. 1 Palasota, Gulf No. 1 Payne, and Gulf No. 1 Peters in the southern end of the field. These were processed and examined for palynomorphs, nannofossils, and foraminifera. Photomicrographs were used for identification of trace fossils.

Ditch cuttings were examined through the interval from the top of the Upper Cretaceous to the top of the Lower Cretaceous. Core samples were taken from the "B" unit, top and bottom of the "C" unit, top and bottom of the "D" unit, between the "D" and "E"

TABLE 2.—KURTEN FIELD CORES UTILIZED IN STUDY

Core	Code	Date Cut	Interval (feet)	Units(s)
Amalgamated Bonanza No. 1 Lloyd	ABL1	8/77	8356-8416 (60)	"C", "D"
			8525-8581 (56)	"E"
Amalgamated Bonanza No. 1 Faulkner	ABF1	3/78	8625-8686 (61)	"D"
Amalgamated Bonanza No. 1 Smith Unit 4	ABS4	1/78	8575-8635 (60)	"C", "D"
Petromark No. 1 Huffman	PH1	12/77	8431-8540 (109)	"B", "C", "D"
Cayuga No. 2 Cobb	CC2	1/78	8828-8863 (35)	"C"

units, top and bottom of the "E" unit, and in the resistive shale below the "E" unit (Fig. 4).

GEOMETRY OF SANDSTONE UNITS

The Kurten Field sandstones have an overall north-south trend extending roughly 11 miles north-south. The trend is approximately 5.5 miles wide. They are eroded down through the "D" unit on the east and south and shale out to the west and northwest. The sandstones can be correlated with wells north and northeast of the field into Madison County. This apparently leaves the field open-ended in the updip direction.

The "A" sandstone unit is the first stringer encountered below the base of the Austin Chalk and varies in thickness from 0 to 10 feet. The "A" unit is not mapped in this study, but it appears to be confined to the northern end of the field. The "A" sand unit shales out to the north and west and was eroded to the south and east (Figs. 5 and 6). On electric logs, the "A" sandstone is a resistive unit with good spontaneous potential which normally deflects further toward the top of the sand. No oil or gas is currently being produced from the "A" unit.

The "B" sandstone unit is developed throughout most of Kurten Field and is productive in the western and northern portions. The "B" net sandstone isopach (Fig. 9) shows a series of elongate, coalesced sandstone pods ranging from 0 to 25 feet in thickness and oriented from northeast to southwest in an area measuring 10 miles by 7.5 miles. The "B" unit shales out to the west and north and is truncated on the southern and eastern limits of the field. The greatest thickness of the "B" unit occurs on the southern end of the field, close to the line of truncation. The "B" unit appears to be a remnant of a much more extensive system that extended to the southwest and was removed by erosion. The "B" unit is shown on stratigraphic cross sections A-A' (Fig. 5) and B-B' (Fig. 6). On logs it generally exhibits a rounded to increasing upward spontaneous potential. The gamma ray log generally increases its deflection upward with a gradational base and an

abrupt top, indicating that the units become cleaner upward. The "B" unit is bounded by marine shale except where it is cut by the unconformity.

The "C" sandstone unit is the best developed and is the most extensive in the field. It has an overall north-south trend, varies in thickness from 0 to 50 feet, and is asymmetrical as shown by the more closely spaced contours which indicate a steep side facing east (Fig. 10). The dimensions of the unit are 5.5 miles by 12.5 miles. Two en echelon sandstone bodies separated by an area of thin sandstone are observed. The "C" unit shales out to the west and is truncated on the southern and eastern limits of the field. It does, however, continue northeast out of Kurten Field toward Madison County for an unmapped distance. The original system probably extended several more miles to the southwest and was removed by erosion. The "C" unit is productive throughout its extent and is the main reservoir in Kurten Field. Production in the "C" unit is largely confined to Brazos County as its permeability becomes progressively poorer to the north. The "C" unit shows both decreasing and increasing upward patterns on spontaneous potential curves, but the gamma ray normally shows an increasing upward character, with a gradational lower boundary and abrupt upper boundary. It appears that the gamma ray is a more reliable indicator of shaliness than the spontaneous potential curve.

The "D" sandstone unit is best developed along the southwest limits of the field. It has a northeast-southwest orientation and is asymmetric as is indicated by the widely spaced contours on the west and the closely spaced contours on the east (Fig. 11). The areal extent of the "D" unit is 8 miles by 11 miles, and it varies in thickness from 0 to 40 feet. It consists of a main sandstone body with smaller pods on the southeast end that are separated by areas of thinner sand development. The unit shales out to the west and northwest and is truncated on the south and east. In the northern and western parts of the field, the "D" unit appears on electric logs as a thin, resistive stringer immediately beneath the "C" unit. Where the unit is

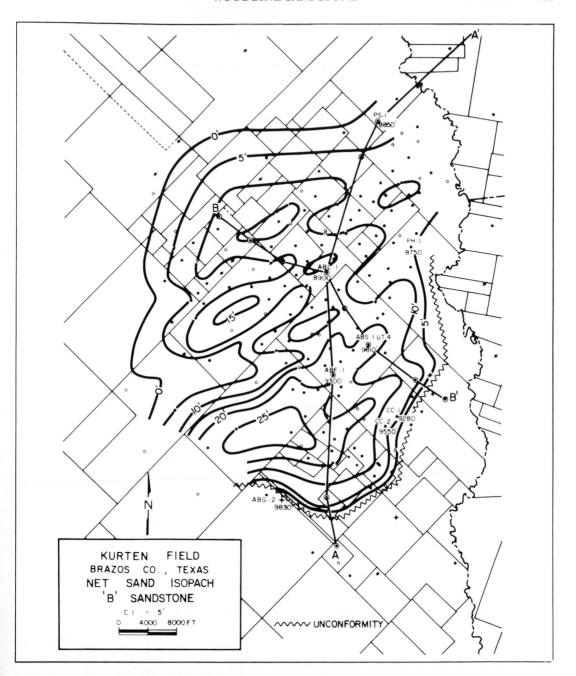

FIG. 9.—Net sandstone isopach, Woodbine "B" sandstone unit, Kurten Field. Locations of cross-sections A-A' (Fig. 5) and B-B' (Fig. 6) and of unconformity along the southeast edge of the area are shown.

developed on the south and east sides of the field, it typically shows a rounded spontaneous potential response. The gamma ray log indicates that the unit is muddy at the base (gradational contact) and becomes sandier upward (abrupt upper contact). Erosion appears to have removed much of the sandstone leaving only a remnant of the original sand body geometry. It is productive on the south and east sides of the field.

The "E" sandstone unit is quite different from the other units as it occurs as a series of one to three discrete stringers in a zone ranging from 100 to 180 feet below the "D" unit and 50 to 80 feet above the

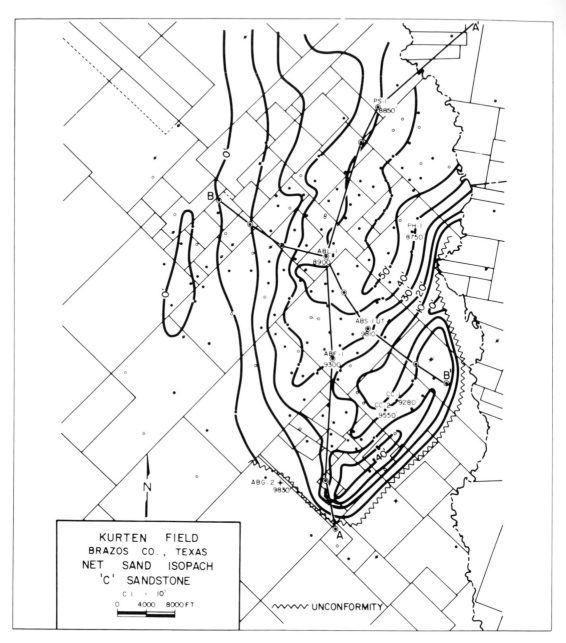

FIG. 10.—Net sandstone isopach, Woodbine "C" sandstone unit, Kurten Field, showing an en echelon pattern of "thicks." Location of cross-sections A-A' (Fig. 5) and B-B' (Fig. 6) and of unconformity along the southwestern edge of the area are shown.

resistive shale. The "E" sandstone stringers are subtle on electric logs, usually showing no spontaneous potential development or, in some cases, a suppressed spontaneous potential response. The gamma ray log is usually reliable in detecting the stringers, but it is often suppressed due to the high clay content. Where the gamma ray is strongly deflected and the resistivity val-

ues are high, 30 ohm-meters or more, the "E" stringer has a symmetrical character with abrupt upper and lower contacts. The "E" sandstone isopach (Fig. 12) was drawn using thickness values measured from the net gamma ray and resistivity deflections on the 1" = 100' scale subsurface logs; different thickness values may be obtained using logs with 5" = 100' scale. The

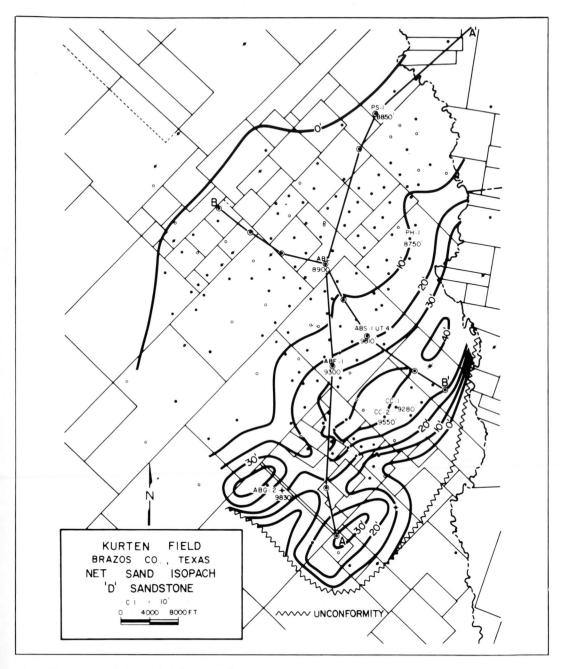

FIG. 11.—Net sandstone isopach, Woodbine "D" sandstone unit, Kurten Field. Location of cross-sections and of unconformity are shown.

"E" isopach shows a series of elongate northeast-southwest trending sandstone pods which thin to the southwest and lie below the unconformity. Although the "E" stringers thicken to the northeast, they do not reach the Harris Delta and are isolated within a marine shale. The unit varies in thickness from 0 to 50 feet and has an areal extent of 9 miles by 9 miles with an overall fan-like geometry. The "E" stringers are productive when present, and their pressures exceed the normally pressured "B", "C", and "D" sandstones by approximately 1000 psi. Individual sandstone units occur as discrete bodies with elongate northeast-southwest orien-

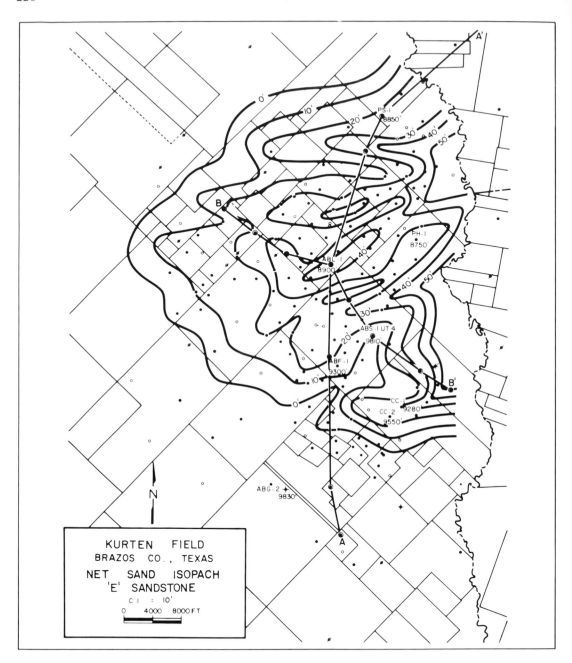

Fig. 12.—Net sandstone isopach, Woodbine "E" sandstone unit, Kurten Field. Location of cross-sections A-A' (Fig. 5) and B-B' (Fig. 6) are shown.

Fig. 13.—Sedimentary structures in Woodbine "C" and "D" units, Amalgamated Bonanza No. 1 Lloyd, Kurten Field. *A,* "C" unit; burrowed (arrow) flaser cross-bedded sandstone; 8356 feet. *B,* "C" unit; burrowed (b), flaser cross-bedded (f) sandstone with sharp basal contact (c) with overlying bedset; 8359 feet. *C,* "C" unit; flaser cross-stratified unit (f) showing stacked lenses of sand, becoming bioturbated (b) upward; 8364 feet. *D,* "C" unit; bedset of massive to indistinctly laminated sandstone (i), distinct contact (c), and flaser cross-bedded sandstone (f); 8366 feet. *E,* "C" unit; bedset of massive to indistinctly laminated sandstone with shale classes (arrow) overlain by flaser cross-beds (f), followed by wavy irregular laminated unit (w) with large burrows (b); 8567-68 feet. *F,* "C" unit; clean flaser cross-bedded sandstone with shale clasts; 8369 feet. *G,* "C"

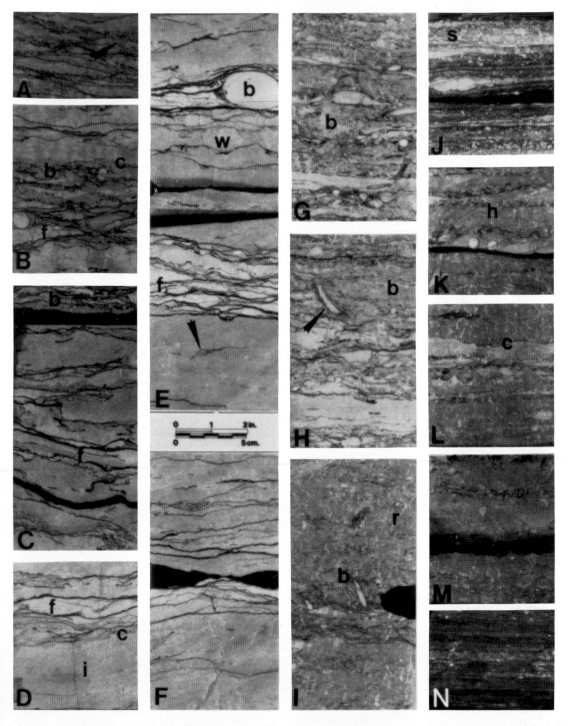

unit; intensely bioturbated (b) clayey, silty sandstone, sedimentary strucutre largely destroyed; 8373 feet. *H*, "C" unit; intensely bioturbated (b) clayey, silty sandstone; inclined shale lined burrow (arrow); 8375 feet. *I*, "C" unit; intensely bioturbated (b) clayey sandstone, with sinuous burrows; 8378 feet. *J*, "C" unit; laminated to bioturbated unit with siltstone-shale bedsets (s); 8392 feet. *K*, "D" unit; wavy laminated, clayey, bioturbated sandstone; 8408 feet. *L*, "D" unit; bioturbated, silty, clayey sandstone with irregular contact (c); 8409 feet. *M*, "D" unit; clayey, bioturbated siltstone; 8410 feet. *N*, Laminated sparsely burrowed marine shale underlying "D" unit; 8411 feet.

tations, approximately paralleling the long axis of the Harris Delta and trending approximately 45° to the overlying sandstones.

SEDIMENTARY STRUCTURES

The sequence of sedimentary structures of the "B", "C", and "D" sandstone units is characterized by what is interpreted to be an increase in depositional energy upward. The sequence begins with a shaley laminated and clayey bioturbated sandstone which grades upward to a wavy, discontinuously laminated, clean sandstone. The "E" unit is characterized by a decrease in depositional energy upward and repeated bedsets of massive and laminated sandstones which grade upward to wavy and undulating laminated shaley siltstones. The best core control is in the "C" unit. On the basis of sedimentary structures and burrowing, it can be subdivided into four sedimentary units (Fig. 13, A-J). A complete vertical sequence is not found in any single core, so it is synthesized from cores from the No. 1 Loyd, the No. 1 Smith Unit 4, and the No. 2 Cobb. A similar sequence of beds is also recognized in the cores of the "B" and "D" units in the No. 1 Faulkner and the No. 1 Huffman. They are thought to represent similar depositional environments. The four sedimentary units in the "C" unit follow in descending order:

#1) Sandstone, clean, burrowed, flaser crossbedded; approximately 9-feet thick.

#2) Sandstone, clean flaser crossbedded; 11 to 16-feet thick.

#3) Clayey, silty sandstone, intensely bioturbated; 12 to 17-feet thick.

#4) Clayey, silty sandstone, laminated, bioturbated; 3 to 16-feet thick.

The laminated, bioturbated sandstone (#4) is well developed in the "C" unit in the Amalgamated Bonanza No. 1 Smith Unit 4 (Fig. 14, F, G, H) and the Cayuga No. 2 Cobb. The lower and upper contacts are gradational over a few feet with marine shale below and clayey intensely bioturbated sandstone above. The unit consists of dark gray, silty, clayey, very fine grain sandstones interbedded with sharp, continuous, slightly wavy, shaley laminae and shale stringers up to 1 inch thick. Thin bedsets occur from 1 to 2 inches thick and consist of very fine grained sandstone and siltstone with small ripple lenses abruptly overlain by shale. Basal contacts of the bedsets are sharp, wavy to slightly irregular, and show little evidence of erosion. Penecontemporaneous contortion is occasionally found (Fig. 14G). The unit contains small, horizontal, cir-

cular, and elliptical burrows, as well as continuous ribbon-like tubes, approximately 0.1 inch in diameter, which are generally confined to the shaley bedding planes. The pattern of alternating bioturbation and lamination implies episodic deposition and is similar to patterns found in turbidity flows (Howard, 1975). Laminated, bioturbated beds are also found in the "D" unit in the No. 1 Smith Unit 4 (Fig. 14, I-N), the No. 1 Huffman, the No. 1 Faulkner, and the No. 1 Lloyd (Fig. 13, K-N). However, in these cores the laminated portions are almost totally destroyed by bioturbation.

An intensely bioturbated gray to dark gray, clayey, silty sandstone (#3) gradationally overlies the laminated sandstone in the "C" unit, and is found in the No. 1 Lloyd (Fig. 13, G, H, I), the No. 1 Smith Unit 4 (Fig. 14, D,E), the No. 2 Cobb, and the No. 1 Huffman. The sedimentary structures are completely destroyed by bioturbation, which consists of both horizontal and vertical burrows that are generally larger (up to 0.5 inch in diameter) than those in the underlying beds. Long, sinuous, unlined burrows approximately 0.1 inch in diameter are abundant. Also, long, lined vertical and slanted (about 45°) burrows with spreiten which resemble *Diplocraterion* (Chamberlain, 1978) are present. A diversity of other burrows occurs in the unit: large, oval, pelleted burrows up to 1 inch in diameter which are probably *Ophiomorpha* (Fig. 14M); irregular oval burrows with concentric clay laminae which resemble *Asterosoma* tubes; small, branching and round unlined burrows which resemble *Chrondrites* (Fig. 14); inclined spreiten filled burrows which are less common and are probably *Rhizocorallium* or *Teichichnus* (Fig. 14N); and small ribbon-like burrows that are abundant throughout the clayey sandstone and resemble *Scalarituba* and *Arenicolites* (Fig. 14, D-G).

A clean, light gray to white, flaser cross-bedded sandstone (#2) gradationally overlies the bioturbated sandstone and represents the highest energy of deposition. It is cored only in the "C" sandstone and is best developed in the No. 1 Lloyd core (Fig. 13, D, E, F). Sets of flaser cross-beds (Reineck and Singh, 1973) alternating with zones of very indistinctly laminated sandstone beds inclined about 20° and containing occasional shale clasts, characterize the sandstone. The indistinctly laminated beds have sharp, irregular, slightly scoured basal contacts (Fig. 13E) and sharp to gradational upper contacts (Fig. 13, D, E). The flaser cross-bedded sandstone contains wavy, discontinuous, slightly irregular, clayey laminae varying in thickness from 0.05 to 0.1 inch. These apparently developed as

---

FIG. 14.—Sedimentary structures in cores of Woodbine "C" and "D" units, Amalgamated Bonanza No. 1 Smith Unit 4, Kurten Field. *A,* "C" unit; clean to burrowed sandstone (b) with wavy shale laminations; 8579 feet. *B,* "C" units; clean burrowed sandstone with sharp basal contact (c) on overlying sand unit, wavy (w) shaley zones probably represent flaser crossbedding; 8580 feet. *C,* "C" unit; clean burrowed sandstone with shale clasts (arrow); 8582 feet. *D,* "C" unit; intensely bioturbated clayey, silty sandstone with small sinuous *Scalarituba*-like burrows (r); 8585 feet. *E,* "C" unit; intensely bioturbated clayey, silty sandstone with vertical and horizontal burrows; 8586 feet. *F,* "C" unit; laminated bioturbated unit showing

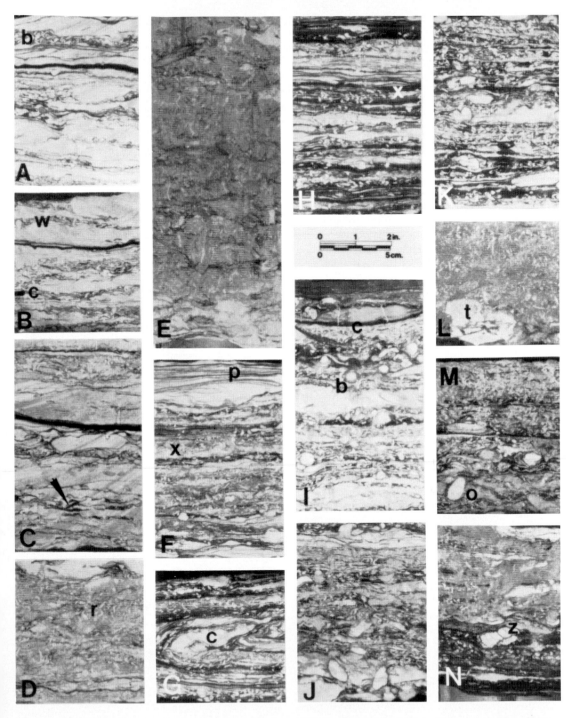

repeating siltstone-shale bedsets (x) and small ripple lenses (p); 8597 feet. *G*, "C" unit; laminated bioturbated unit exhibiting soft sediment deformation (c); 8614 feet. *H*, "C" unit; laminated to bioturbated sets; 8615 feet. *I*, "D" unit; alternating intensely bioturbated and laminated sandstones with sharp, wavy upper contacts (c) and large round horizontal burrows (b); 8620 feet. *J*, "D" unit; intensely bioturbated clayey sandstone with large and small burrows; 8627 feet. *K*, "D" unit; intensely bioturbated to laminated, clayey siltstone; 8631 feet. *L*, "D" unit; clayey, bioturbated sandstone containing a large sandstone clast (t); 8633 feet. *M*, "D" unit; bioturbated to laminated sandstone with pellet-lined *Ophiomorpha*-type (o) burrow; 8634 feet. *N*, "D" unit; bioturbated to laminated siltstone with *Rhizocorallium*-type burrow (z); 8635 feet.

clay drape that filled ripple troughs and outlined ripple lenses (Fig. 13, C, D, F).

Gradationally overlying this sandstone is the slightly more clayey, light gray, burrowed flaser cross-bedded sandstone (#1) (Fig. 13, A, B, C; Fig. 14A). It is best developed in the No. 1 Lloyd core and is similar to the underlying sandstone with the exception of increased bioturbation, increased matrix, and more frequent sets of flaser crossbeds. The sandstone has eight bedsets which consist of indistinctly laminated sandstone followed by flaser cross-bedded sandstone with a mean thickness of 1 foot. Bioturbation consists mostly of small (0.1 to 0.2 inch diameter) unlined, vertical burrows which contain a kaolinite matrix. The sandstone is interpreted to have been deposited during an abandonment or destructional phase when the depositional energy and sediment supply was diminishing. In the No. 2 Cobb core, the sandstones show no evidence of subareal exposure, and the upper contact is gradational with the overlying marine shale.

The "E" unit contrasts with the overlying sandstones and is comprised of about ten repeated bedsets of massive and wavy laminated sandstones without bioturbation. The bedsets become thinner upward and grade over about ten feet to marine shale. The lower contact of the "E" unit is sharp (Fig. 15). The "E" unit bedsets consist of massive to indistinctly laminated sandstones gradationally overlain by distinct wavy laminated siltstone, followed by small scale undulatory to slightly tabular lense sets that are topped by marine shale. The bedsets average about 1 foot in thickness.

PETROGRAPHY

Based on point counts of 59 thin sections, the detrital composition of the Kurten Field cores was determined to average 66% quartz, trace amounts of feldspar, 2% rock fragments (volcanic rock fragments, chert, and polycrystalline quartz), and 28% matrix. Table 3 summarizes the average textural and compositional aspects of each sandstone unit which is further subdivided on the basis of sedimentary structures. For instance, in the Amalgamated Bonanza Lloyd 1 the "C" sand is divided into the clayey bioturbated unit (CB) from which six thin sections were point counted, the clean sand unit (S) from which four thin sections were point counted, and the bioturbated sandy unit (BS) from which four thin sections were point counted. Generally, the Amalgamated Bonanza Lloyd 1 "C" sand quartz content increases upward from 69% to 76% with a corresponding decrease in matrix; grain size also increases upward. The amalgamated Bonanza Smith 1 unit 4 "C" sand increases in quartz content upward from 59% in the clayey laminated unit (CL) to 77% in the bioturbated sand unit (BS). The Petromark Huffman 1 "C" sand likewise increases quartz content upward from 46% to 66%. The "C" unit in the Cayuga Cobb 2 core is bioturbated and does not develop a clear quartz increase upward or a coarsening upward trend of grain size. The only measurements occurring in super

adjacent units in the "D" sand are found in the Amalgamated Bonanza Faulkner 1 core which also shows an increase upward in quartz content from 51% to 66%. The "E" sand measurements are confined to the Amalgamated Bonanza Lloyd 1 core where four thin sections were point counted from two bedsets, one taken from the base and top of each. Each bedset shows a fining upward trend and is illustrated on the left side of Figure 16. The mean quartz content is 75% and is listed on Table 3 along with the grain size means. Compositional and textural trends are correlated with electric and gamma ray log responses and are illustrated in Figures 16 and 17. Although the quartz content and grain size increases upward, the SP curve shows less deflection (becomes more positive) upward, making an environmental interpretation from log response risky.

The quartz grains are largely monocrystalline, all with undulose extinction, and small amounts of polycrystalline and metamorphic quartz. Several types of quartz grains are present and can be classified on the basis of their inclusions. A sample of 600 quartz grains (100 grains per thin section) from the No. 1 Lloyd core was found to contain 60% with few vacuoles, 20% with abundant vacuoles, and 20% with abundant microlites. In thin section, grains sometimes show zoned outer edges, embayments, bubble-wall structure, and rectangular shapes (Pettijohn, Potter, and Siever, 1972).

Rock fragments consist of highly crenulated polycrystalline grains, both with and without vacuoles and inclusions; mosaic non-crenulated polycrystalline grains with vacuoles; and chert. Rock fragments range from 2% to 9% of the detrital composition and are disseminated throughout the sandstone. The chert is rounded to subrounded and occurs as scattered grains which are distinguishable from associated kaolinite by their finer crystalline structure, lower birefringence, and salt and pepper appearance.

Feldspar occurs in trace amounts, usually having plagioclase twinning, sharp cleavage surfaces, and little alteration. No microcline or potassium feldspar was identified, but frequently encountered grains of coarsely crystalline kaolinite with rectangular and boxy shapes are possibly replacements of the potassium feldspars.

Matrix has been identified by scanning electron microscopy, x-ray, and thin section techniques. It consists of approximately 20 to 50% kaolinite, 40 to 60% chlorite (iron rich), and minor amounts of random mixed layer illite-chlorite. Kaolinite occurs as well-crystallized authigenic plates in booklets. It is visible in grains in thin section as low birefringent vermiform booklets and as disseminated plates in pore spaces (Fig. 18C and Fig. 19A). Chlorite occurs as a pore lining, as loose, interlocking blades, and as blades growing on kaolinite plates (Fig. 19C, D). The amount of chlorite appears to increase downward in the sandstone causing the clays to be somewhat segregated. Kaolinite largely occupies pore spaces in the clean

Fig. 15.—Sedimentary structures in Woodbine "E" unit, Amalgamated Bonanza No. 1 Lloyd core, Kurten Field, Texas. *A*, "E" unit, thin-bedded sandstone and silt grading to laminated siltstone and shale above. *B*, "E" unit; laminated siltstone with alternating laminations of sandstone and shale; 8533 feet. *C*, "E" unit; bedset showing indistinct, wavy lamination (a), poorly developed wavy lamination (b), shaley lamination (c), and laminated shale (d); 8534-35 feet. *D*, "E" unit; bedset showing massive sand (a) with clasts (arrow), wavy silty lamination (b), thin, slightly undulating lamination (c), and shale (d); 8536 feet. *E*, "E" unit; massive sand with undulating faintly laminated stringer; 8539 feet. *F*, "E" unit; indistinct wavy slightly undulating laminated sand (b) underlain by thin sandstone stringer with sharp slightly irregular basal contact (arrow), underlain by marine shale; 8542 feet.

TABLE 3.—DATA DETERMINED FROM THIN-SECTIONS

Well	Depth (Ft)	Sand-stone	BDD[a] unit	No. of thin sections	Quartz Size[b]			Detrital Composition[c]					Cement[d]		Permea-[e] bility md	Porosity %
					Mean mm	Max mm	σ ϕ	Qz %	F %	Rx %	Oth %	Mx %	Sil	Cal % of Total		
ABL1	8356-66	C	BS	4	.17	.70	.51	76	0	2	9	13	13	0	2.6	7.4
ABL1	8366-79	C	S	4	.16	.60	.56	77	Tr	2	9	12	15	0	2.5	7.7
ABL1	8379-93	C	CB	6	.12	.40	.64	69	Tr	2	10	19	4	4	.37	10.6
ABL1	8406-10	D	CB	3	.21	.79	.56	67	0	3	1	29	9	0	1.20	6.6
ABL1	8533-43	E	TB	4	.14	.37	.57	75	Tr	3	5	17	11	6	.82	8.6
Means					.16	.57	.57	73	Tr	2	7	18	10	2	1.5	8.2
ABS4	8579-85	C	BS	2	.18	.42	.51	77	0	3	5	15	13	0	.15	7.2
ABS4	8585-01	C	CB	5	.13	.50	.60	67	Tr	2	10	21	11	5	.06	10.2
ABS4	8601-17	C	CL	1	.09	.16	.53	72	0	2	1	24	17	1	.02	9.0
ABS4	8617-29	C	CL	1	.09	.25	.66	59	1	6	11	23	10	0	.52	8.0
ABS4	8629-35	D	CL	2	.19	.57	.72	67	0	2	3	28	7	3	.38	8.3
Means					.13	.38	.57	68	Tr	3	6	22	12	2	.22	8.5
ABF1	8626-39	D	CB	5	.19	.63	.64	66	Tr	2	1	31	7	4	.49	9.5
ABF1	8639-50	D	CL	3	.09	.25	.82	51	0	3	1	45	4	1	.24	8.7
Means					.14	.44	.73	58	Tr	2	1	38	5	2	.37	9.1
PH1	8431-43	B	CB	3	.10	.36	.46	55	0	2	2	40	6	1	.03	9.0
PH1	8471-99	C	CB	3	.12	.34	.50	66	0	2	6	26	5	5	.09	14.0
PH1	8499-23	C	CB	4	.08	.20	.57	46	0	2	1	51	5	0	.11	9.0
PH1	8530-40	D	CB	2	.14	.43	.54	64	0	2	1	32	8	0	.10	8.9
Means					.11	.33	.52	58	0	2	3	37	6	2	.10	10.2
CC2	8842-43	C	S	1	.17	.50	.73	69	0	4	0	27	4	0	--	--
CC2	8843-47	C	S	1	.16	.69	.73	62	0	2	0	36	4	1	--	--
CC2	8847-53	C	S	2	.20	.48	.48	80	0	2	5	12	19	0	.11	11.8
CC2	8853-60	C	CB	3	.15	.65	.58	76	Tr	2	6	16	9	4	.15	12.6
Means				59	.17	.58	.63	72	Tr	2	3	23	9	1	.13	12.1

[a]Bedding unit; TB = thin bedded; CL = clayey, laminated, bioturbated; CB = clayey bioturbated; S = clean sand BS = bioturbated sand

[b]Long axis measurements; σ = PHI deviation, sorting criteria is Folk (1974)

[c]Qz = monocrystalline quartz; F = feldspar; Rx = volcanic rock fragments, polycrystalline quartz and chert; Mx = matrix includes clay, chlorite, kaolinite, montmorillinite, glauconite; Oth = other minerals, heavies, opaques, micas, and porosity

[d]Cement = quartz overgrowths, calcite, amorphous silica

[e]Permeability is logarithmic mean

sandstone zone, and chlorite occurs in the lower bioturbated and laminated zones. Both kaolinite and chlorite have inhibited quartz overgrowths (Fig. 19). In instances where kaolinite appears to have replaced feldspar, recrystallization has resulted in some intercrystalline porosity. The effectiveness of the intercrystalline porosity in the clayey, bioturbated sandstone is easily seen in thin sections which are impregnated with dyed epoxy. In the No. 1 Lloyd core (Fig. 20), measurements show the porosity and oil saturation to be twice as great in the clayey sandstone as in the clean sandstone above. However, the increased oil saturation is not clearly detected by resistivity logs because the dispersed clay tends to mask the resistivity of the oil-bearing zone. Conduction on the large surface area of the dispersed clay plates causes resistivity measurements to be too low (Almon and Schultz, 1979). The kaolinite appears to be associated with the invasion of fresh water. It is known to form in low salinity pore fluids, usually at fairly shallow depths. The requirement for formation of kaolinite is a source of alumina, which could easily be derived from intra-

formational feldspars, volcanics, and incoming meteoric water.

Thin sections display many of the criteria for recognition of diagenetic porosity described by Schmidt and McDonald (1978) which includes: (1) partial dissolution (Fig. 18B), (2) molds (Fig. 18C), (3) floating grains (Fig. 18C), (4) oversized pores (Fig. 18E). (5) elongate pores (Fig. 18D), and (6) corroded grains (Fig. 18E). Much of the porosity developed in the clayey intervals of the "B", "C", and "D" units appears to be diagenetic. Partial dissolution, floating grains, and elongate pore systems comprise the most common criteria recognized; molds and oversized pores are the next most common feature. Dissolution of calcite cement, feldspar grains, and rock fragments has created a pore system ranging from irregular, elongate, and sinuous passageways to large, irregular and boxy pores and cavities that give the reservoir flow characteristics similar to those of carbonate rocks (Schmidt and McDonald, 1978). The average size of the secondary pores in thin section is about 0.13 mm. They are abundant in the intensely bioturbated sand-

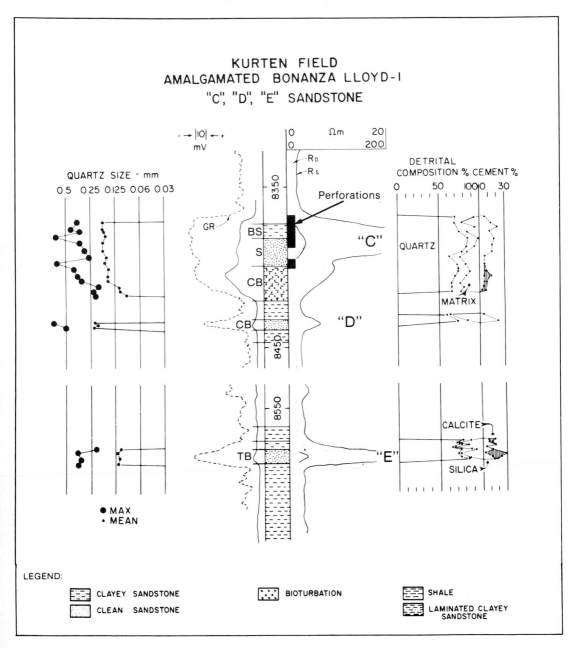

FIG. 16.—Composition and texture, Woodbine "C", "D", and "E" sandstones, Amalgamated Bonanza No. 1 Lloyd, Kurten Field. SP, gamma-ray, and resistivity logs are shown. Cored interval indicated by lithology symbols. Code defines bed types as listed in Table 3, footnote a. Location of well is shown in Figure 9.

stone and do not carry into the overlying clean sandstone or the underlying clayey sandstone. Porosity in the overlying clean sandstone is predominantly occluded by quartz overgrowths and only pin-point widely distributed pores are present. The original porosity was probably 30% or more.

Porosity in the "B", "C", and "D" units has been greatly enhanced by diagenesis after burial, making Kurten Field a diagenetic as well as stratigraphic oil accumulation. In essence, the porosity distribution has been completely reversed by processes operating after burial. The clayey beds are porous and the clean sandstone beds are non-porous. Although it is of great benefit, the diagenetic porosity creates a well completion

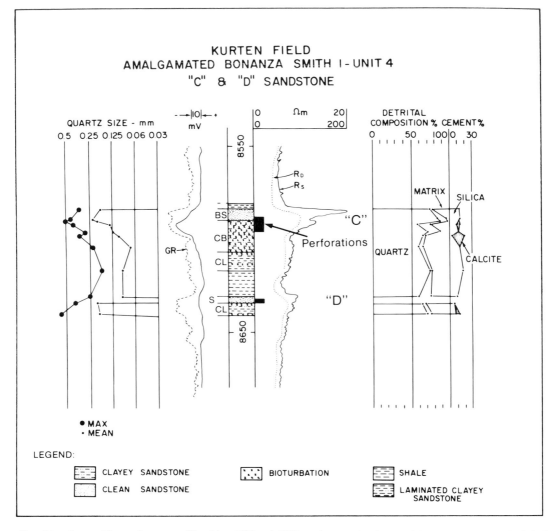

FIG. 17.—Composition and textures, Woodbine "C" and "D" sandstones, Amalgamated Bonanza No. 1 Smith Unit 4, Kurten Field. SP, gamma-ray, and resistivity logs are shown. Cored interval indicated by lithology symbols. Code defines bed types as listed in Table 3, footnote a. Location of well is shown in Figure 9.

problem. The kaolinite plates in the pore spaces are often mobilized during initial flow causing formation damage that may severely curtail production. In addition, acid treatments may be harmful because of the precipitation of insoluble iron hydroxide by the acid reaction with the iron-rich chlorite.

In the clean sandstone beds, glauconite is present as well-formed, disseminated, pale green pellets that are

FIG. 18.—Photomicrographs of thin-sections from cores of Woodbine Sandstone, Amalgamated Bonanza No. 1 Lloyd, Kurten Field, Texas. A, "C" unit; intensely bioturbated zone, quartz grains (q) with few overgrowths loosely packed in poikilitopic calcite cement, overgrowth in left corner (arrow) may be rounded from recycling, plane light; 65x, 8379 feet. B, "C" unit; intensely bioturbated zone, calcite cement (c) stained with alizarin red, quartz grains with few overgrowths, porosity (p) in lower left may be result of dissolution of poikilitopic calcite cement, plane light; 65x, 8383 feet. C, "C" unit; intensely bioturbated zone, oversize pore (p) probably solution mold, with floating quartz grain (fp), and coarsely crystalline kaolinite (k) floating in pore space, plane light; 65x, 8385 feet. D, "C" unit; intensely bioturbated zone, elongate pores (p), floating quartz grain (fg), crystalline kaolinite (k), corroded grain boundary (arrow) in a loose packing arrangement, plane light; 65x, 8388 feet. E, "C" unit; intensely bioturbated facies, corroded quartz grain boundary (arrow) with kaolinite, plane light; 120®, 8388 feet. F, "E" unit; indistinctly laminated "b" division, dark brown grain coatings, probably chlorite (arrow) restricting quartz overgrowths (qo), preserved primary intergranular porosity (p), plane light; 65x, 8540 feet.

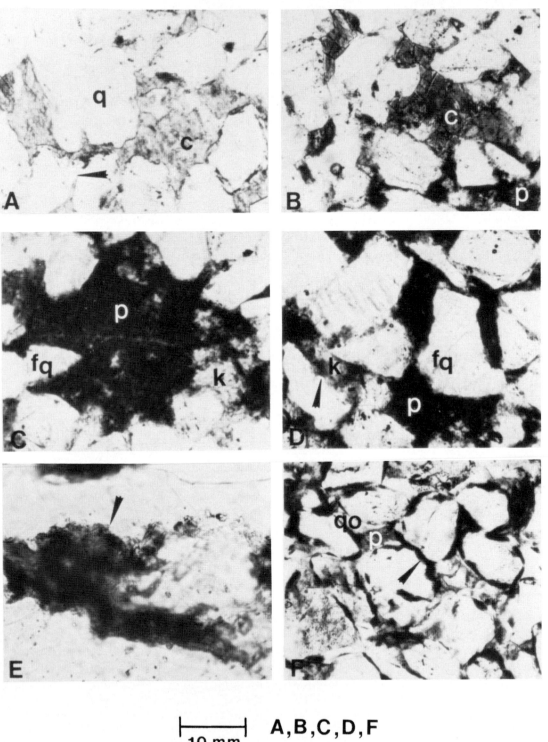

A,B,C,D,F

.10 mm

E

.10 mm

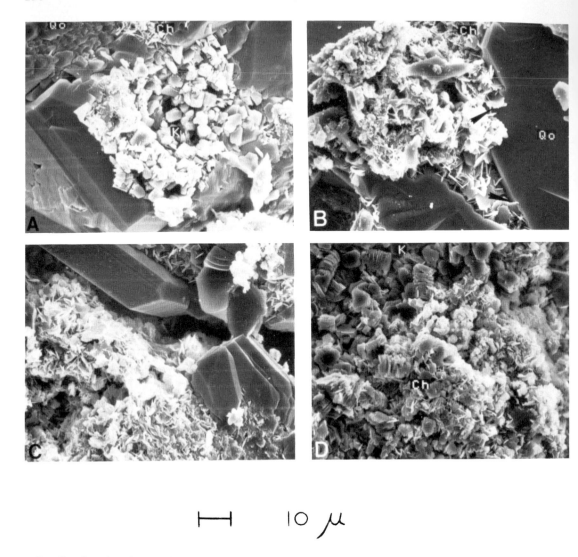

$$\vdash\!\!\!\dashv \qquad 10\ \mu$$

Fig. 19.—Scanning electron microscope photographs of Woodbine Sandstone, Amalgamated Bonanza No. 1 Lloyd, Kurten Field. Photographs courtesy of G. W. Bolger. *A*, "C" unit; pore filled with authigenic kaolinite (KO, small coalescing quartz overgrowths (Qo), and minor amounts of authigenic chlorite (Ch); 1000x, 8356 feet. *B*, "C" unit; pore filled with authigenic chlorite (ch) apparently restricting quartz overgrowths (Qo) with chlorite blades cutting into quartz (arrow); 1000x, 8356 feet. *C*, "E" unit; chlorite pore lining and geometric quartz overgrowth, creating brown coatings on sand grains in Figure 18F; 1000x, 8537 feet. *D*, "E" unit; pore filled with kaolinite (K) and chlorite (Ch) growing on kaolinite books; 1000x, 8540 feet.

cemented by quartz overgrowths. Also present are continuous and discontinuous laminae of dark brown organic-rich clay which forms the clay drape found in the flaser crossbeds.

Other minerals include heavy minerals, opaques, and mica which averages 4% and ranges from 1% to 11% of the detrital composition (Table 3). The heavy minerals are predominantly zircon, tourmaline, garnet, and sphene. Most opaques are pyrite, which occurs in

the highly bioturbated, clayey sandstones as scattered blebs, cubes, and framboids, and are probably formed by reducing conditions created by organic decay.

Cement consists of both silica and calcite and averages 10% of the total composition. Silica is in the form of quartz overgrowths, ranges from 7% in the clayey sandstones to 15% in the clean sandstones, and shows a consistent increase with the decrease in matrix. SEM observation showed that at least part of the quartz overgrowths developed after crystallization of kaolinite because kaolinite plates are found cutting into the overgrowths (Fig. 19A). Chlorite appears to post-date the kaolinite, as it is found to form on kaolinite plates (Fig. 19D). Authigenic chlorite was probably derived from recrystallization of detrital chlorite that was dispersed by burrowing organisms, and from iron and silicates leached from rock fragments within the formation.

Calcite cement is confined to the muddy bioturbated beds (Fig. 16) and is probably derived from the dissolution and reprecipitation of shell material since several molluscan molds were observed in the cores. It occurs as poikilitopic cement around loosely packed grains and as isolated patches (Fig. 18A, B). The association of calcite patches with kaolinite around floating grains and in oversized pores (Fig. 18B) implies that dissolution of calcite provided some of the space in which the kaolinite precipitated. Corroded edges, found on quartz grains (Fig. 18D), probably resulted from the dissolution of calcite that was replacing quartz.

The "E" unit is characterized by intergranular porosity without the extensive diagenetic alteration found in the upper units. Quartz grains in the "E" unit appear to be coated with a dark brown clay rim (Fig. 18F) that has restricted quartz overgrowths and partially preserved primary porosity. The pore throats are less obstructed by kaolinite flakes than those in the upper units. Bulk x-ray diffraction indicates about equal amounts of kaolinite and chlorite. Generally, the "E" unit produces more efficiently than the overlying units due to the lesser amount of diagenetic clays.

The average grain size for the units "B" through "E" is 0.14 mm (fine sand). The sand size in the clayey laminated and clayey bioturbated beds averages 0.13 mm and, in the clean beds, 0.18 mm. The average quartz grain size in the "C" unit is also 0.14 mm. From the bottom to the top of the "C" unit, the average grain size increases from 0.09 mm to 0.18 mm, and the average maximum grain size increases from 0.16 mm to 0.70 mm. The average maximum size of the "C" unit is 0.45 mm (medium sand).

Six hundred sand grains were sampled for rounding and fabric in the No. 1 Lloyd core and were found to be subangular to subrounded with an average of five framework contacts per grain, most of which are point or long contacts without interpenetration. The framework grains are quartz which have 80 to 90% fixed boundaries (bounded edges exceed unbounded edges). In the poikilitopic calcite cemented areas, 32% of the grains are floating and the average is three framework contacts per grain. The loose packing is believed to result from early cementation. The average phi deviation for all sandstones is 0.60 mm, which indicates moderate sorting (Folk, 1974), and ranges from 0.51 mm in the clean sandstones to 0.59 mm in the clayey sandstones. Bioturbation has substantially contributed to the poorer sorting values, and as a result, diminished the possibilities of any grain size data such as sorting to infer depositional processes.

PALEONTOLOGY

Foraminifera, palynology, nannofossils, and trace fossils were examined to aid in the interpretation of the age and depositional environments in Kurten Field.

No age diagnostic foraminifera were observed in the cored intervals. Small *Hedbergella* spp., *Heterohelix* sp., *Ammobaculites* spp., and *Trochammina* spp. were common. In cuttings from wells in the southern end of the field, no *Globotruncana helvetica* (middle Turonian) or Cenomanian age foraminifera were present.

Nannofossils were generally not present in the cored intervals. *Lithastrinus floralis* was present in a few core samples and ditch cuttings, giving an early Coniacian to late Turonian age to these sandstones. In the Gulf No. 1 Palasota, Cenomanian age nannofossils and planktonic foraminifera were present 90 feet above the Early Cretaceous. The fossils are possibly reworked for they were not observed in nearby wells.

Palynomorphs indicated an upper Woodbine flora with possibly some reworked Early Cretaceous. Most palynomorphs were poorly preserved or had a burned appearance.

Bioturbation was common within the sandy and silty intervals. A type of *Teichichnus*, small *Ophiomorpha*, *Zoophycos*, *Asterosoma*, and oblique burrows were recognized. This trace fossil assemblage is one generally found in foreshore to offshore environments. Irregular U-shaped burrows were present in sandy muds and muddy sands believed to be deposited in quiet waters.

The "B" unit was essentially barren of fauna with the exception of a few *Ammobaculites* species.

The "C" unit had rare planktonic foraminifera, common arenaceous foraminifera, and *Inoceramus* at the top and bottom.

The "D" unit had arenaceous foraminifera and Woodbine palynomorphs at the top and common compressed *Ammobaculites* sp. and poorly preserved broken planktonic foraminifera at the base. Below the "D" unit in the shale section, common compressed *Trochammina* spp. and rare planktonic species were present.

The "E" unit had common planktonic and *Trochammina* species.

Within the resistive shale unit, planktonic foraminifera were common; *Inoceramus*, mollusc impressions, and arenaceous foraminifera were present; and pyritization was common.

The age of the Woodbine in Kurten Field is early Coniacian to late Turonian. The assemblage indicates that the "E" unit was deposited in offshore marine waters with periods of sediment reworking by bioturbation. An arenaceous fauna developed in low energy areas and was surrounded by quiet water marine shales containing planktonic foraminifera.

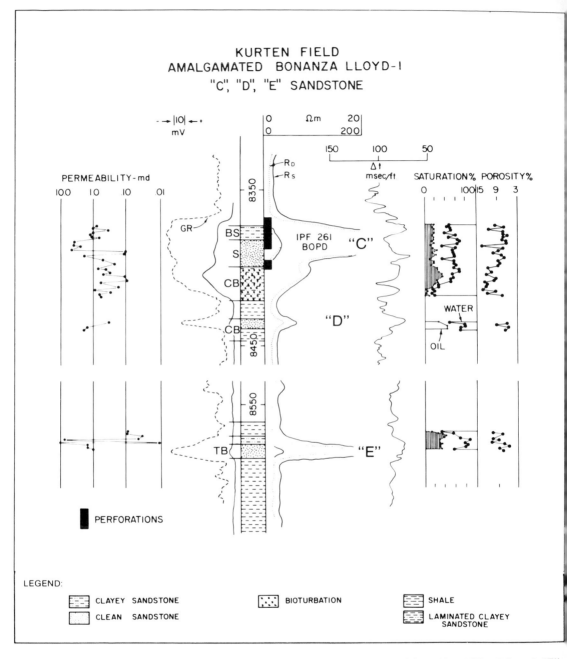

Fig. 20.—Core porosity, porosity log, permeability and oil and water saturations of Woodbine "C", "D", and "E" sandstones, Amalgamated Bonanza No. 1 Lloyd, Kurten Field. Cored interval shown by lithology symbols. Code defines bed types as listed in Table 3, footnote a. Location of well is shown in Figure 9.

RESERVOIR CHARACTERISTICS

The Kurten Field reservoirs are characterized by low porosity, low permeability, and high clay content. Reservoir properties are plotted in Figures 20 and 21 which show porosity, permeability, fluid saturation, and porosity log response. The average porosity measured in the cores is 9.3% and the average permeability is 0.55 millidarcy (Table 3). Porosity ranges from 6 to 14%, and permeability ranges from less than 0.01 millidarcy to 2.6 millidarcys. In the clean sandstone, the average porosity is 8.4% and ranges up to 11%; permeability averages 1.5 millidarcies and ranges up to 40

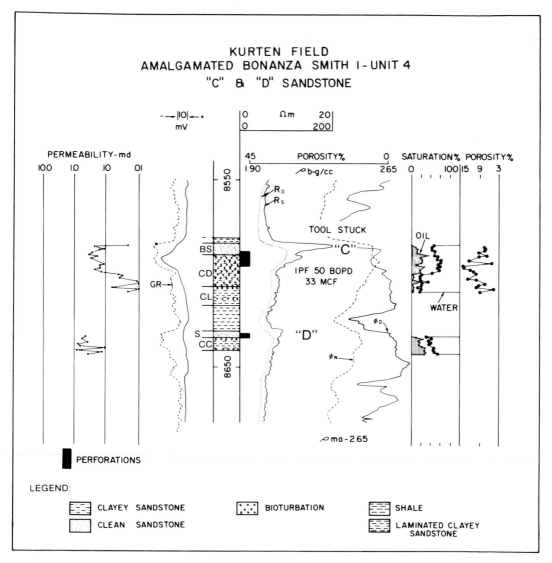

FIG. 21.—Core porosity, log porosity, permeability and oil and water saturations of Woodbine "C" and "D" sandstones, Amalgamated Bonanza No. 1 Smith Unit 4, Kurten Field. Cored interval shown by lithology symbols. Code defines bed types as listed in Table 3, footnote a. Location of well is shown in Figure 9.

millidarcies. The clayey bioturbated and clayey laminated sandstones average 9.8% porosity, which is 30% higher than the clean sandstone, and the permeability is 0.28 millidarcy.

The mean oil saturation for all the sandstones in the cored intervals is 14.3%, while the mean oil saturation in the clayey sandstone is 17.2% (Fig. 20). In the No. 1 Lloyd core, the clayey sandstone contains 23.6% oil saturation, while the clean sandstone contains only 17.6%. The "E" unit measurements are comparable to the "B", "C", and "D" units and average 28% oil saturation, 0.82 millidarcy permeability, and 8.6% porosity (Table 3). In a similar fashion, the Smith 1 Unit 4 has 18.17% oil saturation in the clayey bioturbated

sand and only 12% in the clean sand (Fig. 21).

The average production per well in the field is about 133 barrels of oil per day with 108 MCF of gas per day and 5 barrels of water. There are over one hundred producing wells in the field. The average API gravity is 38.5°. The "C" unit normally produces an average of 37.8° API, the "D" unit 44° API, and "E" unit 40° API. The average gas oil ratio is 757 cubic feet per barrel. The original reservoir pressure in the "C" unit was measured at 3836 psi in the Amalgamated Bonanza No. 1 Lang in 1977, which reflects a normal pressure gradient of 0.45 psi per foot. The original reservoir pressure in the "E" unit was measured at 4838 psi in the Amalgamated Bonanza No. 1 Lang and

TABLE 4.—PERCENT OF OIL SATURATION IN THE WOODBINE SANDSTONE CLAYEY AND CLEAN UNITS, KURTEN FIELD

	MEAN OIL SATURATION %				
Well Name	ABL1	ABS4	PH1	CC2	ABF1
Clean Sandstone	17.59	12.60	—	12.68	—
Clayey Bioturbated Sandstone	23.66	18.17	7.68	15.78	28.72
Means	20.62	15.38	7.68	14.23	28.72

2707 psi in the Amalgamated Bonanza No. 3 DRB, which reflects an overpressure of about 850 psi and a pressure gradient of 0.55 psi per foot. From October 1976, when the field began production, to June 1980, the field produced over 5 million barrels of oil and 3,000 mm MCF of gas. The ultimate recovery is estimated at over 75 million barrels of oil. Enchanced recovery techniques could increase the yield to over 100 million barrels of oil.

INTERPRETATION OF PALEOENVIRONMENTS

The interpretation of the depositional environment at Kurten Field is based on sand body geometry, composition, texture, sedimentary structures, paleontology, and regional setting. The elongate, asymmetrical nature of the "B", "C", and "D" units, illustrated in Figures 9, 10, and 11, coupled with the regional setting west of the Harris Delta, imply submarine bar (sand ridge) deposition. Both mean and maximum grain size increase upward which is typical of many bar sequences. The thinner areas that separate the thicker pods of sandstone may represent channels or gaps cut by currents passing obliquely over the sand ridges. A similar process is suggested by Caston (1972) and illustrated by Walker (1979). A diagram showing the asymmetrical profiles and vertical sequence of sedimentary structures in the Kurten sand bars is presented in Figure 22. The slight variation in body morphology from the ridge-like "D" unit to the pod-like "B" unit may reflect changing circulation patterns and current velocities resulting from a rise in sea level. A rising sea level is suggested by the apparent onlap of the "A" and "B" units onto the Harris Delta in Madison County, northeast of the field. Changes in sand body morphology relative to current strength are documented by Kenyon (1970) in an example from the continental shelf near England.

Both the microfauna and the diversity and intensity of bioturbation points to a marine environment with good circulation, normal salinity, good oxygenation, and stability where sedimentation was out-paced by the biological processes. *Rhizocorallium*, *Asterosoma*, and *Diplocraterion* are indicators of a marine offshore environment (Chamberlain, 1978). The initial environment of deposition appears to have been localized in water sufficiently deep to be out of the reach of normal

tides and wave action. As low relief developed in areas of sand deposition, a favorable substrate for burrowing developed which was apparently elevated into a more oxygenated zone. Continued growth accompanied by the deflection of longshore currents began to elongate the sand bodies while building and winnowing their central portions. As the ridges reached fair-weather wave-base the biological processes were overtaken by the depositional processes and the clean sand flaser cross-bedded unit was developed. Although they are not generally environmentally diagnostic, flaser cross beds have been described from the wave dominated sublittoral facies (de Raaf *et al.*, 1977). The bioturbated sand unit at the top of the sequence is believed to represent an abandonment phase due to a combination of a deepening of the water and a reduction in sediment supply.

The thin-bedded facies of the "E" unit is characterized by repeating bedsets that have an ordered sequence of sedimentary structures. The bedset sequence consists of a poorly laminated to massive sand overlain by a wavy to undulatory laminated sand which is topped by a thinly interlaminated sandstone and shale. The sequence demonstrates a decrease in flow regime upward and is suggestive of deposition from turbidity flows. However, many of the characteristics of classical turbidity flows are lacking, such as fluid escape structures or dish structures. The wavy laminated sandstone could be deposited from low flow regime currents, and the thinly interlaminated sand and shale beds could be suspension deposits. The lack of bioturbation in the "E" unit indicates deposition in an environment relatively hostile to benthic organisms, possibly at the middle or outer shelf. The sharp to irregular basal contacts of the beds indicate mild scouring of the shelf floor. The mechanism of deposition is interpreted to be sand sheet flows that moved periodically across the shelf and were generated by storm action. Shale units separating the storm events represent periods of normal deposition. To a certain extent, the wavy, oscillatory laminae may have resulted from wave action associated with times of increased water agitation. A similar mode of deposition was described for thin-bedded sandstone units in the Shannon of the Wyoming Upper Cretaceous by Spearing (1976). On the modern shelf, Hayes (1967) documented sand beds up to 9 cm. thick

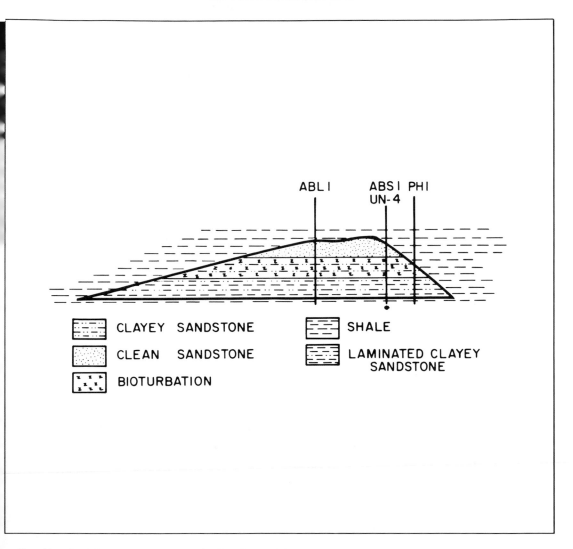

Fɪɢ. 22.—Sand-ridge model interpreted for Woodbine "C" sandstone at Kurten Field. Cored wells are shown in their inferred relative positions within the bar. Asymmetrical geometry of ridge is shown.

at depths of up to 120 feet off Padre Island on the Texas Gulf Coast after hurricane Carla in 1961. These may be a modern analog. These sedimentary features probably exist in the zones of extensive bioturbation in the "B", "C", and "D" sands but were destroyed. Sheet flows and possibly shelf turbidites must have been the transport mechanism originally for placing the Kurten Sands. Finally, a significant point associated with the sheet flow deposits is that they may indicate additional sand accumulations down dip from the bar system.

Figure 23 is a schematic diagram of the depositional patterns. The Harris Delta is shown as a braided distributary channel network, with interchannel marsh areas. Progradation is to the southwest. The delta is thought to have been reworked by marine currents that

came from the east flank of the Sabine Uplift, paralleled the Angelina-Caldwell Flexure, and turned north after passing the delta front. This would account for the sparse sand distribution in Walker County, south of the delta, and the trend of strandline and bar sands found in Madison and Leon Counties, north of the delta. The delta seems to have prograded to the southwest along the updip side of the Angelina-Caldwell Flexure parallel to the Lower Cretaceous shelf edge which was about 30 miles to the south.

The thinning of the Woodbine, south and west of Kurten Field (Fig. 24), and the truncation of sandstone units, south and east of Kurten Field, demonstrate uplift and erosion in the vicinity of the shelf edge. The isopach map (Fig. 24) is drawn between the BAC

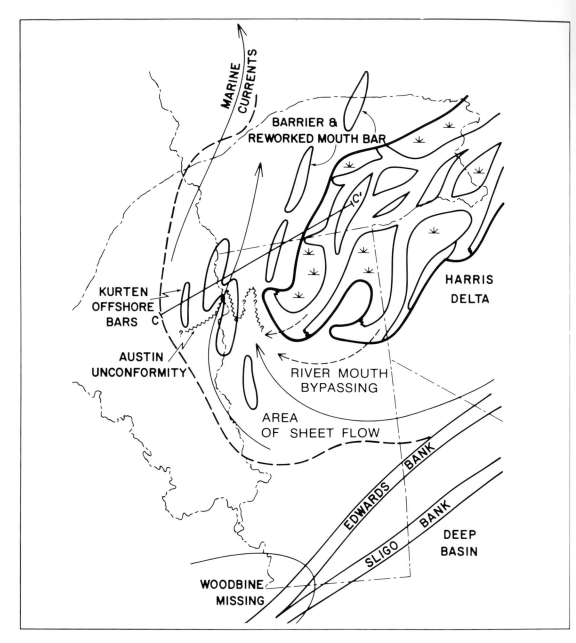

FIG. 23.—Depositional model interpreted for Woodbine sandstones at Kurten Field. Shown are the southwest prograding Harris Delta, river mouth by-passing, and bars modified by storm and tidal currents.

marker and the top of the Lower Cretaceous using well logs and regional seismic lines. The Woodbine and Buda are missing in the northwestern corner of Waller County in the Shell No. 1 Chapman; therefore, the Austin Chalk rests directly on the Georgetown Formation. An elongate thin in the Woodbine isopach marks an area of erosion along the Lower Cretaceous shelf edge through Washington County, southern Grimes County, and possibly into western Montgomery County. The Woodbine section thickens rapidly to the south of this area.

INTERPRETATION OF DIAGENESIS

Erosion of the Woodbine resulting from uplift along the Lower Cretaceous shelf edge with concurrent invasion of fresh water is believed to have strongly influ-

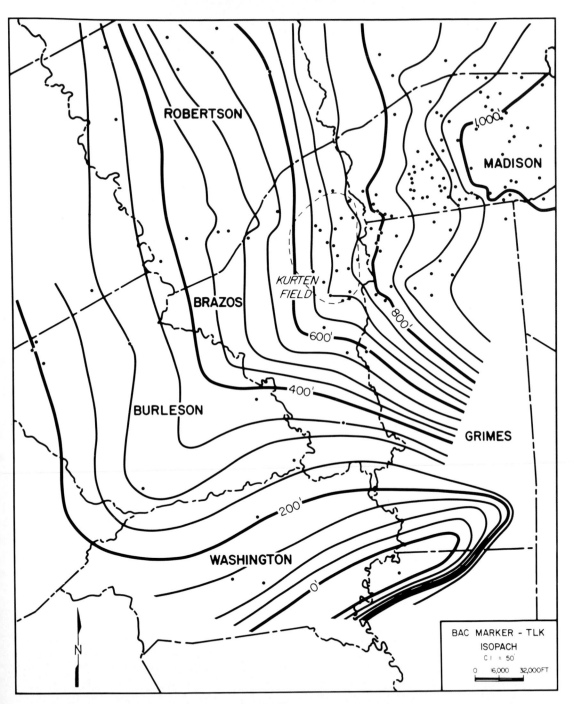

FIG. 24.—Regional isopach of interval between BAC marker and the top of the Lower Cretaceous. Thinning of the Woodbine is shown to the south of Kurten Field; post-Woodbine erosion along the Mesozoic shelf-edge is inferred in the southern part of the map area.

enced the diagenetic processes at Kurten Field. The diagenesis developed in the following sequence:

1. Development of poikilitopic calcite cement derived from shell material;

2. Invasion of fresher meteoric water at the time of the uplift which created diagenetic porosity by dissolving calcite cement, feldspar, and volcanic clastics, while precipitating kaolinite;

3. Development of authigenic chlorite and chlorite coatings with deeper burial;
4. Selective development of quartz overgrowths; and,
5. Migration of oil into the sandstones.

The sequence of events from the time of deposition to the present is schematically summarized in Figure 25. Initially sand was deposited in the marine shelf and worked into bars, continued sedimentation and subsidence buried the sand units, and carbonate cementation took place. Uplift of the shelf occurred during the Austin unconformity along which meteoric diagenesis enhanced the porosity and permeability in the clayey units, but had little effect on the clean sand units. Subsidence followed and the Kurten sands were once again buried. The generation and migration of oil occurred. Gulfward structural rotation followed leaving the oil trapped in the down dip reaches of the sandstone. The best production at Kurten is found in the southern end of the field which is structurally low, but is more closely associated with the unconformity. It appears that the invasion of fresh water at the unconformity was geographically limited and, farther away from the truncation, the porosity is much less enhanced (Fig. 25). Hypothetical situations of trapping oil in structurally unfavorable positions have been described by Wilson (1977) and termed "frozen in" hydrocarbon traps. His concept calls for the diagenetic sealing of hydrocarbons on positive areas that were subsequently rotated to structurally low positions. The process is somewhat different, but the effect is essentially the same at Kurten where the oil, to some degree, is "frozen in" on the southern end of the field, and the limit of diagenesis defines the northern end of production.

DISCUSSION OF SHELF-SAND SEDIMENTATION

Shelf sand sedimentation has received considerable attention in recent years. Off (1963) described linear sand ridges associated with tide-dominated deltas: Houbolt (1968) described sand ridges in the North Sea generated by tidal currents; and Caston (1972) described a model for development of sand ridges in tidally worked areas. A current summary of shelf sedimentation and processes in shallow siliciclastic seas is presented by Johnson (1978).

Originally, it was believed that a progressive decrease in grain size took place with increasing distance from clastic shorelines, but it was found from bottom sampling that the continental shelves were actually a mosaic of sandy, silt, and shaley areas (Shepard, 1932). The bodies of sand found on the shelf were explained as relic sediments deposited at lower stands of sea level (Emery, 1968). Sedimentological features found on the shelf were theorized to develop from reworking of relic sand by marine currents, and the reworked sands were termed "palimpsest" sediments by Swift and others (1973). Studies of the North Sea and the northeast Atlantic Shelf advanced the idea that sed-

iments were not only relic, but dynamic processes were actually introducing new sand to the shelf (Curray, 1964; Swift, et al., 1979). Two types of shelf environment were catagorized: the tide-dominated shelf and the storm-dominated shelf, each with its own set of attributes (Walker, 1979).

Tide-dominated shelves produce large-scale linear ridges and sand waves and occur in association with enclosed seas, blind gulfs, and restricted bays (Ball, 1967; Brenner, 1980). They usually involve considerable volumes of sediment.

On storm-dominated shelves, seasonal weather changes strongly influence deposition (Johnson, 1978). The preservation potential of beds placed on the shelf below storm wave base is very good. The volumes of sediment moved by storms is probably less than that moved by continual tidal processes, but storms probably place more new sediment into the system.

Moving sediment from the shoreline to the shelf involves breaking through the "littoral energy fence" (Swift, 1967) which can be accomplished by a variety of processes including river mouth bypassing, shoreface bypassing (Brenner, 1980) and storm surge turbidity flows. River mouth bypassing occurs during flood stage and during ebb tide flow, when the flow exiting the delta is strong enough to carry clastics past the mouth bar (Brenner, 1980). Shoreface bypassing occurs when seas transgress shorelines and the sand is reworked into shelf sand bodies. Storm surge flows occur when storm-piled water in shore areas flows back into the sea carrying sediments (Hayes, 1967). With these processes fairly well-documented, it is reasonable to expect numerous offshore bars to occur in the rock record. Offshore bars are described by several authors and summaries are presented by Brenner (1980), Walker (1979), and Johnson (1978).

During the Mesozoic, the interior of North America was flooded by epicontinental seas, which were periodically open to the north, south, or both directions, allowing the possibility of widespread occurrences of marine bars. In the Jurassic (Oxfordian) of Montana, sand bars are found which are characterized by well-sorted coarsening upward vertical sequences and gradational bases. The lower bar is bioturbated, grades upward to wave-rippled sandstone with shale partings, and is capped by trough crossbedded sandstone (Brenner and Davies, 1974). The bars are 12 miles wide, 3 miles long, and 65 feet thick. These are interpreted as shallow marine shoals molded by wave action.

Bars, described by Spearing (1976) in the upper Cretaceous of Wyoming, are characterized by bioturbated shale, grading upward to thin-bedded bioturbated sandstone, and overlain by crossbedded clean sandstone with clay drapes and clasts. The sand units are 18 miles wide, 30 miles long, and 50 feet thick. They are believed to form by the action of storm waves and tidal currents that shape the sand into discrete bodies on the shelf. The sand bodies appear to build by vertical

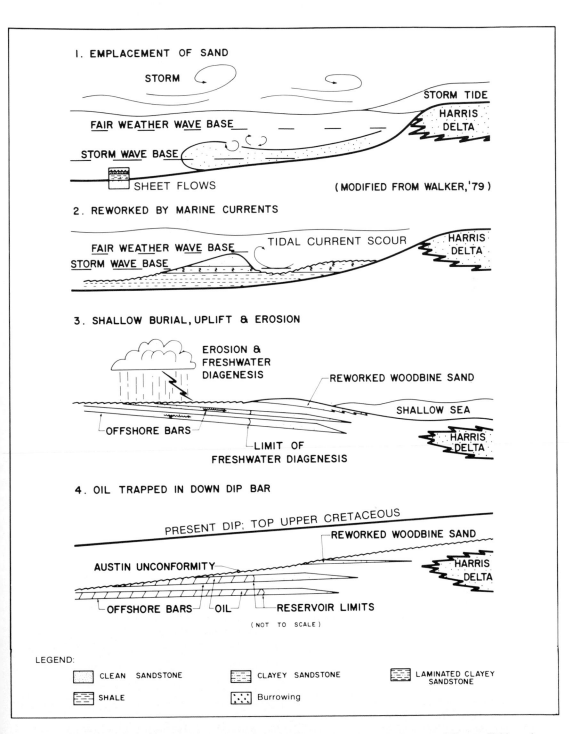

1. EMPLACEMENT OF SAND

STORM

STORM TIDE

HARRIS DELTA

FAIR WEATHER WAVE BASE

STORM WAVE BASE

SHEET FLOWS

(MODIFIED FROM WALKER, '79)

2. REWORKED BY MARINE CURRENTS

FAIR WEATHER WAVE BASE

TIDAL CURRENT SCOUR

HARRIS DELTA

STORM WAVE BASE

3. SHALLOW BURIAL, UPLIFT & EROSION

EROSION & FRESHWATER DIAGENESIS

REWORKED WOODBINE SAND

SHALLOW SEA

OFFSHORE BARS

LIMIT OF FRESHWATER DIAGENESIS

HARRIS DELTA

4. OIL TRAPPED IN DOWN DIP BAR

PRESENT DIP; TOP UPPER CRETACEOUS

REWORKED WOODBINE SAND

AUSTIN UNCONFORMITY

HARRIS DELTA

OFFSHORE BARS   OIL   RESERVOIR LIMITS

(NOT TO SCALE)

LEGEND:

CLEAN SANDSTONE   CLAYEY SANDSTONE   LAMINATED CLAYEY SANDSTONE

SHALE   Burrowing

FIG. 25.—Inferred sequence of depositional and diagenetic events leading to development of Kurten Field sandstone reservoirs.

stacking of thin beds deposited from sheets of moving sand that eventually may have reached wave base where they were winnowed by currents in a longshore direction.

The Upper Cretaceous Sussex Sandstone of the Powder River Basin is characterized by an upward graded sequence of shale with ripple lenses, ripple-bedded and thinly laminated sandstone, and a medium- to coarse-grained pebbly sandstone (Berg, 1975). The sandstone bodies are 25 miles long, 1.5 miles wide, and 25 miles thick and are interpreted to have been deposited in 100 to 200 feet of water on the Cretaceous shelf. The grain size in the bars described by Berg (1975) implies a high level of current energy acting on the shelf.

The Upper Cretaceous Hygiene Sandstone member of the Pierre Shale in the Denver Basin is interpreted as an offshore bar (Porter, 1976). The Hygiene Sandstone is characterized by a coarsening-upward sequence of burrowed, laminated beds grading to ripples and large-scale crossbeds. The bioturbation is very similar to that found at Kurten.

Upper Cretaceous sand bars are also described by Cotter (1975) in the Mancos Shale of east central Utah. The bars are series of coarsening-upward sandstone beds, up to 10 feet thick, that trend southwest-northeast. They are believed to have been formed from sediment moved by storms and shaped by longshore currents. The clayey, laminated bedsets found in the lower reaches of the sand bars compare with the thin bedded "E" sand unit at Kurten field. These bedsets are believed to be similar to the thin-bedded features described by Spearing (1976) in the lower portion of the Shannon Sandstone which was interpreted to be deposited from moving sand sheets.

Based on the Kurten field study and descriptions in the literature, it appears that shelf sands bodies are fairly common in the rock record, and that they are often characterized by a distinctive succession of sedimentary structures. The complete order of bedding in ascending sequence is thin-bedded laminated fine sand and silt interbedded with shale, bioturbated clayey, sand, clean sand with tubular or undulating cross-beds or flaser cross-beds, and slightly bioturbated sand that grades upward to shale. This sequence records the bar history from initiation to abandonment. Additionally, thin-bedded sheet flow deposits frequently underlie bar deposits and may emplace sand downdip from established bar sequences.

CONCLUSIONS

Based on the foregoing discussion, a number of conclusions can be drawn from the Kurten Field study concerning the impact of continental shelf deposition on the geology and economics of hydrocarbon exploration. They are listed below:

1) Kurten sandstones are offshore bars created by shelf processes.

2) Offshore bars are numerous in the geologic record.

3) Offshore bars are often not detectable on conventional seismic lines.

4) Offshore bars on continental shelves have a characteristic order of bedding and their presence can be inferred through paleogeographic reconstructions.

5) Stratigraphically trapped oil in offshore bars is probably not uncommon.

6) Diagenetic porosity associated with uplift can significantly enhance petroleum reservoirs.

7) Diagenetic enhancement of porosity associated with uplift on the craton is probably not uncommon.

These items are especially important to the exploration geologist who, today, is faced with the ever increasing challenge to discover new hydrocarbon reserves. Since the ancient continental shelves and epeiric seas contain thin, widespread sandstone reservoirs that are not associated with structure or readily seen on seismic lines, many fields similar to Kurten remain to be discovered. These accumulations may well be a major portion of the undiscovered reserves remaining in the United States.

ACKNOWLEDGEMENTS

We wish to acknowledge J. Greco, M. Kontz, B. Statham, J. Benton, H. O. Arnold, and S. Cage for their financial and managerial support, and Pat Roberts and her staff for the technical illustrations, Gulf Exploration and Production Co., Houston, Texas; staff of Gulf Research and Development Co., Houston, Texas, for their expertise: George Bolger, SEM clay analysis; Mark Nations, palynology; Mike Keppler and Warren Cooper, nannofossils; John Duncan foraminifera; Dr. John Warme, Department of Geology, Colorado School of Mines, Golden, Colorado, for consultation concerning trace fossils in the cores; and Dr. R. R. Berg, Department of Geology, Texas A&M University, College Station, Texas, for technical advice. We wish also to acknowledge Seiscom Delta, Houston, Texas, for release of the seismic line through Kurten Field.

REFERENCES

ALMON, W. R., AND A. L. SCHULTZ, 1979, Electric log detection of diagenetically altered reservoirs and diagenetic traps: Gulf Coast Assoc. Geol. Societies Trans., v. 29, p. 1–10.

BALL, M. M., 1967, Carbonate sand bodies of Florida and the Bahamas: Jour. Sed. Petrology, v. 37, p. 556–591.

BERG, R. R., 1975, Depositional environment of Upper Cretaceous Sussex Sandstone, House Creek Field, Wyoming: Am. Assoc. Petroleum Geologists Bull., v. 59, p. 2099–2110.

BRENNER, R. L., 1980, Construction of process-response models for ancient epicontinental seaway depositional systems using partial analogs: Am. Assoc. Petroleum Geologists Bull., v. 64, p. 1223–1244.

_____, AND D. K. DAVIES, 1974, Oxfordian sedimentation in western interior United States: Am. Assoc. Petroleum Geologists Bull., v. 58, p. 407–428.

CASTON, V. N. D., 1972, Linear sand banks in the southern North Sea: Sedimentology, v. 18, p. 69–73.

CHAMBERLAIN, C. K., 1978, Recognition of trace fossils in cores, in P. B. Basan (ed.), Trace Fossil Concepts: Soc. Econ. Paleontologists and Mineralogists Short Course No. 5, p. 119–166.

COOPER, M. R., 1977, Eustacy during the Cretaceous: Its implications and importance: Palaeogeogr., Palaeoclimatol., Palaeoecol., v. 22, p. 1–60.

COTTER, M. R., 1975, Late Cretaceous sedimentation in a low energy coastal zone: The Ferron Sandstone of Utah: Jour. Sed. Petrology, v. 95, p. 664–685.

CURRAY, J. R., 1964, Transgressions and regressions, in R. L. Miller (ed.), Papers in Marine Geology: McMillan, New York, p. 176–203.

de RAAF, J. F. M., H. G. READING, AND R. G. WALKER, 1977, Wave generated structures and sequences from a shallow marine succession, Lower Carboniferous, County Cork, Ireland: Sedimentology, v. 24, p. 451–483.

EATON, R. W., 1956, Resume of the subsurface geology of northeast Texas with emphasis on salt structures: Gulf Coast Assoc. Geol. Societies Trans., v. 6, p. 79–84.

EMERY, K. O., 1968, Relic sediments on continental shelves of the world: Am. Assoc. Petroleum Geologists Bull., v. 52, p. 445–464.

FOLK, R. L., 1974, Petrology of Sedimentary Rocks: Hemphill, Austin, Texas, 182 p.

GRANATA, W. H., JR., 1963, Cretaceous stratigraphy and structureal development of the Sabine Uplift area, Texas and Louisiana, in L. A. Herrmann (ed.), Report on Selected North Louisiana and South Arkansas Oil and Gas Fields and Regional Geology: Shreveport Geol. Soc. Ref. Vol. 5, p. 50–96.

HAYES, M. O., 1967, Hurricanes as geological agents: Case studies of hurricanes Carla, 1961, and Cindy, 1963: Univ. Texas Bur. Econ. Geol. Rep. Invest. 61, 56 p.

HOUBOLT, J. J. H. C., 1968, Recent sediments in the southern Bight of the North Sea: Geol. en Mijnbouw, v. 47, p. 245–273.

HOWARD, J. D., 1975, The sedimentological significance of trace fossils, in R. W. Frey (ed.), The Study of Trace Fossils: Springer-Verlag, New York, p. 131–146.

HUNTER, B. E., AND D. K. DAVIES, 1979, Distribution of volcanic sediments in the Gulf Province—Significance to petroleum geology: Gulf Coast Assoc. Geol. Societies Trans., v. 29, p. 147–155.

JOHNSON, H. D., 1978, Shallow siliclastic seas, in H. G. Reading (ed.), Sedimentary Environments and Facies: Elsevier, New York, p. 207–258.

KENYON, N. H., 1970, Sand ribbons of European tidal seas: Marine Geol, v. 9, p. 25–39.

MARTIN, R., 1966, Paleogeomorphology and its application to exploration for oil and gas (with examples from western Canada): Am. Assoc. Petroleum Geologists Bull., v. 50, p. 2227–2231.

NICHOLS, P. H., 1964, The remaining frontiers for exploration in northeast Texas: Gulf Coast Assoc. Geol. Societies Trans., v. 14, p. 7–22.

_____, G. E. PETERSON, AND C. E. WUESTNER, 1968, Summary of subsurface geology of northeast Texas, in B. W. Beebe and B. F. Curtis (eds.), Natural Gases of North America: Am. Assoc. Petroleum Geologists Mem. 9, p. 982–1004.

OFF, T., 1963, Rhythmic linear sand bodies caused by tidal currents: Am. Assoc. Petroleum Geologists Bull., v. 47, p. 324–341.

OLIVER, W. B., 1971, Depositional systems in the Woodbine Formation (Upper Cretaceous) northeast Texas: Univ. Texas Bur. Econ. Geol. Rep. Invest. 73, 28 p.

PETTIJOHN, F. J., P. E. POTTER, AND R. SIEVER, 1973, Sand and Sandstone: Springer-Verlag, New York, 618 p.

PORTER, K. W., 1976, Marine shelf model, Hygiene Member of Pierre Shale, Upper Cretaceous, Denver Basin, Colorado, in Studies in Colorado Field Geology: Colorado School of Mines Prof. Contrib. 8, p. 251–265.

REINECK, H. E., AND I. B. SINGH, 1973, Depositional sedimentary environments: Springer-Verlag, New York, 439 p.

SCHMIDT, V., AND D. A. McDONALD, 1978, Secondary reservoir porosity in the course of sandstone diagenesis: Am. Assoc. Petroleum Geologists Continuing Education Course Notes Series 12, 125 p.

SHEPARD, F. P., 1932, Sediments on continental shelves: Geol. Soc. America Bull., v. 43, p. 1017–1034.

SPEARING, D. R., 1976, Upper Cretaceous Shannon Sandstone: An offshore shallow-marine sand body: Wyoming Geol. Assoc., Guidebook 28th Ann. Field Conf., p. 65–72.

SWIFT, D. J. P., 1976, Continental shelf sedimentation, in D. J. Stanley and D. J. P. Swift (eds.), Marine Sediment Transport and Environmental Management: John Wiley and Sons, New York, p. 311–350.

_____, D. B. DUANE, AND T. F. McKINNEY, 1973, Ridge and swale topography of the Middle Atlantic Bight, North America: Secular response to the Holocene hydraulic regime: Marine Geol., v. 15, p. 227–297.

VAIL, P. R., R. M. MITCHUM, JR., AND S. THOMPSON, III, 1977, Global cycles of relative changes of sea level, in C. E. Payton (ed.), Seismic Stratigraphy—Applications to Hydrocarbon Exploration: Am. Assoc. Petroleum Geologists Mem. 26, p. 83-98.

WALKER, R. G., 1979, Shallow marine sands, in R. G. Walker (ed.), Facies Models: Geol. Assoc. Canada Reprint Series 1, p. 75-90.

WILSON, H. H., 1977, "Frozen-in" hydrocarbon accumulations or diagenetic traps—Exploration targets: Am. Assoc. Petroleum Geologists Bull., v. 61, p. 483-491.

HIGH-ENERGY SHELF DEPOSIT: EARLY PROTEROZOIC WISHART FORMATION, NORTHEASTERN CANADA

BRUCE M. SIMONSON
Department of Geology, Oberlin College, Oberlin, Ohio 44074

ABSTRACT

The Wishart Formation of the Early Proterozoic Labrador trough has the sedimentary structures of a high-energy shelf deposit. It is divisible into six facies designated A through F: (A) very fine to fine-grained sandstone in cm-thick, laminated to rippled layers and lenses; (B) fine-grained sandstone in decimeter-thick layers displaying internal flat lamination to hummocky cross-stratification and capped by a veneer of ripples; (C) fine to coarse-grained sandstone with dm-thick trough cross-stratification and biomodal paleocurrents; (D) thin, laterally persistent beds of coarse conglomerate in which the larger clasts are all slabs of penecontemporaneously-cemented sandstone; (E) crudely stratified graywacke containing pseudonodules and soft-sediment folds; and (F) stromatolitic to thinly laminated dolomite.

Ninety percent of the aggregate thickness of the Wishart Formation consists of facies A, B, and C. For about half of that thickness, they are grouped into upward-coarsening cycles similar to ones in various Phanerozoic marine shelf deposits; elsewhere, they appear to be randomly interstratified. Facies A and B, which form the lower parts of the cycles, are interpreted as storm deposits which accumulated in deeper areas of the shelf. Facies C, which forms the upper parts of the cycles, is interpreted as tidal sands deposited in shallower areas. The cycles average 12 m in thickness and are attributed to shelf aggradation during pulses of more rapid sedimentation. All of the cycles are capped by erosional surfaces. Many are overlain by a thin lag conglomerate (facies D) formed by the winnowing of partially cemented sands during episodes of sediment starvation. Where cycles are absent, facies shifts presumably reflect long-term and possibly random variations in tidal currents, storms, or other factors.

Facies E and F are not incorporated into cycles. Facies E formed by either slumping or foundering of previously-deposited strata (probably facies A). The dolomites of facies F indicate that siliciclastic sedimentation was outpaced by carbonate accumulation for a time. Meager evidence suggests that some of the carbonates are cryptalgal, but given the Early Proterozoic age of the Wishart, they could have been deposited anywhere from the intertidal zone to the lower limit of the photic zone. Hence, facies E and F add little to the paleoenvironmental picture already gleaned from facies A, B, and C.

The characteristics of the Wishart can all be explained in terms of processes taking place on modern continental shelves, supporting a uniformitarian interpretation of Early Proterozoic shelf environments. However, this analysis also serves to highlight certain shortcomings in our understanding of the stratigraphic record those processes leave. In particular, it cannot be determined with the available evidence (which is considerable) whether or not the Wishart in the Schefferville area was deposited near a coastline. Likewise, the reasons for the variability of the sandstones attributed to storm deposition are obscure. Lastly, the interpretation of the Wishart as a high-energy shelf deposit supports a marine origin for the superjacent Sokoman Formation, one of the largest of the cherty iron formations.

INTRODUCTION

Geological Setting

The Wishart Formation was deposited in the Labrador trough, an Early Proterozoic (Aphebian) geosyncline in northeastern Canada (Fig. 1a). The trough extends for more than 800 km in a north-northwesterly direction and is up to 100 km wide (Gross, 1968, Fig. 1). The Wishart is restricted to the western, miogeosynclinal half of the trough, but persists along strike for most of its length (Dimroth et al., 1970). The only penetrative deformation of Proterozoic rocks in the area studied (the vicinity of Schefferville, Quebec) took place during the Hudsonian Orogeny about 1.6 billion years ago (Dimroth, 1970, p. 2734) and comprised overthrust faults with tectonic transport towards the southwest. The Wishart is exposed repeatedly in successive thrust sheets as long strike ridges trending northwest-southeast (Fig. 1b). Slaty cleavage is present only locally (notably in fold hinge zones), and the met-

amorphic grade of the rocks is zeolite grade or lower (Zajac, 1974, Fig. 14).

The Wishart lies immediately below the largest Precambrian cherty iron formation of the Canadian Shield, the Sokoman Formation. The contact between the Wishart and Sokoman is abrupt in the Schefferville area, but the following evidence indicates that it is nevertheless comformable: (1) lack of angular discordance between the Wishart and Sokoman in the Schefferville area (Zajac, 1974, p. 15); (2) minor intercalations of chert and possibly ferruginous shale in the uppermost few meters of the Wishart (Simonson, 1982a, p. 58); and (3) the Wishart and the Sokoman both interfinger with volcanic rocks of the Nimish Subgroup southeast of the Schefferville area (Evans, 1978). The Sokoman is one of a number of large, Early Proterozoic iron formations with abundant granular or arenaceous textures. Others occur in the Lake Superior area (Bayley and James, 1973) and in the Nabberu basin of Western Australia (Hall and Goode, 1978).

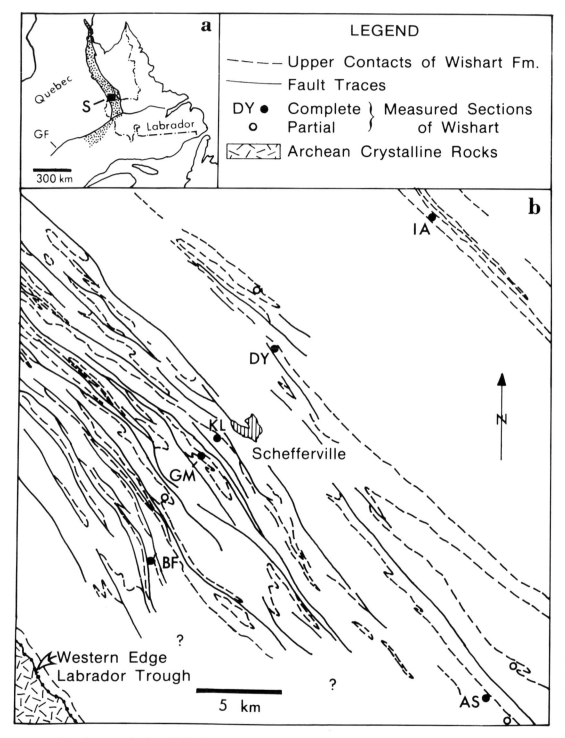

FIG. 1.—Location maps. A, sites of Labrador trough (stipple area), Schefferville area (black square labeled S), and Grenville Front (line labeled GF) in northeastern Canada; B, geological sketch map of Schefferville area (after Fig. 6 of Zajac, 1974) showing fault traces, upper boundary of Wishart Formation in successive thrust sheets, and measured section locations; most interruptions in strike belts and fault traces are due to lakes not shown on this map.

Most of these large, arenaceous iron formations rest with apparent conformity on units rich in quartz sand which are similar in many ways to the Wishart Formation, and which various workers have independently interpreted as shelf to nearshore deposits (e.g. James, 1954; Morey, 1973; Van Schmus, 1976; Hall and Goode, 1978).

The Wishart Formation rests on different stratigraphic units in different parts of the Schefferville area (Zajac, 1974, p. 13 and Fig. 3). At the extreme western edge of the Labrador trough, the Wishart rests unconformably on the Ashuanipi Complex which consists of feldspathic gneisses and granitic rocks of Archean age (Harrison *et al.*, 1972, p. 4). In the central and western portions of the Schefferville area, the Wishart rests successively on three formations named the Fleming, Denault, and Dolly. The Fleming Chert Breccia is an enigmatic unit whose origin is still in dispute (Zajac, 1974, p. 13; Simonson, 1982b). The Denault in the Schefferville area is predominantly a basinal dolomite in which most of the sedimentary structures point to deposition by suspension settle-out, turbidity currents, and debris flows (Dimroth, 1971, p. 1440). Most commonly, the Wishart rests on the Dolly Formation, which consists mainly of slaty siltstones with thin, even stratification. Dimroth (1971, p. 1448-1449) regards the Dolly as a basinal to "prograded sublittoral deltaic sequence." Contact relations between the Wishart and the underlying formations are generally disconformable along the western fringe of the trough, but become conformable moving northeastwards across the Schefferville area (Dimroth *et al.*, 1970, p. 54; Harrison *et al.*, 1972, p. 12).

The Wishart Formation ranges in thickness between 20.6 and 64.7 m (Fig. 2). Throughout the Schefferville area, it is mainly resistant, quartz-rich sandstone. Shale is a widespread constituent, but minor volumetrically and restricted mainly to thin partings between sandstone layers. Fine extra-basinal and coarse intrabasinal conglomerates form decimeter-scale beds, but they are much fewer in number than the shaly partings. Overall, the Wishart becomes progressively thinner, the abundance and coarseness of conglomerate beds increase, and the amount of shale decreases approaching the southwestern edge of the trough.

Based on distinctive associations of rock types and primary sedimentary structures, the Wishart Formation is subdivided into six facies (Fig. 2). Facies A, B, and C make up about 91% of the aggregate thickness of the measured sections of the Wishart. These three facies are organized into upward-coarsening cycles for about 51% of their aggregate thickness. Elsewhere, facies A, B, and C generally lack recognizable cycles, although a few, thin, upward-fining packages are present. Facies D is a genetically important component of the upward-coarsening cycles, but contributes little to their aggregate thickness. Facies E and F are not integral parts of the upward-coarsening cycles.

Petrography

Most sandstones in the Wishart Formation are protoquartzites (Fig. 3), classified according to Pettijohn (1957, Table 48). Subarkoses and true arkoses also occur, but true orthoquartzites are very rare or absent. The feldspars are predominantly perthitic microclines, consistent with the derivation of the detritus from the Ashuanipi basement complex to the west of the Labrador trough. This is corroborated by deformation textures in the quartz grains, and zircon and tourmaline being the only common heavy minerals. Most individual sandstones are well sorted, and mean grain size ranges from very fine to very coarse. Cements, where present, consist of authigenic overgrowths (Fig. 3c), or, more rarely, chert (Fig. 3b). The finer-grained sandstones tend to have less cement, more angular grains, a higher percentage of feldspar, and more interstitial chlorite than the coarser ones (contrast Figs. 3a and 3c). Feldspar is most abundant, however, in the conglomeratic beds of the Wishart.

DESCRIPTIONS OF FACIES AND UPWARD-COARSENING CYCLES

In this section, the six facies of the Wishart Formation are described in succession, focusing primarily on their characteristic sedimentary structures. A brief description of the stratigraphic packaging of facies A, B, C, and D is also included (after the description of facies D). In the section which follows this one, the processes responsible for forming the structures in each facies are discussed. The interpretations of the various facies are then synthesized into a unified model of deposition on the Wishart shelf in the next to last section of the paper. The final section is a brief summary of the broader implications of this study of the Wishart.

Facies A

By definition, most of the aggregate thickness of any interval of facies A consists of very fine- to fine-grained sandstone in layers and lenses which are about one to two centimeters thick and separated by shaly partings (Fig. 4a). The sandstone layers typically pinch and swell, yet persist laterally for tens of centimeters. Internally, the layers are thinly laminated. Laminae within a given layer tend to be concordant with the bottom surface and truncated above at very low angles by either the top surface of the layer or other laminae within the layer. Sandstone lenses extending for less than 10 cm parallel to bedding also occur, although they are less abundant than the laterally persistent layers. Most of the lenses have internal cross-laminae with tangential bases and maximum dips of 20°. Up to five laterally equivalent lenses may be approximately equally spaced along discrete bedding planes.

Most bedding plane exposures in facies A exhibit one of several different kinds of ripple marks (although

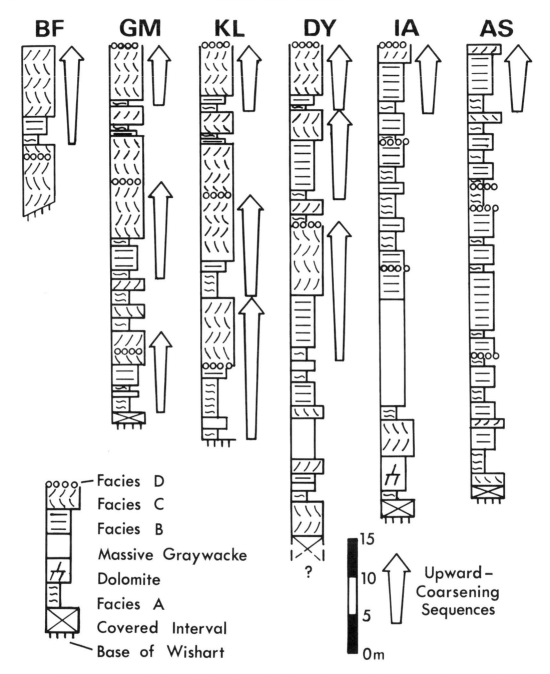

FIG. 2.—Schematic measured sections of Wishart Formation. The top of each section is the contact between the Wishart and the overlying Sokoman Iron Formation. Upward-coarsening sequences are indicated. Facies are described in the text.

such exposures are rare). Some of the marks are identical in shape to certain well-known types of modern ripples. For example, small linguoids are present on a few bedding planes, and symmetrical, straight-crested oscillation ripples occur on others (Fig. 4b). The latter have either sharp or rounded crests and spacings of about 10 cm. However, the most frequently observed type of bed form does not have a normal ripple morphology. It comprises equant, rounded domes and intervening hollows. Each dome or hollow is about 10 cm in diameter, has about 1 to 2 cm of relief, and generally lacks any marked asymmetry. Based on the common presence of undeformed ripple cross-laminae within the domes, these structures appear to be ripples that have been modified in some way.

The shaly partings which separate sandstone layers and lenses from one another rarely exceed a millimeter in thickness, yet they are rarely spaced more than a centimeter or two apart. The partings are fissile, dark green, and consist of equal parts of quartzofeldspathic silt and chloritic matrix. Most partings undulate sinusoidally, bifurcating in cross section in places (Fig. 4a), and a minority are erosionally truncated. The thin partings are crinkled on a fine scale, probably as a result of distortion during compaction and/or tectonic deformation. Flat pebbles of mudstone occur in a few

of the intervening sandstones, and a few of the fine-grained layers exceed 1 centimeter in thickness. The pebbles and thicker layers are both similar to the thin partings petrographically and both lack internal lamination.

Decimeter-scale beds of sandstone are the only important component of facies A, aside from the cm-scale sandstone layers and the shaly partings which form the majority of it. The dm-scale beds are very similar to the cm-scale layers; indeed, the two appear to be end members of a continuum. However, in addition to being thicker, the dm-scale beds appear to be slightly coarser than the cm-scale sandstones on average. Identical dm-scale sandstone beds are described more fully below, because such beds are the main component of facies B.

Fig. 3.—Photomicrographs of Wishart sandstones, all fields of view are approximately 1.1 mm by 0.9 mm. *A*, very fine to fine-grained sandstone rich in matrix, uppermost bed of facies B in section KL. *B*, medium-grained sandstone with chert cement from a few centimeters below top of measured section DY. *C*, fine to medium-grained sandstone cemented with syntaxial overgrowths from the middle bed of facies C in section KL; dust rings outlining detrital cores are visible in many grains. *D*, medium-grained sandstone intraclast with chert cement (upper right) in sharp contact with finer-grained, interstitial sandstone (lower left) lacking visible cement; edge of intraclast runs diagonally across photo; from bed of facies D closest to middle of section AS.

FIG. 4.—Predominant sedimentary structures of facies A, B, and C. *A*, typical exposure of cm-scale sandstones of facies A; each dark, recessed line is a shaly parting; from uppermost bed in section DY; scale bar is 3 cm long. *B*, straight- and sharp-crested, symmetrical ripple marks on dipslope in facies A near middle of section DY; scale bar 20 cm long. *C*, typical exposure of decimeter-scale fine sandstones of facies B; dark lines again indicate shaly partings; from middle of section AS; scale bar 20 cm long. *D*, most of thick interval of facies B in upper half of section DY; hammer in center is about 30 cm tall. *E*, trough cross-stratified medium sandstone in bed of facies C in middle of section DY; scale bar is 10 cm long. *F*, paleocurrent rose of cross-strata from the two thick intervals of facies C in upper half of section DY; note predominance of paleoflow towards north-northwest and south-southeast.

Facies B

By definition, most of the aggregate thickness of any interval of facies B consists of fine sandstone in layers one to several decimeters thick that are thinly laminated internally (Fig. 4c). The basal contacts of these layers are sharp and virtually flat, save for local scours one to several centimeters deep into the underlying strata. The scours are filled with either flat-laminated or massive sandstone, but their shapes in plan are not known because no good exposures of the soles of such beds were encountered. In almost every decimeter-scale layer of sandstone, most of the bed is thinly and homogeneously laminated. The laminae are generally nearly horizontal, never inclined at angles exceeding 10°, and truncated in places along low-angle erosion surfaces. The laminated zone is in turn overlain in most dm-scale beds by up to several centimeters of ripple cross-laminated sandstone. The contact between the ripples and the subhorizontally laminated sandstone below is erosional and consists of small trough-like scours, which are lined with a shaly parting in some beds. Bedding plane exposures of cm-scale beds capped with ripple cross-laminated sandstone display the same varieties of ripple marks observed on the thinner layers of sandstone characteristic of Facies A.

A minority of these decimeter-scale sandstone beds exhibit the structure Harms (1975, p. 87) calls hummocky cross-stratification. The characteristics displayed by the cross-strata which indicate that they are hummocky are as follows: (1) Where sets of planar to gently-curved laminae are truncated erosionally by other sets, the truncation surfaces and the cross-strata both dip more gently than the angle of repose. (2) Cross-strata and truncation surfaces look the same in mutually perpendicular cross sections. (3) Sets of cross-strata thicken laterally in a uniform, fan-like manner, especially over the shoulders and into the bottoms of scour hollows. (4) Nonplanar cross-strata display both concave-up and convex-up curvature. (5) Most cross-strata resting on truncation surfaces are comformable with same. All four of the criteria Harms used to identify hummocky cross-stratification are included amongst these. However, planar lamination is proportionately more abundant and both the curvature and the dips of the cross-strata are less pronounced in the Wishart than they are in other recently-described examples of hummocky cross-stratification (e.g., Bourgeois, 1980).

Despite their flat appearance (Fig. 4d), I could not correlate individual decimeter-scale beds between replicate measured sections, even where they were separated by less than ten meters along strike. Hence, although the beds look continuous at first glance, they must be lenticular on a scale of meters. The lateral disappearance of at least some of these beds appears to have been caused by progressive down-cutting of the ripples veneering their upper surfaces. Therefore, the original lateral persistence of the decimeter-scale sandstone beds cannot be determined with certainty.

Departures from the dm-scale sandstone beds in intervals of facies B consist almost entirely of strata which, in isolation, would be indistinguishable from those of facies A. The distinction between facies A and B therefore rests on the fact that the dm-scale beds form the majority of intervals of facies B, whereas they are in the minority in facies A. This distinction, although subtle, is important: it serves to highlight depositional trends through time, particularly the upward-coarsening cycles (Fig. 2).

Facies C

Facies C consists mainly of cross-stratified, very well-sorted, fine- to coarse-grained sandstones. Most of the cross-strata are from 1 to 5 decimeters thick and gently curved (Fig. 4e), although a subsidiary number are nearly planar in cross-section. Sets of cross-strata persist laterally for a few decimeters to several meters, pinching out where the flat to gently curved erosional surfaces which bound them converge. Reactivation surfaces are present within some of the sets. The cross-strata of facies C resemble herringbone cross-strata in that adjacent sets frequently dip in opposite directions, yielding paleocurrent roses with two dominant modes almost 180° apart (Fig. 4f).[1] However, herringbone cross-strata are generally thought of as having planar-tabular foresets that dip at high angles. (<30°). Inasmuch as the cross-strata of facies C generally have maximum dips of only 15° to 20°, and the foresets are mostly trough-shaped instead of tabular-planar, they are not conventional herringbone cross-strata. Cross-strata of the "coarse sand complex sand wave facies" of Klein (1970) are the closest morphological analogs reported from modern sediments to those of facies C (compare Fig. 4e to Klein's Fig. 33b). The comparison is strengthened by the fact that the paleocurrent roses for these two occurrences are also similar (compare Fig. 4f with Fig. 18, lower left corner, of Klein, 1970).

Departures from well-sorted sandstone with dm-scale cross-stratification are uncommon in facies C and take one of two forms. One consists of thin layers of finer-grained, ripple cross-laminated sandstone, commonly with shaly partings. Such layers occur between the thicker sets of cross-strata and do not intertongue

---

[1]Throughout the field area, all large exposures of facies C lie parallel to structural strike. In order to prevent this from biasing the data, I used the following procedure to measure paleocurrents: (1) a small area with exceptionally good exposures of facies C was selected; (2) the apparent dips of all layers visible within sandstones in that area were measured in two cross-sections which were nearly mutually perpendicular; (3) using a stereonet, the true dip of each layer was determined from the two apparent dips, then rotated to compensate for structural tilting; and (4) the azimuths of the true dips of all layers in a given area were plotted in a rose diagram (e.g., Fig. 4f), save for layers with dips less than 5°. This procedure was applied to three separate areas in three different measured sections, and all of them revealed two major components of paleoflow which were northwesterly and southeasterly.

with the overlying or underlying foresets. Laterally, these finer-grained layers are truncated erosionally by the basal scour surfaces of sets of larger cross-strata. Such intercalations are rare in facies C.

The second departure comprises sandy, fine conglomerate in beds about 20 cm thick on average. The larger pebbles in these beds rarely exceed 1 cm in diameter and consist of silty shale, microcline, and/or black chert. Small numbers of sandstone intraclasts (to be described in facies D) are also present in places. The beds have sharp basal contacts, a few of which display evidence of load-type soft-sediment deformation involving the underlying sandstone. Internal stratification is generally absent, although some beds display a single, decimeter-thick set of cross-strata. The upper contacts of the beds are gradational, commonly consisting of alternating coarse and fine laminae in-

clined at moderate angles. Several of the coarser beds are topped by symmetrical megaripple marks, with average spacings of up to 80 cm and average heights of as much as 15 cm. Most of these conglomerates can be traced along strike for a few hundred meters at a constant stratigraphic level.

Facies D

Facies D is unique in that each "interval" consists of a single bed of sandy conglomerate. Furthermore, the conglomerates contain the coarsest clasts in the Wishart Formation, namely boulders up to 85 cm across of very well-sorted, chert-cemented sandstone (Fig. 5a). Pebbles and cobbles of such sandstone are the main constituents of these beds. They originated as intraclasts, because they are texturally identical to in situ sandstones (especially those of facies C) which are

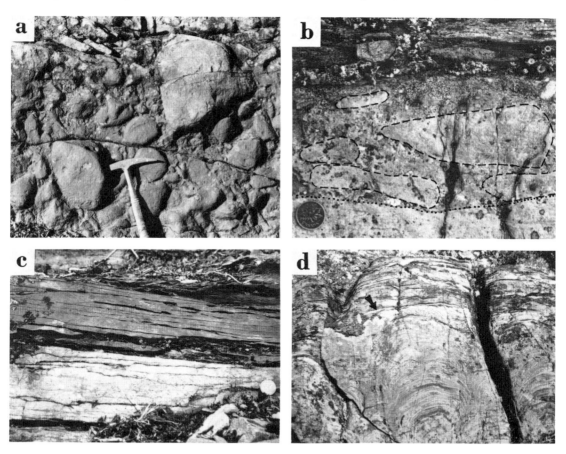

FIG. 5.—Predominant lithologies of facies D, E, and F. A, plan view of bed of facies D exposed on dipslope at top of section GM; each oval object is individual sandstone intraclast, the largest of which is thicker than the average thickness of the entire bed; hammer is 30 cm tall. B, cross-sectional view of bed of facies D closest to middle of section AS; bed rests on flat-laminated sandstone of facies B (contact indicated by dotted line), is overlain by shaly strata of facies A (dark zone at top), and contains flat, intraformational pebbles and cobbles of sandstone (largest ones outlined by dashed lines); white, cm-size clasts are among the coarsest grain of microcline found in the entire Wishart Formation; penny for scale in lower left. C, typical exposure of unstratified graywacke (facies E); dark lines are stylolite-like seams; from thick bed near base of section DY; penny for scale in lower right. D, dolomite from facies F in section IA; layer of stromatolites truncated sharply by thinly-laminated dolomite at arrow; approximately 40 cm from base of photo to truncation surface.

cemented with chert (compare Figs. 3b and 3d). The sandstone intraclasts are typically so closely spaced that they form a nearly continuous carpet of gravel, yet the carpet is rarely much more than one clast thick (Fig. 5b). The intraclasts are very well rounded, most are tabular or discoidal, and they contain internal laminae or cross-strata which are sharply truncated along their edges. The spaces between these large sandstone slabs are generally occupied by poorly sorted coarse sandstone to fine conglomerate which is very similar petrographically to the conglomeratic rocks in facies C. More rarely, the sediment associated with the sandstone intraclasts consists mainly of medium to coarse sand-size peloids of black chert. Chemical cement is generally absent from the sediment between the sandstone intraclasts where it is siliciclastic, but chert cements are commonly abundant where chert sands are present.

The basal contacts of beds of facies D are erosional surfaces which truncate all sedimentary structures in the substrate (most notably cross-strata). Although these contacts are broadly conformable with bedding, they have local irregularities with relief of up to half a meter. The slopes on the sides of such irregularities commonly exceed the angle of repose, and in some places they are even vertical or form overhangs. The layers in the rocks underlying such irregularities show no signs of soft-sediment deformation (in contrast to the basal contacts of conglomeratic beds in facies C); instead, they contain abundant chert cement.

The upper contacts of beds of facies D are also sharp and irregular, but otherwise they are quite different in character from the basal contacts. The upper contacts are not erosional and are always draped with at least a thin veneer of shaly sediment (Fig. 5b). The irregularities in this case result from the tendency of the larger sandstone intraclasts to protrude above the general level of the associated sediments, thus forming a bumpy surface which was draped with mud. This is in marked contrast to the conglomeratic beds in facies C whose upper contacts are generally gradational from fine conglomerate into medium- or coarse-grained sandstone.

The two coarsest beds of facies D, one at the top of the Wishart Formation and the other near the middle, can be traced or correlated with confidence for distances of several kilometers (and possible tens of kilometers) both along and across strike. Although locally they are devoid of large sandstone intraclasts and resemble the conglomeratic rocks of facies C, they are distinctly different entities. They are the only beds in the Wishart with abundant cobbles and boulders of chert-cemented sandstone. Consequently, they are the best marker beds in the Wishart.

Upward-coarsening Cycles and Other Stratigraphic Patterns

About 46% of the aggregate thickness of the measured sections of the Wishart Formation is incorporated into twelve upward-coarsening cycles which average 11.9 m in thickness (Fig. 2). From bottom to top, almost all of the cycles display a systematic succession from facies A to B to C. The relative proportions of these three facies vary considerably; facies A is consistently the thinnest, but either facies B or C can be the thickest. In some cycles, basal facies A rests sharply on coarser beds of facies C (e.g., uppermost cycles in measured sections GM and KL, Fig. 2). Other cycles have no precise starting points, passing downwards gradationally into randomly interstratified facies A and B (e.g. lowermost cycle in section DY). Within the cycles, the upward transitions from facies A to B are completely gradational; the decimeter-scale sandstone beds simply increase in thickness and abundance up-section, and the changeover occurs where they become dominant. Transitions from facies B to C are also gradational, but more abrupt. Typically, a few decimeter-scale sets of cross-strata appear high in facies B, then increase rapidly in abundance and, to a lesser extent, thickness upsection until they coalesce to form rocks typical of facies C. Such transitions take place within a thickness of only one to two meters. In addition, most of the finer intercalations in facies C are stratigraphically close to such transition zones.

At the top of every upward-coarsening cycle is an erosion surface truncating facies C and overlain by finer sediment and/or a layer of facies D conglomerate. Some beds of facies D also occur in association with facies A and B where they seem to be randomly interstratified (e.g. measured section AS, Fig. 2). However, only one bed of facies D occurs within the confines of a well-defined upward-coarsening cycle (basal cycles, measured section GM and Kl, Fig. 2). This bed is unique among facies D conglomerates in that it has abundant cobbles and pebbles of black chert. Hence, it must differ in origin from the others, but it is not clear what the difference was (Simonson, 1982a, p. 268).

Available evidence suggests that the cycles and most individual facies intervals persist laterally for kilometers, but not tens of kilometers. For example, the vertical sequences of facies in measured sections KL and GM (which are currently 1.3 km apart, Fig. 1b) are similar (Fig. 2), whereas only a few beds in sections KL and DY (which are 6 km apart) are clearly correlative. Major thrust faults intervene between both pairs of sections (Fig. 1b), so they must have originally been even further apart than they are today. Unfortunately, the measured sections are too few in number and the deformational history of the region is too poorly known for the three-dimensional shapes of individual cycles or facies intervals to be determined at present.

The strata of facies A, B, C, and D which are not incorporated into upward-coarsening cycles exhibit one of three types of stratigraphic organization. The commonest of the three consists of random interstratification, particularly of facies A and B, about which little more can be said. The second type of organization is rarer and is exemplified by several upward-coarsening sequences in measured sections IA and AS which differ from those already described in being thinner and less dramatic. In the thinner upward-coarsening sequences, facies A passes gradationally upsection into facies B which in turn ends in an erosion surface. Some of these sequences are capped by facies D, but facies C is absent. The third and rarest type of stratigraphic organization consists of upward-fining sequences. These are thinner than the normal upward-coarsening cycles and, from bottom to top, each one consists of a stack of facies C, B, and A. Within these sequences, the facies have gradational contacts and appear to be more variable along strike than they are in the upward-coarsening cycles. Scattered pebbles may occur at several levels within upward-fining sequences, but beds of facies D are absent. Because of their rarity, I did not study the upward-fining sequences in detail.

Facies E

The rocks of facies E are predominantly graywackes (sandstones rich in matrix; definition of Pettijohn, 1957, Table 48) lacking good stratification. Petrographically, they consist of very fine to fine quartzofeldspathic sand set in a chloritic paste. The only stratification they exhibit consists of paper-thin, finely crinkled partings that are parallel to general bedding and spaced about 0.5 to 1.0 cm apart (Fig. 5c). Pseudo-nodules and soft-sediment folds up to 20 cm long are sparsely disseminated throughout the graywacke. They consist of well-sorted, very fine to fine sandstone (like that of facies A and B) and display excellent lamination internally. The long dimensions of the pseudo-nodules and the axes of the folds are parallel to general bedding. Even rarer are blocks up to 55 cm long which are isolated in the graywacke and have undisturbed internal stratification. The stratification in the largest blocks is similar to that of facies A, except that it is tilted with respect to general bedding.

Uninterrupted sequences of facies E can be at least 6 m thick (and possibly twice that). Nevertheless, layers of facies E are not as persistent laterally as other facies. Layers of facies E in some measured sections appear to be laterally equivalent to beds of facies A and possibly facies B in other sections. The upper and lower contacts of layers of facies E are sharp and horizontal with one exception. The latter exhibits vertical tongues of graywacke containing pseudonodules which penetrate downwards more than 40 cm from a layer of facies E into underlying sandstones of facies B. Scattered soft-sediment deformation structures (mainly pseudo-nodules) also occur in many of the strata adjacent to layers of facies E. The timing of the formation of facies E relative to the deformation of the surrounding beds can rarely be established. However, in one exceptional case, soft-sediment deformation structures are clearly truncated erosionally at the base of a layer of facies E.

Facies F

Facies F is unique in the Wishart Formation in that it consists mostly of carbonate (dolomite, possibly ankeritic) rather than siliciclastic sediment. I only observed two such occurrences in the Schefferville area, one in measured section IA and the other in a partial section 2.5 km north-northeast of measured section AS (Fig. 1b). Both occurrences are near the base of the Wishart and display the same internal stratigraphy. From the bottom up, both consist of siliciclastic sediments, stromatolitic dolomite, flat-layered dolomite, and finally siliciclastic sediments again. Despite their similarity, these two occurrences are more than 27 km apart and neither one can be traced along strike for more than a few hundred meters.

The stromatolites form a single row of columns whose lateral and vertical dimensions both range from a few to 40 cm (Fig. 5d). Although they consist mainly of carbonate, most of the stromatolites contain a small amount of finely disseminated siliciclastic sand. Internally, their laminae are thin, hemispheroidal, pinch out on the sides of the columns, and have synoptic relief of 12 cm or less. Fenestrae and carbonate peloids also occur in some stromatolites, but they are rare. The fenestrae are either within thicker laminae or between laminae. They are elongate parallel to layering and filled with coarser crystals of quartz and/or carbonate. The sediment between the stromatolites can either be nearly pure carbonate or contain abundant fine to coarse siliciclastic sand. The intercolumnar sediment is always chaotic and unstratifed.

In the flat-layered dolomite, micritic laminae 5 to 30 microns thick alternate with thicker, more irregular laminae. The latter contain rare pockets of well-sorted, very fine to medium sand-size carbonate peloids and siliciclastic grains. A few of these pockets are in the form of thin lenses with tapering terminae much like the low-relief starved ripples figured by Hardie and Ginsburg (1977, Fig. 36). The micritic laminae are more persistent laterally, although locally they are discontinuous. Fenestrae like those found in the stromatolites are abundant in these rocks. The flat-layered dolomite rests directly on the stromatolitic layer. The contact is horizontal and erosional, cutting across the stromatolites and intercolumnar sediment indiscriminately.

Both occurrences of flat-layered dolomite also contain a single layer of flat pebble conglomerate. The conglomerate is composed of a combination of tabular carbonate intraclasts up to 7 cm long and poorly sorted (but sandy) dolomitic and siliciclastic sediment. The intraclasts are imbricated where they are closely spaced.

Microscopically, all of the dolomite consists of minute crystals in tightly-interlocking mosaics. The mosaics are locally equigranular, but the average crystal size varies between 5 to 80 microns. Despite the pockets of carbonate peloids and siliciclastic sand grains, clastic textures are quite rare overall. Alternations of mosaics with different crystal sizes generally give definition to the laminae, commonly accentuated by variations in their content of siliciclastic silt and sand. No argillaceous detritus was detected petrographically in any of the dolomite; however, no insoluble residues were examined.

INTERPRETATIONS OF FACIES

Only shallow marine processes will be considered in the interpretations which follow, because several independent lines of evidence indicate that the Wishart Formation is a shallow marine deposit. Foremost among these are the large areal extent of the Wishart, the complexity of its paleocurrent patterns (interpreted below as the product of tidal currents), and the near-total absence of mudcracks[2] or other indicators of subaerial exposure. The abundance of wave-generated structures in the Wishart rules out deposition in deeper marine environments.

Facies A and B will be discussed in reverse order in this section for the following reasons. Both facies consist of two components: (1) decimeter-scale beds of fine sandstone, and (2) centimeter-scale layers and lenses of finer-grained sandstone separated by shaly partings. The decimeter-scale beds have well-studied analogs in Phanerozoic shallow marine deposits, whereas few, if any, analogs for the centimeter-scale layers have received as much attention. Interpretations of the former can be extended to the latter in the Wishart because the two appear to be end members of a continuum. Consequently, facies B will be discussed first (it contains the greater abundance of decimeter-scale sandstone beds) followed by facies A (in which centimeter-scale beds predominate).

Facies B

The following lines of evidence suggest that the decimeter-scale sandstone beds characteristic of facies B were deposited rapidly during high-energy events: (1) They consist largely of horizontally-laminated to hummocky cross-stratified fine sandstone, suggesting that they were deposited by unidirectional and/or oscillatory currents in the transitional to upper flow regimes (Allen, 1970, Figs. 2.6 and 5.8). (2) They closely resemble Phanerozoic beds whose rapid deposition is well established on the basis of trace fossil evidence (e.g., Howard, 1972; Goldring and Bridges, 1973). (3) The intervening strata are rippled and shaly, both indications that lower-energy conditions prevailed between times when the decimeter-scale beds were being deposited.

---

[2]Howell (1954, p. 46) asserted that desiccation cracks are common in the Wishart Formation. I examined numerous bedding planes with shaly veneers through three field seasons, and I found only one possible example.

Most of the decimeter-scale beds have a thin, capping layer of rippled sandstone with a basal contact showing evidence of scour. Therefore, the beds must have generally been deposited in two distinct steps separated by an episode or erosion. Shaly partings lining these scours in some of the beds further indicate that a lull in current activity sometimes preceded deposition of the ripples. The sequence of sub-horizontal lamination overlain by ripples is found in clasical turbidites, but evidence of an erosional hiatus between the Bouma b and c is units rarely, if ever, observed. Moreover, the decimeter-scale sandstone beds in the Wishart are not even and continous, and many are capped by ripples formed by waves rather than currents. Therefore, the dm-scale beds in the Wishart could not have been deposited as turbidites, unless the rippled parts of the beds were formed by subsequent events and/or other processes.

Storm systems traversing a shelf environment are believed to be the most plausible explanation for deposition of the decimeter-scale sandstone beds. Sands with sub-horizontal lamination which are interpreted as storm deposits occur on modern continental shelves (Howard and Reineck, 1972; Kumar and Sanders, 1976). Similar beds in a number of Phanerozoic shelf sequences have also been interpreted as storm layers (Harms, 1975, p. 87-88; Hamblin and Walker, 1979, p. 1681; Cant, 1980, p. 128-129). Sand transport onto modern shelves during storms has been attributed to both turbidity currents (Hayes, 1967) and bottom return flow (Morton, 1981); the laminated, high-energy parts of the decimeter-scale beds may have been deposited from either or both. The suprajacent ripples on such beds could be the result of reworking during fair weather, waning storm, and/or minor storm conditions. The sediment incorporated into the ripples could have been brought in by the major storms and/or shelf sediment dispersal systems active at a slower but more consistent pace.

Almost all of the remainder of facies B consists of thinner layers of sandstone and intervening shaly partings similar to those which are the dominant component of facies A. These are probably a combination of fair-weather and milder storm deposits, as explained in the following paragraphs.

Facies A

The centimeter-scale sandstone layers which are the main constituent of facies A show internal lamination and external geometries which are very similar to those of the decimeter-scale beds characteristic of facies B. This close resemblance bespeaks a common origin. Therefore, each cm-scale sandstone layer is also interpreted as a storm deposit, although clearly less sand was deposited during each storm and/or more was removed or reorganized between storms than during the formation of a decimeter-scale bed. The presence of scattered, decimeter-scale beds indicates that thicker storm sands were deposited and preserved from time to time during formation of facies A. However, deposits from lower-energy events were formed and preserved more frequently, because rippled sandstones and shaly partings are more abundant than in facies B.

The origin of the dome and hollow structures which are the most abundant type of ripple mark in facies A (and B) is not known. They may represent normal current ripples which were rounded off by ebbing tidal currents. Ebb-modified current ripples are well-known from modern environments (Reineck and Singh, 1975, p. 369), and evidence reviewed in the next paragraph suggests that tidal currents played a large role in deposition of the Wishart Formation. Alternatively, they may represent some obscure type of interference wave ripple. Whatever their origin, facies A probably formed via a combination of the following processes: (1) storm-induced currents which deposited thin, sub-horizontally laminated sands; (2) gentler (fair weather, mild storm, and possibly tidal) currents and waves which formed ripples; and (3) settling out of suspended mud during times of reduced current and wave activity to form shaly partings.

Facies C

The ubiquity of decimeter-scale cross-strata indicates that the sandstones of facies C were originally deposited as large, migrating bed forms. The predominance of trough shapes further shows that the bed forms were short-crested megaripples, or dunes (rather than sand waves) in the terminology of Southard (1975, p. 24-25). The two opposed modes of the rose diagrams (e.g., Fig. 4f) indicate that paleocurrents flowed predominantly from the northwest and southeast, which is consistent with the widespread occurrence of herringbone-like structures in outcrops of facies C. The only modern environments where superimposed dunes migrate repeatedly in opposite directions are (1) high-energy, tide-swept shelves (Reineck and Singh, 1975, Fig. 466), and (2) marine embayments with amplified tides such as the Bay of Fundy (Klein, 1970). The morphological similarity between the cross-strata of facies C and certain cross-strata in the Bay of Fundy has already been noted. Consequently, the sandstones of facies C were most probably deposited in a tide-swept environment.

Estimating the depth of water in which facies C was deposited is very difficult. The attributes of the cross-strata offer one of the few approaches to this question. Some of the dunes were more than 50 cm high (and possibly much higher) based on the maximum heights of the preserved foresets. Most foresets are curved, which suggests that water depths were many times greater than the bed form height (Jopling, 1965). Together, these suggest a minimum water depth of many meters (10?). However, foreset curvature can also be caused by the simultaneous action of waves and currents (Harms, 1969), which casts doubt on this reasoning. The apparent absence of desiccation cracks also

suggests subtidal deposition, but it is not very reliable because shale is so rare in facies C. The special ripple types peculiar to intertidal zones (e.g., double-crested, ladderback, and truncated wave ripples) are also absent from facies C, but again ripple marks of any sort are rare. Finally, it should be noted that the structures which appear to be the closest modern analog to facies C cross-strata are located in an intertidal, not subtidal, environment (Klein, 1970, his Fig. 33b). Thus, while it seems likely that the sandstones of facies C originated in subtidal environments, deposition deep in the intertidal zone of a coastal or shoal area with a large tidal range cannot be ruled out categorically. In any case, the environment was one in which both ebb and flood tidal dunes were formed, then partially preserved in intimate association with one another.

The interbedded finer- and coarser-grained sediments show that additional processes helped shape facies C. The ripple cross-laminated, shaly layers must represent lower-energy conditions. The ripples were evidently not active at the same time as the dunes because they occur between the thicker sets of cross-strata, yet the two do not interfinger. Perhaps they were deposited during neap tides. The paucity of such finer-grained beds may be due simply to poor preservation, a view supported by their erosional truncation. The conglomeratic beds, on the other hand, must represent higher-energy conditions. Their sharp bases, gradational tops, foundered basal contacts, and lateral persistence indicate that they were deposited quickly and simultaneously over broad areas. Hence, they are probably storm deposits, too. This interpretation is strengthened by the large oscillation ripple marks atop some of these layers, which indicate that high-energy waves played a role in their deposition.

Facies D

Beds of facies D formed as lag deposits based on the following lines of evidence. The sandstone clasts which are the main component of the conglomerates are intraformational in origin, because of their close petrographic similarities to in situ sandstones of the Wishart Formation (particularly those of facies C) and the lack of any other potential source beds. The largest sandstone intrclasts are almost two orders of magnitude larger than any demonstrably extrabasinal detritus (e.g., fine pebbles of microcline vs. boulders of sandstone). Therefore, long-distance transport of the coarser clasts as traction bedload is highly unlikely. The thinness of the beds of facies D precludes deposition via any type of mass movement, because there is simply not enough matrix to support such large clasts. Finally, the intraclasts are believed to have formed via penecontemporaneous silica cementation of some of the sands (Simonson, 1981), and early cementation is most easily accomplished during times when sedimentation rates are low. Low sedimentation rates would also promote the formation of a surface lag gravel in a high-energy shelf environment.

At first glance, local derivation and minimal transport of the sandstone intraclasts seem to be incompatible with their excellent rounding, especially since there is good evidence that the penecontemporaneous cements were rigid precipitates of silica (Simonson, 1982a, p. 140-141). However, this poses no problem if the intraclasts originated as concretions, which tend to be highly rounded (e.g., Pettijohn, 1975, p. 464 and Fig. 12-1). If the sandstone intraclasts in the Wishart formed as ellipsoidal concretions surrounded by more friable sands, they could have been very well rounded when first exhumed, in which case no rounding during transport would have been necessary. This interpretation is supported by the fact that the occurrences of chert cements in the sandstones of other facies tend to be highly localized.

Deposits akin to facies D occur in several modern environments. Morgan and Treadwell (1954), van Straaten (1957), and Garrison *et al.* (1969) have described large, tabular clasts of sandstone formed by penecontemporaneous cementation in sediments from three different coasts. The modern sandstone intraclasts are cemented with calcium carbonate instead of chert, but they closely resemble those of the Wishart Formation in their sizes and shapes, and in having been winnowed from penecontemporaneously cemented sands to form a lag gravel. Morgan and Treadwell (1954) found that cementation was taking place in intertidal beach sands, but in the other two examples, subtidal cementation was inferred. Most of the in situ chert cements in the Wishart are in facies C, which suggests that they formed in subtidal environments.

Facies E

The abundance of mud and the lack of stratification in facies E can be explained in two ways: either (1) it was deposited as debris flows, or (2) it formed from interlayered sands and muds that were homogenized in situ by soft-sediment deformation. Choosing between these two hypotheses is difficult because the characteristics of facies E can be interpreted as products of either process. For example, the pseudo-nodules and soft-sediment folds displayed by facies E could have formed in situ from original bedding, but alternatively, they could have been transported by debris flows. Likewise, the partings which define bedding in facies E could have originated in situ in a manner similar to dish structures (Lowe and LoPiccolo, 1974), or they could be akin to the "slump cleavage" described by Corbett (1973). Lastly, some beds of facies E display crude stratification which could have either developed as a result of shearing in a debris flow or been inherited from original bedding.

Whatever its origin, facies E must have formed from sediments similar to those of facies A for the following reasons: (1) facies A and E consist of the same types of fine sediment (although E contains a higher proportion of shale); (2) the few intact sedimentary structures within facies E match those of facies A; and (3) one bed of facies E in one measured section is stratigraphically equivalent to a bed of facies A in another section nearby. The formation of facies E from beds of A by slumping and/or soft-sediment deformation has little or no paleoenvironmental significance inasmuch as both of these processes occur in wide variety of subaerial and subqueous environments.

Facies F

The dolomite stromatolites of facies F probably formed by the trapping of fine carbonate detritus on cyanophyte mats rather than by either inorganic or biochemical carbonate precipitation for the following reasons: (1) Laminae pinch out on the margins of the stromatolites (Fig. 5d), whereas overturned laminae are commonplace in inorganically-precipitated stromatolitic structures (e.g., Smoot, 1978, Fig. 12). (2) The Wishart stromatolites have detrital textures in places, whereas inorganically-precipitated stromatolitic structures (e.g., speleothems) display crystalline rather than detrital textures (e.g., Bjaerke and Dypvik, 1977, Fig. 6; Kendall and Broughton, 1978). (3) Fenestrae in stromatolitic structures formed by inorganic or biochemical precipitation are generally elongated normal to layering (e.g., Monty and Hardie, 1976, Fig. 11; Smoot, 1978, Fig. 13), whereas the fenestrae in the Wishart stromatolites are elongated parallel to layering. Nevertheless, the number of stromatolites examined was small enough so that all inferences about their origin must remain tentative.

The flat-layered dolomites display three distinct types of stratification, each of which must reflect a different depositional process. Infrequent clastic textures and probable ripple cross sections suggest that laminae of the most abundant type are current-laid deposits. However, in the absence of cross-laminae or ripple marks, it is impossible to be more specific. Strata of the second type (thin micritic laminae) have characteristics which are very similar to Bahamian cryptalgal laminae (Hardie and Ginsburg, 1977, p. 109). However, the sample population is again too small to be certain that this comparison is widely applicable. The third stratification type (layers of flat pebble conglomerate) was deposited by high-energy currents which eroded the other types of dolomitic sediments, based on the large size, imbrication and intraformational origin of the clasts they contain.

The sedimentary structures of facies F place no new constraints on the paleoenvironments of the Wishart Formation. The stromatolites and flat laminae tentatively interpreted as cryptalgal suggest that the dolomite was deposited in water no more than 10 to 15 m deep (the lower limit of abundant algal growth according to Wilson, 1975, p. 3). However, water depths of 100 m or more have been inferred for some ancient algal stromatolites (Playford *et al.*, 1976, p. 553-557; Hoffman, 1974), and deposition in such deep water cannot be ruled out for the Wishart stromatolites with the evidence at hand. Almost the entirety of modern continental shelves are within that limit; hence, it is conceivable that facies F could have been deposited anywhere from the intertidal zone to the shelf edge.

Why was a thin sequence of fairly pure dolomite deposited in an area where siliciclastic deposition predominated? Two possibilities exist: (1) the normal influx of siliciclastic sediment into the basin was interrupted, permitting carbonate to accumulate by default, or (2) an abnormal abundance of carbonate flooded the basin and diluted the siliciclastic sediment for a time. The paucity of siliciclastic mud in the dolomite, coupled with the presence of siliciclastic sand, argues against mere dilution and supports the first of these two interpretations. Two possible sources exist for the dolomite. One source consists of the thick accumulations of Denault dolomite north (Baragar, 1967, p. 33-40) and east (Donaldson, 1966, p. 23-33; Wardle, 1979, p. 11-12) of the Schefferville area. These are largely platform deposits which may be time equivalent to the Wishart in part. If so, dolomite mud (or its precursor) produced on these platforms could have been transported into the Schefferville area in suspension. The other potential source of dolomite is the water of the Wishart basin itself. It is unclear whether one or both of these sources contributed carbonate to the sediments of the Wishart Formation.

MODEL OF DEPOSITION ON WISHART SHELF

Using the interpretations presented in the preceding section, the Wishart Formation is inferred to be a high-energy shelf deposit which accumulated via two main processes. One process, episodic storm deposition, is represented by facies A and B, which make up about 49% of the aggregate thickness of the Wishart. These two facies differ somewhat in the sedimentary structures they display. However, the difference is one of degree rather than kind, and it probably reflects changes in water depth and/or average storm intensity through time. The second major process, deposition from normal tidal currents, is represented by facies C, which makes up about 41% of the aggregate thickness of the measured sections. The sands of facies C were probably deposited in large-scale bed forms, based on their demonstrable similarity to such deposits in recent environments. They almost certainly were not deposited in tidal inlets. A typical inlet filling sequence displays a sharp, erosional basal contact overlain by a lag gravel as well as an upward decrease in grain size (Kumar and Sanders, 1974, p. 505; Barwis, 1978, Fig. 15); these features are absent from intervals of facies C.

Within the Wishart Formation, facies A, B, and C do not follow one another in random succession. For about half of their aggregate thickness, they are grouped into upward-coarsening cycles in which the finer-grained storm deposits are superseded gradationally by the coarser-grained tidal deposits. It is almost as common to find the two are randomly interstratified. Gradational transitions upwards from tidal to storm deposits are quite exceptional. If sedimentation had been controlled by random variations in such factors as tidal currents, storms, and sediment supply rate, the facies would presumably have been stacked in random order, or at least with equal thicknesses of upward-fining and upward-coarsening sequences. Inasmuch as this is not observed in the Wishart, a mechanism linking the deposition of facies A and B with that of facies C is required. The preferred hypothesis is that, at certain times, the shelf aggraded rapidly as a result of rapid sedimentation, and that the upward-coarsening cycles formed during such aggradational pulses. The arguments and evidence supporting this hypothesis are summarized in the following paragraphs.

Shoaling during times of rapid sediment influx offers a simple and repeatable mechanism for creating the upward-coarsening cycles of the Wishart Formation as follows. Normal tidal currents would be more likely to rework sediments on the shallower parts of the shelf. Consequently, sediments deposited by storms would stand a better chance of being preserved in the deeper areas. Inasmuch as rapid sedimentation would cause the shelf to become shallower, it could result in an upward-coarsening transition from storm to tidal deposits in a given area. It should be noted here that the average grain size of the Wishart decreases from southwest to northeast across the Schefferville area, suggesting that the shelf deepened towards the northeast. The abundance of facies A and B relative to C increases towards the northeast (Fig. 2, sections BF through IA), which independently suggests that facies A and B were deposited in deeper shelf environments than facies C. Overall, the stratigraphic organization of facies A, B, C, and D in the Wishart probably reflects a combination of long-term and possibly random variations in storms, tidal currents, or other factors, and the shoaling episodes outlined above.

The mechanism proposed for the creation of the upward-coarsening cycles is somewhat analogous to that of delta lobe formation. Delta lobes grow during times of rapid sediment influx, giving rise to upward-coarsening sequences. The lobes cease to grow, then become truncated by erosional surfaces and capped with thin lag deposits during times of sediment starvation. Such lag deposits commonly include sandstone intraclasts (Morgan and Treadwell, 1954; Oomkens, 1970, p. 210). Moreover, Coleman and Gagliano (1964, p. 71) maintain that such "bounding. . . sediments form the major stratigraphic markers in deltaic sequences" because of their lateral continuity. The same characteristics are displayed by beds of facies D, which commonly cap the upward-coarsening cycles in the Wishart.

Departures from storm and tidal deposition were infrequent and took one of three forms, resulting in three additional facies. Facies F represents a unique switch from siliclastic to carbonate sedimentation, but not necessarily a switch in the physical environment of deposition. Facies E represents local soft-sediment deformation and/or slumping on the shelf. Facies D in-

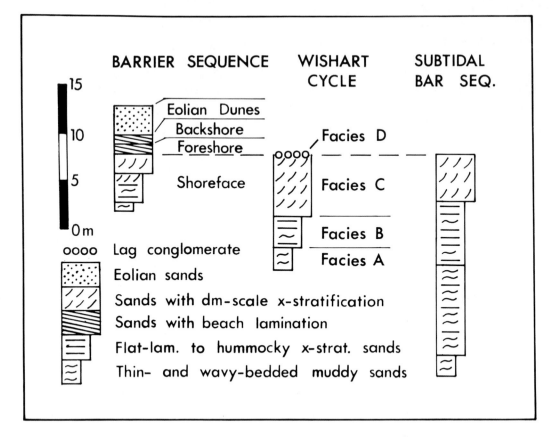

Fɪɢ. 6.—Comparison of three schematic upward-coarsening sequences, one generated by prograding barrier island (after Fig. 24 of Reinson, 1979), one from the Wishart Formation, and one from Mesozoic deposits of the Western Interior of North America and interpreted as an aggrading subtidal bar sequence (after Fig. 16 of Brenner and Davies, 1974); dashed horizontal line connects base of foreshore (i.e., top of subtidal deposits) in barrier sequence with upper contacts of Wishart and Mesozoic sequences.

cludes a scant 1% of the aggregate thickness of the measured sections, yet it appears to record two major and several minor episodes of sediment starvation, winnowing, and penecontemporaneous silica cementation. Lag gravels were formed as a result of these processes, and the uppermost bed of facies D coincides with the near-total cessation of siliciclastic deposition on the Wishart shelf.

One final question needs to be addressed: Is the Wishart Formation in the Schefferville area a proximal or a distal shelf deposit? In other words, was it deposited in close proximity to a shoreline at any time? The data at hand are inconclusive, because the upward-coarsening cycles of the Wishart show similarities with Phanerozoic sequences formed both on distal shelves and via coastal progradation (Fig. 6). On the one hand, the average grain size of the Wishart is significantly coarser than the best-documented upward-coarsening sequences from distal shelf environments (Brenner and Davies, 1974, Figs. 16, 17; Spearing, 1975, Figs. 6-7). Moreover, the modified ripples of facies A and B

and the possible desiccation cracks likewise suggest shallow water deposition. However, shallow water cannot be equated with coastal proximity. Sandy, high-energy environments with water as shallow as 4 m exist on modern shelves well over 100 km from the nearest shoreline (Stewart and Jordan, 1964). Furthermore, no unequivocally intertidal or supratidal deposits must have been eroded from the tops of most of all of them. This is a distinct possibility, in as much as supratidal deposits have been identified in the Wishart. If the cycles formed by coastal progradation, the supratidal deposits must have been eroded from the tops of most of all of them. This is a distinct possibility, inasmuch as supratidal deposits have the lowest preservation potential of any in the coastal zone (Fischer, 1961, p. 1662). In summary, it is not possible to resolve the question of coastal proximity with the data at hand. If any coastal deposits are to be found in the Wishart, they are probably located near the western margin of the Labrador trough, given the decrease in grain size towards the northeast.

CONCLUSIONS

A uniformitarian interpretation of Early Proterozoic shelves is supported by the fact that the sedimentary facies and upward-coarsening cycles of the Wishart Formation are similar in many respects to those of their Phanerozoic counterparts (Fig. 6). The main ways in which the Wishart differs from, e.g., the Mesozoic shelf deposits described by Brenner and Davies (1974), Spearing (1975), and in this volume are as follows: (1) shale is much less abundant, (2) trace fossils are absent, and (3) the paleocurrent patterns are different. The paucity of mud in the Wishart can be ascribed to a higher average energy level (probably stronger tidal currents) which permitted the sediments to be winnowed more throughly. The absence of trace fossils simply reflects the lack of metazoa in the Early Proterozoic. The contrast in paleocurrent patterns probably reflects a larger tidal range on the Wishart shelf relative to the Mesozoic ones. Tidal sands with sedimentary structures comparable to those of facies C have nevertheless been reported from modern environments. In short, no radical differences in tidal circulation, the nature of chemical weathering, or any other physical or chemical process in the Early Proterozoic need be invoked to explain the characteristics of the Wishart Formation. Of particular interest is the presence of hummocky cross-stratification in the Wishart, one of the oldest occurrences of this structure yet reported. It suggests that storms periodically swept across Precambrian shelves much as they did throughout the Phanerozoic (and do today).

This analysis of the Wishart Formation also brings into focus how little we know about the stratigraphic record left by certain shelf processes. For example, it is not clear whether a shoreline was ever present in or near the Schefferville area during deposition of the Wishart. Given our current knowledge of shelf sedimentology, it is exceedingly difficult to determine how far from a shoreline particular shelf sediments were deposited, unless there is good faunal evidence, or isochronous datum planes to facilitate paleogeographic reconstructions. Facies A and B provide another example of the limits to our interpretations of ancient shelf sediments. The differences between the centimeter- and decimeter-scale sandstone beds they contain are attributed to differences in storm intensity, but it is by no means clear in what way(s) the storms actually varied. Factors which could have varied include (1) storm size, (2) storm position relative to the site of deposition, and (3) the speed at which storms traveled across the shelf. Unfortunately, it is probably impossible to specify how variations in these or other storm parameters are reflected in the sediment deposited by a single storm. More work on modern storm deposits along the lines of Hayes' (1967) classic study will be necessary before the sedimentary record of storm-dominated shelves can be interpreted in detail.

Finally, the interpretation of the Wishart Formation places constraints on the depositional environment of the overlying Sokoman Formation, an extensive cherty iron formation. Inasmuch as the contact between the two formations is believed to be conformable, the basal strata of the Sokoman must have been deposited in an environment adjacent or related to a marine shelf. Throughout most of the Schefferville area, the lowermost beds of the Sokoman are monotonous, thinly laminated slates high in iron and silica (Gross, 1968, p. 24; Harrison et al., 1972, p. 12-13) with stray, thin beds of tuffaceous detritus that exhibit excellent normal grading (Zajac, 1974, Fig. 21). These beds clearly escaped reworking by strong currents and waves. Hence, they probably accumulated in deeper water than the sediments of the Wishart. Inasmuch as the Wishart is a shelf deposit, it follows that the basal strata of the Sokoman were deposited in a slope to basinal marine environment.

ACKNOWLEDGEMENTS

This paper is based on part of a PhD dissertation completed at the Johns Hopkins University, Department of Earth and Planetary Sciences, under the guidance of Drs. Lawrence A. Hardie and Francis J. Pettijohn. Grants in aid of research from Sigma Xi, the Geological Society of America, and the Department of Earth and Planetary Sciences (The Johns Hopkins University), and NSF Dissertation Improvement Grant EAR 77-08137 (L.A. Hardie, principal investigator) all helped to support the fieldwork. L.A. Hardie kindly loaned me a field vehicle, and Om Garg, Frank Nicholson, and Douglas Barr gave me invaluable logistical assistance in the Schefferville area. Charles Minero served as my field assistant during the summer of 1977. Discussions with many people contributed to this study; foremost among them were I.S. Zajac, Joseph P. Smoot, and my thesis advisers. D.W. Houseknecht and C.J. Stuart critically reviewed a draft of this manuscript and suggested many improvements, as did R.W. Tillman and unnamed colleagues of his. Veronica Kusznir typed several drafts of the manuscript. My thanks to them all.

REFERENCES

ALLEN, J.R.L., 1970, Physical Processes of Sedimentation: George Allen & Unwin, Ltd., 248 p.

BARWIS, J.H., 1978, Sedimentology of some South Carolina tidal-creek point bars, and a comparison with their fluvial counterparts, in A. D. Miall, (ed.), Fluvial Sedimentology: Canadian Soc. Petroleum Geologists Mem. 5, p. 129-160.

BARAGAR, W.R.A., 1967, Wakuach Lake map-area, Quebec and Labrador: Geol. Surv. Canada Mem. 344, 174 p.

BAYLEY, R.W., AND H.L. JAMES, 1973, Precambrian iron-formations of the United States: Econ. Geology, v. 68, p. 934-955.

BJAERKE, T., AND H. DYPVIK, 1977, Quarternary 'stromatolitic' limestone of subglacial origin from Scandinavia: Jour. Sed. Petrology, v. 47, p. 1321-1327.

BOURGEOIS, J., 1980, A transgressive shelf sequence exhibiting hummocky stratification: The Cape Sebastian Sandstone (Upper Cretaceous), southwestern Oregon: Jour. Sed. Petrology, v. 50, p. 681-702.

BRENNER, R.L., AND D.K. DAVIES, 1974, Oxfordian sedimentation in western interior United States: Amer. Assoc. Petroleum Geologists Bull, v. 58, p. 407-428.

CANT, D.J., 1980, Storm-dominated shallow marine sediments of the Arasaig Group (Silurian-Devonian) of Nova Scotia: Canadian Jour. Earth Science, v. 17, p. 120-131.

COLEMAN, J.M., AND S.M. GAGLIANO, 1964, Cyclic sedimentation in the Mississippi River deltaic plain: Gulf Coast Assoc. Geol. Soc. Trans., v. 14, p. 67-80.

CORBETT, K.D., 1973, Open-cast slump sheets and their relationship to sandstone beds in an Upper Cambrian flysch sequence, Tasmania: Jour. Sed. Petrology, v. 43, p. 147-159.

DIMROTH, E., 1970, Evolution of the Labrador Geosyncline: Geol. Soc. America Bull., v. 81, p. 2717-2742.

_____,1971, The Attikamagen-Ferriman transition in part of the Central Labrador Trough: Canadian Jour. Earth Science, v. 8, p. 1432-1454.

_____,W.R.A. BARAGAR, R. BERGERON, AND G.D. JACKSON, 1970, The filling of the Circum-Ungava Geosyncline: Geol. Surv. Canada Paper 70-40, p. 45-142.

DONALDSON, J.A., 1966, Marion Lake map-area, Quebec-Newfoundland: Geol. Surv. Canada Mem. 338, 85 p.

EVANS, J.L., 1978, The geology and geochemistry of the Nimish Subgroup, Dyke Lake area, Labrador: Newfoundland Dept. Mines and Energy, Mineral Develop. Div., Rept. 78-4, 39 p.

FISCHER, A.G., 1961, Stratigraphic record of transgressing seas in light of sedimentation on Atlantic Coast of New Jersey: Am. Assoc. Petroleum Geologists Bull., v. 45, p. 1656-1666.

GARRISON, R.E., J.L. LUTERNAUER, E.V. GRILL, R.D. MacDONALD, AND J.W. MURRAY, 1969, Early diagenetic cementation of Recent sands, Fraser River delta, British Columbia: Sedimentology, v. 12, p. 27-46.

GOLDRING, R., AND P. BRIDGES, 1973, Sublittoral sheet sandstones: Jour. Sed. Petrology, v. 43, p. 736-747.

GROSS, G.A., 1968, Geology of iron deposits in Canada, v. III, Iron ranges of the Labrador geosyncline: Geol. Surv. Canada Econ. Geol. Report 22, 179 p.

HALL, W.D.M., AND A.D.T. GOODE, 1978, The Early Proterozoic Nabberu Basin and associated iron formations of Western Australia: Precambrian Res., v. 7, p. 129-184.

HAMBLIN, A.P., AND R.G. WALKER, 1979, Storm-dominated shallow marine deposits: the Fernie-Kootenay (Jurassic) transition, southern Rocky Mountains: Canadian Jour. Earth Science, v. 16, p. 1673-1690.

HARDIE, L.A., AND R.N. GINSBURG, 1977, Layering: The origin and environmental significance of lamination and thin bedding, in L.A. Hardie, (ed.), Sedimentation on the Modern Carbonate Tidal Flats of Northwest Andros Island, Bahamas; Johns Hopkins Univ. Studies in Geology, no. 22, p. 50-123.

HARMS, J.C., 1969, Hydraulic significance of some sand ripples, Geol. Soc. America Bull., v. 80, p. 363-396.

_____,1975, Stratification and sequences in prograding shoreline deposits, in J.C. Harms, J.B. Southard, D.R. Spearing, and R.G. Walker (eds.), Depositional Environments as Interpreted from Primary Sedimentary Structures and Stratification Sequences: Soc. Econ. Paleontologists and Mineralogists Short Course Notes, no. 2, Dallas, p. 81-102.

HARRISON, J.M., J.E. HOWELL, AND W.F. FAHRIG, 1972, A geological cross-section of the Labrador Miogeosyncline near Schefferville, Quebec: Geol. Surv. Canada Paper 70-37, 34 p.

HAYES, M.O., 1967, Hurricanes as geological agents: Case studies of hurricanes Carla, 1961, and Cindy, 1963: Univ. Texas, Bur. Econ. Geology, Rept. Inv. 61, 54 p.

HOFFMAN, P., 1974, Shallow and deepwater stromatolites in Lower Proterozoic platform-to-basin facies change, Great Slave Lake, Canada: Am. Assoc. Petroleum Geologists Bull., v. 58, p. 856-867.

HOWARD, J.D., 1972, Trace fossils as criteria for recognizing shorelines in the stratigraphic record, in J.K. Rigby and W.K. Hamblin, (eds.), Recognition of Ancient Sedimentary Environments: Soc. Econ. Paleontologists and Mineralogists Spec. Pub. 16, p. 215-225.

_____,AND H.E. REINECK, 1972, Georgia coastal region, Sapelo Island, U.S.A.: sedimentology and biology, IV, Physical and biogenic sedimentary structures of the nearshore shelf: Senckenberg Marit., v. 4, p. 81-123.

HOWELL, J.E., 1954, Silicification in the Knob Lake Group of the Labrador Iron Belt: Unpublished PhD Dissertation, Univ. Wisconsin, Madison, Wisconsin, 81 p.

JAMES, H.L., 1954, Sedimentary facies of iron-formation: Econ. Geology, v. 49, p. 235-293.

JOPLING, A.V., 1965, Hydraulic factors controlling the shape of laminae in laboratory deltas: Jour. Sed. Petrology, v. 35, p. 777-791.

KENDALL, A.C., AND P.L. BROUGHTON, 1978, Origin of fabrics in speleothems composed of columnar calcite crystals: Jour. Sed. Petrology, v. 48, p. 519-538.

KLEIN, G. DEV., 1970, Depositional dynamics of intertidal sand bars: Jour. Sed. Petrology, v. 40, p. 1095-1127.

KUMAR, N., AND J.E. SANDERS, 1974, Inlet sequence: A vertical succession of sedimentary structures and textures created by the lateral migration of tidal inlets: Sedimentology, v. 21, p. 491-532.

_____,AND_____, 1976, Characteristics of shoreface storm deposits: modern and ancient examples: Jour. Sed. Petrology, v. 46, p. 145-162.

LOWE, D.R., AND R.D. LOPICCOLO, 1974, The characteristics and origins of dish and pillar structures: Jour. Sed. Petrology, v. 44, p. 484-501.

MONTY, C.L.V., AND L.A. HARDIE, 1976, The geological significance of the freshwater blue-green algal calcareous marsh, *in* M.R. Walter, (ed.), Stromatolites: Elsevier, Amsterdam, p. 447-477.

MOREY, G.B., 1973, Stratigraphic framework of Middle Precambrian rocks in Minnesota, *in* G.M. Young, (ed.), Huronian Stratigraphy and Sedimentation: Geol. Assoc. Canada Spec. Paper 12, p. 211-249.

MORGAN, J.P., AND R.C. TREADWELL, 1954, Cemented sandstone slabs of the Chandeleur Islands, Louisiana: Jour. Sed. Petrology, v. 24, p. 71-75.

MORTON, R.A., 1981, Formation of storm deposits by wind-forced currents in the Gulf of Mexico and North Sea, *in* S.-D. Nio, R.T.E. Shüttenhelm, and Tj. C.E. van Weering, (eds.), Holocene Marine Sedimentation in the North Sea Basin: Intern. Assoc. Sed. Spec. Pub. 5, p. 385-396.

OOMKENS, E., 1970, Depositional sequences and sand distribution in the postglacial Rhone delta complex, *in* J.P. Morgan, (ed.), Deltaic Sedimentation: Soc. Econ. Paleontologists and Mineralogists Spec. Pub. 15, p. 198-212.

PETTIJOHN, F.J., 1957, Sedimentary Rocks (2nd ed.): Harper and Bros., New York, 718 p.

_____, 1975, Sedimentary Rocks (3rd ed.): Harper and Row, New York, 628 p.

PLAYFORD, P.E., A.E. COCKBAIN, E.C. DRUCE, AND J.L. WRAY, 1976, Devonian stromatolites from the Canning Basin, Western Australia, *in* M.R. Walter, (ed.), Stromatolites: Elsevier, Amsterdam, p. 543-563.

REINECK, H.E., AND I. SINGH, 1975, Depositional Sedimentary Environments: Springer-Verlag, New York, 439 p.

REINSON, G.E. 1979, Barrier island systems, *in* R.G. Walker (ed.), Facies Models: Geoscience Canada Reprint Series 1, p. 57-74.

SIMONSON, B.M., 1981, Cherts in Wishart Formation (Aphebian) of Labrador: example of rapid shallow-water silica sedimentation [abs.] Am. Assoc. Petroleum Geologists Bull., v. 62, p. 992-993.

_____, 1982a, Sedimentology of Precambrian Iron-Formations with Special Reference to the Sokoman Formation and Associated Deposits of Northeastern Canada: Unpublished PhD Dissertation, The Johns Hopkins University, Baltimore, Maryland, 359 p.

_____, 1982b, Fleming Chert Breccia: A probable hot spring plumbing system in the 2 billion year-old Labrador trough [abs.]: Geol. Soc. Amer. Abstracts with Programs, v. 14, p. 618.

SMOOT, J.P., 1978, Origin of the carbonate sediments in the Wilkins Peak Member of the lacustrine Green River Formation (Eocene), Wyoming, U.S.A., *in* A. Matter and M.E. Tucker, (eds.), Modern and Ancient Lake Sediments: Intern. Assoc. Sedimentologists Spec. Pub. 2, p. 109-127.

SOUTHARD, J.B., 1975, Bed configurations: *in* J.C. Harms, J.B. Southard, D.R. Spearing, and R.G. Walker, Depositional Environments as Interpreted from Primary Sedimentary Structures and Stratification Sequences: Soc. Econ. Paleontologists and Mineralogists Short Course Notes, no. 2, p. 5-43.

SPEARING, D.R., 1975, Shallow marine sands, *in* J.C. Harms, J.B. Southard, D.R. Spearing, and R.G. Walker, Depositional Environments as Interpreted from Primary Sedimentary Structures and Stratification Sequences: Soc. Econ. Paleontologists and Mineralogists Short Course Notes, no. 2, p. 103-132.

STEWART, H.B., AND G.F. JORDAN, 1964, Underwater sand ridges on Georges Shoal: *in* R.L. Miller, (ed.), Papers in Marine Geology: Macmillan, New York, p. 102-114.

VAN SCHMUS, W.R., 1976, Early and Middle Proterozoic history of the Great Lakes area, North America: Phil. Trans. Royal Soc. London A., v. 280, p. 605-628.

VAN STRAATEN, L.M.J.U., 1957, Recent sandstones on the coasts of the Netherlands and of the Rhone delta: Geol. en Mijnbouw (new series), v. 19, p. 196-213.

WARDLE, R.J., 1979, Geology of the eastern margin of the Labrador trough: Newfoundland Dept. of Mines and Energy, Mineral Develop. Div. Rept. 78-9, 22 p.

WILSON, J.L., 1975, Carbonate Facies in Geologic History: Springer-Verlag, New York, 471 p.

ZAJAC, I.S., 1974, The stratigraphy and mineralogy of the Sokoman Formation in the Knob Lake area, Quebec and Newfoundland: Geol. Surv. Canada Bull. 220, 159 p.